D0215141

Fiber Optics

Communication and Other Applications

HENRY ZANGER

CYNTHIA ZANGER

Merrill, an Imprint of
Macmillan Publishing Company
New York

Collier Macmillan Canada, Inc.
Toronto

Maxwell Macmillan International Publishing Group
New York Oxford Singapore Sydney

Administrative Editor: David Garza
Production Editor: Rex Davidson
Art Coordinator: Vincent A. Smith
Cover Photo: Stephen Feld
Text Designer: Denise L. Shaw
Cover Designer: Russ Maselli

This book was set in Century Schoolbook.

Merrill is an imprint of Macmillan Publishing Company.

Printed in the United States of America.

Collier Macmillan Canada, Inc.

International Standard Book Number 0-675-20944-7

Library of Congress Catalog Card Number: 90-62185

Printing: 1 2 3 4 5 6 7 8 9 Year: 1 2 3 4

Preface

In the last decade we have seen the introduction of a new communication medium: the glass (or plastic) fiber with light as the data carrier. Although the principles of guiding light in a glass (or other similar "pipe") is not new, only in the last few years has this technology become popular. There is even serious consideration of using glass fibers to replace the telephone lines to the home. This rapid development is motivated by economics. Indeed, presently fiber optic communication is cost competitive with the conventional copper transmission lines.

In addition to their use in communication, optical fibers have applications in a variety of other areas. They are used in medical applications, industry, local area communication, computer communication, and more.

In spite of the enormous expansion in the use of fiber optic technology, presently there is a shortage of texts that cover the subject in a simple, straightforward fashion, without resorting to sophisticated physics and mathematics. Although most texts use such techniques as Maxwell's equations with boundary conditions and other advanced analysis techniques, this text follows a simplified approach, relying mostly on geometric ray-tracing methods. In specific areas where ray tracing is not applicable, the results of other analytical methods are explained without derivation.

In general, this text takes a practical approach, yet it gives the reader the theoretical background necessary for further study. It is intended for the 2- and 4-year engineering technology curricula, as well as for the electrical engineer who has a general interest in fiber optics or is looking for a change in specialty.

The material covered includes, in Chapter 1, a historical note and principles of general fiber optics. Chapter 2 is a basic review of the physics of light as it applies to fiber optics. Chapter 3 discusses principles of fiber optics in more detail and is followed in Chapter 4 by coverage of the characteristics of fiber optic transmission, particularly as the characteristics relate to communication. Chapter 5 describes a variety of fiber structures: step-index, graded-index, single-mode, and more. Chapter 6 discusses analog and digital fiber optic communication systems, showing how the fiber characteristics (discussed in Chapter 4) affect system performance. Chapter 7 gives a more expanded description of communication technology. Chapter 8 deals with the practical aspects of how to connect fibers (with a splice or a connector) and the losses encountered, as well as a number of practical fiber optic components such as couplers and switches. Chapters 9 and 10 discuss light sources and detectors, respectively, including the electronic circuitry used to transmit and receive optical data. Chapter 11 discusses a number of typical fiber optic communication systems. Chapter 12 rounds out the text with a discussion of a number of noncommunication applications of fiber optics and a discussion of the future directions of fiber optic technology.

It is worth repeating that the major thrust of this book is an introduction to fiber optic technology, and as such, the mathematics prerequisites are kept to a minimum. Basic algebra, geometry, and trigonometry combined with some background in electronics are the only prerequisites.

As an aid to the reader, chapter objectives appear at the beginning of each chapter and give the highlights of the chapter. A summary and glossary at the end of each chapter provides a short review of the material covered. A large number of illustrative examples throughout are designed to clarify the analysis, while the questions and problems in most chapters allow the reader to practice what has been learned. The extensive use of manufacturer's data sheets gives the reader almost hands-on experience with the devices and systems discussed.

The coverage offered is relatively broad, allowing the instructor flexibility in choosing specific areas for the course. Chapters 1 and 2, the introduction and review chapters, respectively, can be used to provide background material for weaker students. In addition, a course oriented toward electro-optics might consist of the following chapters:

Chapter 3	Principles of Fiber Optics
Chapter 4	Fiber Characteristics
Chapter 5	Optical Fibers
Chapter 9	Optical Sources for Communication
Chapter 10	Optical Detectors
Chapter 12	Applications (optional)

The course oriented toward communication might consist of Chapters 1–8, with Chapter 11 as an option.

The technical level and writing style used are intended to make the text student oriented; it is readable and readily understandable.

While researching and writing the text, we received advice and assistance from numerous people in industry and academia. A number of reviewers gave their input, influencing the final form of the text. We thank the following reviewers: **Naqi Achter,** *DeVry Institute of Technology,* Chicago, Illinois; **Archie Campbell,** *ITT Technical Institute,* West Covina, California; **Gary Carter,** *Nashville State Technical Institute,* Nashville, Tennessee; **Ray Davidson,** *Texas State Technical Institute,* Waco, Texas; **Ralph Folger,** *Hudson Valley Community College,* Troy, New York; **Leon Heselton,** *Mohawk Valley Community College,* Utica, New York; **Steve Hixson,** *TRW Corp.*; **Ron Moody,** *Pima Community College,* Tucson, Arizona; **William Maxwell,** *Nashville State Technical Institute,* Nashville, Tennessee; **Raymond McNamee,** *Triton College,* River Grove, Illinois; **Harry Partin,** *Hinds Junior College,* Raymond, Mississippi; and **Howard Yoder,** *Detroit Engineering Institute,* Detroit, Michigan. In particular, we owe thanks to the Merrill editors who so diligently worked to obtain the reviews and to put us in direct contact with some of the reviewers.

Henry Zanger

Cynthia Zanger

Contents

1
Introduction to Fiber Optics

CHAPTER OBJECTIVES

The purpose of this chapter is to give you an overview of fiber optic technology, its history, and its applications. It discusses the advantages of using fiber optics. In particular, it emphasizes the low-loss and high data-carrying capacity (high bandwidth) of the fiber. In terms of applications, this chapter includes communication as a common application, as well as its medical and industrial uses.

1–1 FIBER OPTICS

Until about a decade ago, most electronic communication was carried by copper cables, whether twisted pairs, coaxial cables, or copper waveguides. Communication was accomplished by sending electrical signals through the copper cables or waveguides. In recent years, a new medium has been introduced: optical fibers. In optical fiber communication, light signals replace the electrical signals. (The terms light, light signals, and optical signals will be used interchangeably. All refer to both visible and invisible light [such as **infrared**[1] and **far infrared**].) Although a distinction exists between light signals and electronic communication signals, both fall into the category of electromagnetic waves. The coaxial cable, used for cable television, carries electromagnetic waves that are **modulated** (modified to carry the television

1. The Summary and Glossary for each chapter contain definitions of boldface terms.

signal) in much the same way that the optical fiber carries modulated light waves. As you will see later, even though both methods—copper cable and optical fibers—of communication use electromagnetic waves, the practical differences are substantial.

An optical fiber is a transparent rod, usually made of glass or clear plastic, through which light may propagate. The light signal travels through the rod (sometimes called a light pipe) from the transmitter to the receiver and can easily be detected at the receiving end of the rod, provided the losses in the fiber are not excessive. The structure of the modern fiber consists of an optical rod **core** coated with a **cladding.** The rod and the cladding have different optical characteristics.

Figure 1–1 shows an unclad glass rod and a clad rod through which the light travels. With the unclad rod, only a small portion of the light energy is kept inside; most of the light leaks to the surroundings. The clad fiber is a much more efficient light **carrier.** The losses of the light energy as it travels through the fiber are much smaller for the clad fiber than for the unclad one.

The problems of light loss in the fiber and the distortion of light pulses introduced by the fiber were two of the past difficulties of this technology. The low loss and reduced pulse distortion in modern fibers have become two of the major advantages of fiber optic transmission. These advantages will be discussed in more detail later.

The transmission of light via a cable, or fiber, has a wide variety of applications other than communication. Fiber optic technique can be used in indoor lighting, with fibers conducting sunlight from the outdoors. In the medical field, well-focused light energy is used in surgical procedures. Other applications in computers, measurements, and quality control are mentioned in Section 1–4.

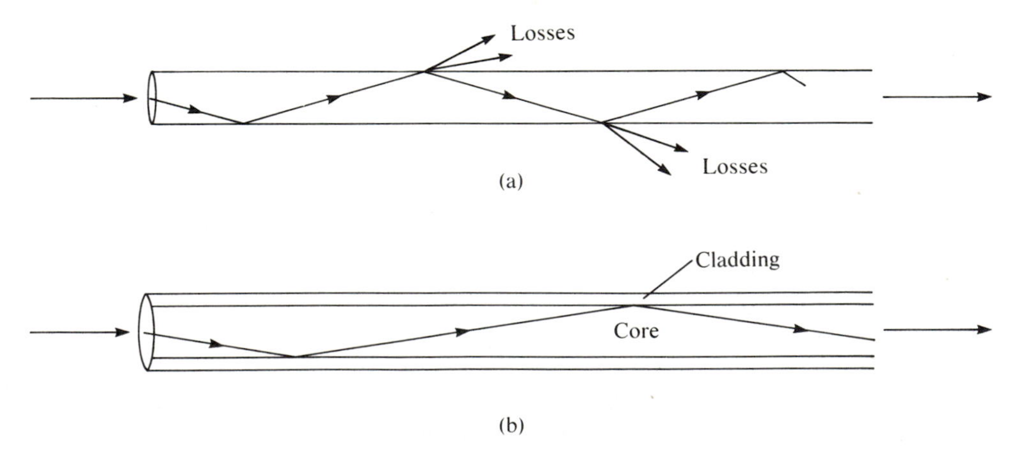

FIGURE 1–1 Light guides. (a) Simple glass rod. (b) Glass rod and cladding with different refraction qualities.

Some of the important advantages of fiber optic communication are listed here:

1. Transmission loss is low. Long-distance transmission is possible without the need to amplify and retransmit the signal along the way.
2. Fiber is lighter and less bulky than equivalent copper cable.
3. More information can be carried by each fiber than by equivalent copper cables (higher data rate).
4. There is no electrical connection between the sender and the receiver (complete electrical isolation).
5. There is no interference in the transmission of light from electrical disturbances (lightning) or electrical noise. (Electromagnetic waves generated by electrical appliances cannot interfere with the light signal.)
6. The fiber itself can better withstand environmental conditions such as salt, pollution, and radiation with no resulting corrosion and minimal nuclear radiation effects, so it is more reliable.
7. Transmission is more secure and private. "Listening in" is nearly impossible.
8. The overall cost of a fiber optic communication system is already lower than that of an equivalent cable communication system. In the near future, fiber optic communication systems should become more economical than most other types of communication systems.

The use of optical fiber, particularly in communication, is projected to increase rapidly. Estimates (to nearest $100 million) of the total U.S. fiber optic market, prepared by the Kessler Marketing Intelligence Organization, are shown in Table 1–1. Expenditures on telephone services are expected to account for the largest portion of these funds.

TABLE 1–1
Estimated Total U.S. Fiber Optic Market (in billion dollars)

Year	Market
1987	0.7
1990	1.3
1991	1.8
1992	2.9

1–2 HISTORICAL NOTE

The use of light as a means of communication is not new. Fire was used as a communication signal early in human history. Morse code communication used light reflected by mirrors, particularly in ship-to-ship signaling. As early as 1860, Alexander Graham Bell demonstrated voice transmission, using mirrors that were vibrated by the sound waves of the voice so that the light reflected from the mirrors was modulated by the sound. The modulated light at the receiving end was focused on a selenium plate. The resistance of the plate and thus the current through the plate varied with the changes in light intensity. The current changes were used to drive a device similar to the modern speaker.

All these methods depended on weather conditions (visibility). They were usable for short distances and for direct line of sight applications only. With the invention of the **laser** in 1960, interest in light communication intensified. (The laser is a high-intensity light source that can easily be modulated, producing an inherently well focused light beam.) Even with lasers, open-air light communication was limited to short distances and good weather.

The first attempts at long-distance transmission of light through glass fibers were made in 1966. Because of extensive impurities in the glass, the light losses in the fiber were high. Transmission was still limited to short distances. In addition, the size of the available lasers made it difficult to couple the light energy into the small fibers efficiently.

With the development of the semiconductor laser (diode laser) and the light-emitting diode and, later, the introduction of high-purity fiber, fiber communication came of age. Transmission along distances of many miles, without reamplifying the light signal, became commonplace.

The history of the development of fiber optic technology centers on applications in the communication industry and on government-sponsored research, which is focused almost exclusively on communication. The major advances have taken place relatively recently, in the 1970s and 1980s. The general theory of light propagation, however, developed over a long period of time. Important milestones are listed here.

1621 Willebrord Snell formulated his law, which deals with the behavior of light as it crosses from one material to another.

1870 John Tyndall demonstrated light transmission in a water stream. The light bent and followed the water stream. This behavior was the first indication that light could propagate through a medium, along a curved path as well as a straight line.

1897 John William Strutt, the third Baron Rayleigh, formulated some of the basic laws governing light propagation.

1900 Max Planck developed the theory of radiation in discrete amounts (later called photons) and Planck's constant h, relating electron and photon energy.

1905 Albert Einstein proposed the photon theory, explaining photoelectric effects.

1930 Willis Lamb, Jr., experimented with light guided in a glass fiber.

1951 A group of researchers in the United States demonstrated the transmission of an image through a bundle of glass fibers.

1953 Narinder Singh Kapany developed fibers with cladding. These fibers greatly improved transmission characteristics.
1960 Theodore Maiman demonstrated the first laser.
1962 Theodore Maiman invented the semiconductor laser.
1966 Charles Kao and Charles Hockman proposed the use of fiber in long-distance transmission.
1970 Robert Maurer and a group at Corning Glass Company produced low-loss fibers (loss under 20 dB/km).
1980 American Telephone and Telegraph (AT&T) began major fiber optic communication link between Boston, Massachusetts, and Richmond, Virginia.
1981 Corning Glass Company commercially introduced single-mode fibers with high **bandwidth** and low loss.
1983 AT&T, MCI, and others installed long-distance fiber optic communication links, using single-mode fibers.

1–3 RECENT DEVELOPMENTS

In the early stages of development, fiber communication promised extremely high **data rates,** which would allow large masses of data to be transmitted quickly. It also had the potential for transmission over long distances without the need to amplify and retransmit along the way. Recent developments have exceeded the hopes of those involved with the technology.

In the last few years, fibers with losses of 0.2 decibels per kilometer (dB/km) have been produced (a power loss of less than 4.5% for a 1-km length). This development means that transmission over a distance of more than 100 km (60 mi) is now possible, depending on the data rate. New, high-intensity, long-life semiconductor lasers may increase this distance.

The bandwidth of the fiber optic communication system, which determines the maximum data rate, depends on the major components of the system (Figure 1–2). Both the light source on the sending end and the light detector on the receiving end must be capable of operating at the system data rate. The circuitry that drives the light source and the circuitry that amplifies and processes the detected light must both have suitable high-frequency response. The fiber itself must not distort the high-speed light pulses used in the data transmission.

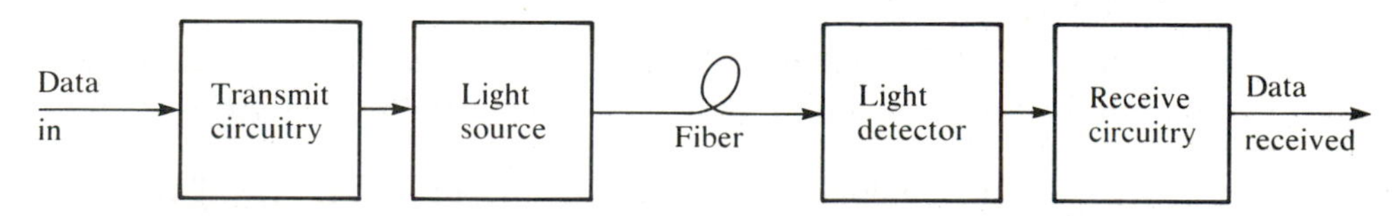

FIGURE 1–2 Fiber optic communication system.

Recently, circuits have been developed that can operate above 1 GHz (1000 MHz). (The relation between hertz (cycles per second) and bits per second can be simplified to 1 Hz is equivalent to either 1 b/s or 2 b/s, depending on transmission details.) Light detectors and light sources that have a similar frequency response are now available. Single-mode fibers (see Section 5–4) that can operate at extremely high data rates have been made. These developments have led a number of communication companies to experiment with 2-GHz light communication systems. A system operating at 56 Gb/s (56,000 Mb/s) over a span of 72 km has been demonstrated by the Standard Electric Lorenz Company of West Germany. The efforts to produce fibers that can carry data at very high rates and exhibit low losses are motivated by the cost savings that the fibers bring.

If a fiber can operate at a data rate of, say, 200 Mb/s, it can carry more than 3000 separate voice communications. The method used in such a system is called **multiplexing.** It allows multiple, independent data sources to be transmitted through a single fiber (or carrier). Since a twisted pair of copper wires typically carry only one voice channel, there is a 3000 to 1 improvement in data-carrying capacity.

The enormous bandwidth of optical fibers results in a great reduction in the size and weight of data cables. For example, a shipboard advanced radar system requires about 750 ft of coaxial cable, with a weight of 7 tons and a diameter of 18 in. These cables can be replaced by an optical fiber weighing 40 lb and measuring 1 in. in diameter.

The typical copper, high data rate system, which uses coaxial cables, requires a **repeater station** approximately every 1.8 km. The repeater consists of a receiver, an amplifier, and a transmitter. All of these are relatively expensive and tend to fail much more often than the cable itself. With very low loss fibers, the repeaters may be located more than 50 km apart, an improvement of 30 to 1 in terms of distances, and reduce overall cost by requiring fewer repeaters (excluding the improvement that comes from the much higher data rate that the fiber can carry). As was noted, fibers have now been operated at 2 Gb/s (2000 MHz/s) and above.

1–4 APPLICATIONS

You have already noted some of the communication applications of light transmission in fibers. The following list highlights specific data relating to a few of these applications.

- In Montana in 1978, Cablecom General installed one of the first fiber cables in a cable television application. It consisted of a 3-km span of cable.
- In 1979, General Telephone Company of Florida built one of the first underwater fiber optic communication systems. Also in 1979, the U.S.

Department of Energy installed a fiber optic cable in an underground nuclear testing facility.

- In 1981, the Continental Telephone Company of Virginia installed a 17-km fiber cable that operated at 45 Mb/s.
- In 1982, The Kawai Telephone Company installed a 36-fiber cable that operated at 45 Mb/s over an unrepeatered span of 32 km.
- In 1982, MCI Telecommunications installed a fiber link in New York that operated at 135 Mb/s.
- In 1983, the Central Telephone Company installed a fiber optic link connecting various industrial plants in a **local area network.** Also in 1983, the Continental Telephone Company used fiber communication in a subscriber loop; that is, it connected fibers directly to the telephone company customers. And in that same year, MCI Telecommunications installed one of the first high data rate, single-mode fiber systems. It operates at 405 Mb/s, and the cable contains both single-mode and multimode fibers. (To lower costs and avoid delays in obtaining rights of way for laying the cables, particularly across state lines, many companies leased rights of way from railroad companies. Then they laid the fiber optic cable next to the railroad tracks.) This application has been further extended to connect peripheral devices such as printers and large memory systems to central computers. The light weight and high data rate of the fiber link makes installation easy and cost effective.
- As early as the mid-1970s, optical fibers were used in a number of medical applications. Kidney stones have been pulverized by well-focused energy carried by fibers. Similarly, cancer cells have been treated by localized heat introduced by fibers connected to relatively high power lasers.
- In 1987, F & P Fiber Optics AG of Switzerland introduced a **micro-endoscope** (model FU7 150). It is used to inspect visually internal organs such as intestines (colonoscopy). The diameter of this **endoscope** is 1.5–2 mm, much smaller than that for conventional endoscopes, and it is up to 260 mm long. A similar device has also been used experimentally in angiograms to view various blood vessels, and high-pulsed laser energy has been used to open blocked vessels.
- Optical fibers are used in the remote measurement of temperature and pressure. The use of optical fibers has been preferred, particularly when the measurements involved highly radioactive environments such as the area of a nuclear explosion. Optical fibers are unaffected by nuclear radiation or by electrical transients that usually accompany nuclear explosions.
- Many fiber optic applications involve the optical transmission of images. This application is useful to inspect industrial parts to which there is no free access. A thin fiber can be maneuvered close to the area to be inspected and used to carry the image to a convenient display.

- For about 10 years, a special high-intensity fiber light has been used as a replacement for the dentist's spotlight. The fiber carrying the light is attached to the drill to illuminate the work area.
- Another promising area of application is the use of fiber optics for shipboard communication because of its light weight, bandwidth, and immunity from electromagnetic interference.

SUMMARY AND GLOSSARY

The terms and definitions given here should serve as a review of this chapter. As you read this section, try to relate the terms to the material that you have just read. Make sure that you are thoroughly familiar with these terms.

Bandwidth. The frequency range (the highest frequency for systems with a direct current response) that the system can handle with little amplitude distortion and low loss. (Signals with frequencies outside the bandwidth will suffer relatively high loss.)

Carrier. An electromagnetic wave that can be modulated (see "modulation") to carry information.

Cladding. The outer coating (not the structural covering), usually made of glass or plastic, of an optical fiber.

Core. The center portion, usually made of glass or plastic, of an optical fiber.

Data rate. The highest rate of digital binary information that the system can handle (related to bandwidth). (The unit of digital information is the "bit.")

Endoscope. An instrument used to inspect (view) the inside of internal organs such as intestines.

Far infrared. Electromagnetic waves with frequencies below the infrared (well below the visible range).

Fiber loss. The loss of energy of light as it travels along the fiber, expressed in decibels per kilometer. (The decibel is a logarithmic unit relating the input power to the output power [input to the fiber and output from the fiber].)

Infrared. Electromagnetic waves with frequencies just below visible red. (It is subdivided into near infrared [highest frequency], middle infrared, and far infrared.)

Local area network. A communication system that interconnects computers or other digital systems, located within a relatively short distance of 1 or 2 mi.

Laser. *L*ight *A*mplification by *S*timulated *E*mission of *R*adiation. Refers to a device that produces light, typically of a single color (nearly a single frequency). (Some such devices are the gas laser, the ruby laser, the semiconductor laser, and the diode laser.)

Microendoscope. An endoscope (see "endoscope") with a diameter of 1.5–2 mm.

Modulation. Modulation applied to a carrier (see "carrier") refers to the modification in some way by the information to be transmitted (carried) by the carrier. The modification can be applied to the amplitude of the carrier, the frequency of the carrier, or the phase of the carrier.

Multiplexing. Techniques used to transmit data from many sources via a single channel (a single fiber or wire). (The sources may be analog or digital.)

Refraction. A phenomenon that causes light to change its direction as it travels from one material to another (described by Snell's law of refraction).

Repeater station. A device or system used to reshape (amplify) the signal along a transmission cable. (It consists of a receiver, an amplifier, and a transmitter. The signal is received, reshaped, amplified, and retransmitted along the cable; typically, analog signals are amplified and digital signals are reshaped.)

QUESTIONS

1. What is the basic structure of a modern optical fiber?
2. The carrier in electronic communications is an electromagnetic wave (or current). What is it in fiber optic communication?
3. Which of the advantages of fiber optic transmission in data communications do you think are most important? Give reasons.
4. Some missiles are guided by data carried by optical fibers that are attached to the missiles. Could regular copper wires be used? Explain.
5. During the early stages of fiber optic communications, why were data transmitted only over short distances?
6. What two developments are largely responsible for today's improved fiber optic communication systems?
7. The most advanced fiber optic systems operate at approximately 20 Gb/s. If a voice transmission requires 64 kb/s, how many voice channels can be carried by a single fiber?
8. What are two medical applications of optical fibers?
9. What are two inspection applications of fibers?
10. What features of the optical fiber are the major reasons for the low cost of the fiber optic communication system? Discuss these features.
11. It is said that the standard telephone line can be easily tapped. Is this true for the fiber optic line? Explain.
12. Today, the preferred underwater communication lines are fiber optic cables. Why?
13. Can you think of a home or office application of fiber optic technology?

2 Physics of Light

CHAPTER OBJECTIVES

This chapter discusses principles of physics relating to the propagation of light. It aims to give you an understanding of the behavior of the electromagnetic wave and its spectrum, light being a part of this spectrum. You will learn about the velocity of the electromagnetic wave (light being an electromagnetic wave) in free space (or vacuum), denoted by c, as well as in different materials. The velocity is related to the refractive index of the material.

The laws governing reflection, refraction, and diffraction are described in some detail. In the last part of Section 2–3–5, you will study various causes of light loss (such as absorption and scattering) in materials.

2–1 INTRODUCTION

The propagation of light can be analyzed in detail using **electromagnetic wave** theory. Simplified analysis uses the **ray-tracing** method. This method examines the propagation direction, ignoring the electric and magnetic fields involved. Light falls in the general category of electromagnetic waves, much like radio waves. Light, however, has a much higher frequency than radio waves. Visible light, for example, covers the range between 0.43×10^{15} and 0.75×10^{15} Hz. Fiber optic communication uses light in the frequency range from 0.2×10^{13} to 0.37×10^{13} Hz, called the infrared range. In contrast, radio frequencies, including radar, television, and radio, cover the range from 0.5×10^{6} Hz to 5×10^{10} Hz.

The idea of representing electromagnetic waves as rays propagating in straight lines is valid only for very high frequencies. (More precisely, it is valid when the wavelength [see Section 2–2] is small relative to the structure carrying the wave.) Light, including infrared radiation, falls into the category of rays; however, not all of the characteristics of light can be analyzed or explained with the ray-tracing approach.

Another theory relating to light propagation and generation is the **quantum theory of light,** also referred to as the **photon theory.** This approach views light as the propagation of packets of energy (analogous to basic atomic particles) called **photons.** The energy contained in each photon is related to the frequency of the light by the equation

$$E_p = h \times f \qquad \textbf{(2–1)}$$

where E_p is the energy of the photon, h is **Planck's constant** ($h = 6.626 \times 10^{-34}$ joule-seconds [J-s]), and f is the frequency. This theory will be employed in analyzing and explaining light generation and detection. It is particularly helpful in describing the transformation of light into electron current and vice versa.

Electromagnetic wave propagation is usually described in terms of frequency, wavelength, and **phase velocity.** To facilitate the discussion, the units of measure used are listed here.

micron (μm)	10^{-6} m
nanometer (nm)	10^{-9} m
angstrom (Å)	10^{-10} m
kilohertz (kHz)	10^{3} m
megahertz (MHz)	10^{6} Hz
gigahertz (GHz)	10^{9} Hz
terahertz (THz)	10^{12} Hz

2–2 ELECTROMAGNETIC WAVES

2–2–1 Principles

When an alternating current passes through a wire, it sets up an electromagnetic field that propagates away from the wire. As its name implies, the electromagnetic field consists of an electric field and a magnetic field. These fields are at right angles to each other and to the direction of propagation. Figure 2–1(a) shows the electric field *(E)* in the *y* direction, the magnetic field *(H)* in the *x* direction, and the propagation in the *z* direction. (The conventional notation for the *x-y-z* orthogonal coordinate system is used.) This type of field is called a **transverse electromagnetic field.** Since the wave is induced by a sinusoidal current, the field increases, decreases, and reverses polarity with

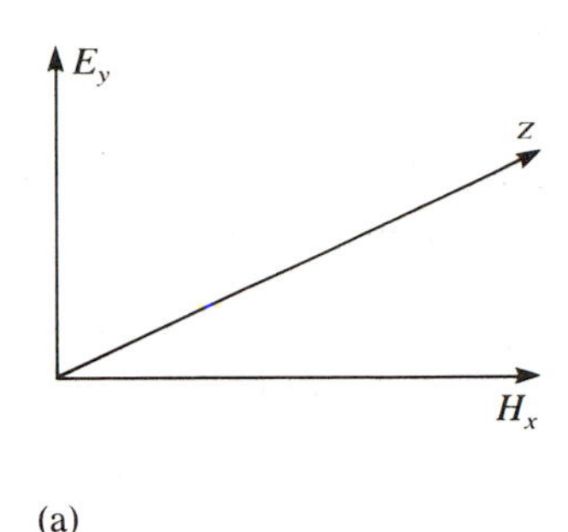

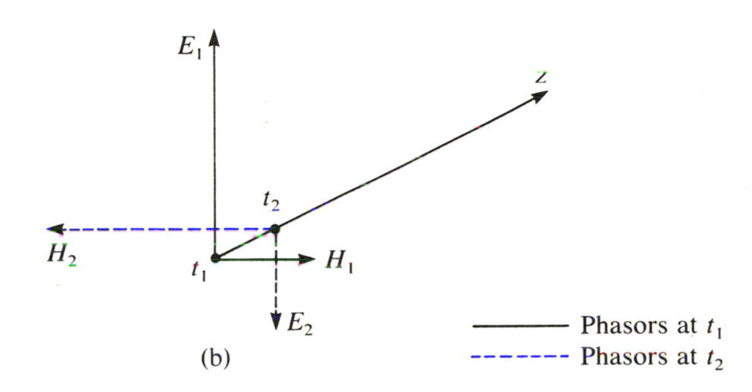

FIGURE 2–1 Electromagnetic field representation. (a) E-H amplitude phasors. (b) E-H instantaneous phasors (at times t_1 and t_2).

time; the field intensity varies sinusoidally with time. In Figure 2–1(b), the electric and magnetic fields are shown at different points in time. The line along the z axis between t_1 and t_2 represents the distance that the wave has traveled between time t_1 and t_2.

The propagation velocity of the electromagnetic waves in a vacuum is denoted by c, where

$$c = 300 \times 10^6 \text{ m/s} \tag{2–2}$$

This velocity is independent of frequency. The velocity of propagation in a medium (rather than a vacuum) depends on the particular material and is always less than c.

Most light consists of a group of frequencies rather than a wave with a single frequency. The light of a typical incandescent lamp—visible light—contains frequencies ranging from blue light (approximately 0.6×10^{15} Hz) to red (0.4×10^{15} Hz) and lower. Because light consists of a group or range of frequencies the behavior of light requires a more careful definition of the parameters used to describe it. It also directly affects the performance of fiber optic systems. The propagation velocity just mentioned has to be redefined as the group velocity (the velocity of a group of waves with different frequencies) not simply the velocity of a single wave.

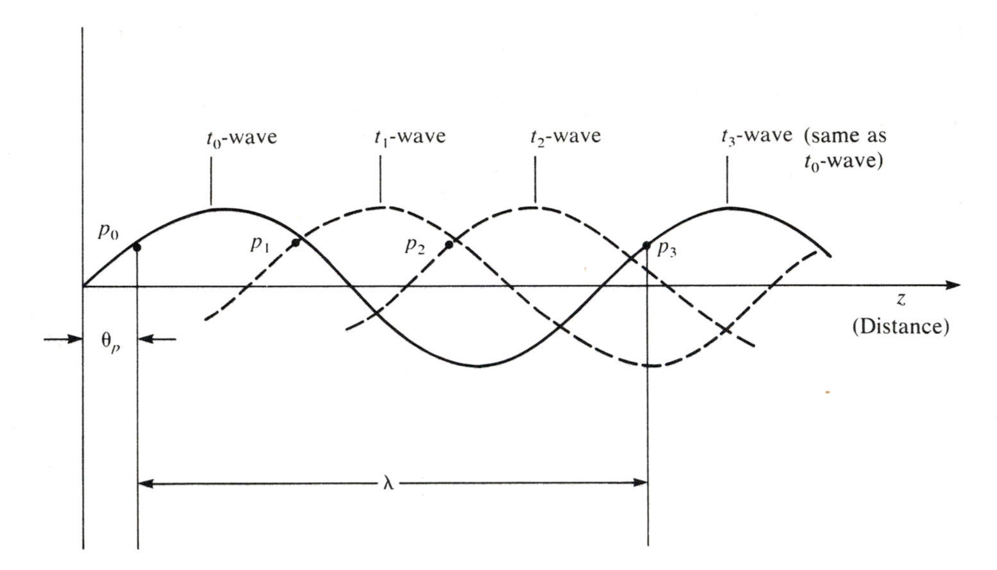

FIGURE 2–2 Electromagnetic wave propagation.

2–2–2 Discussion of Terms

Wavelength and Phase Velocity. Figure 2–2 shows a wave propagating along the z axis, starting at time t_0. The horizontal axis is distance not time. The waveforms t_1-wave, t_2-wave, t_3-wave in Figure 2–2 are sinusoidal waveforms in the z direction, that is, along the axis of propagation. At time t_0, the wave t_0-wave looks like a sine wave (with z as the variable not t) with zero-phase shift. At time t_1, the t_1-wave is a sine wave that has advanced along z. At times t_2 and t_3, the wave is further advanced along the z axis. As the wave moves, you can look at a point on the wave p and see it move along the z axis: p_0 at time t_0, p_1 and time t_1, and so forth. The t_3-wave is the same as the t_0-wave. The time t_0 to t_3 equals one period T $(T = 1/f)$ of the propagating sine wave. The distance the wave (or a point on the wave such as p) has traveled in time T is called the **wavelength** λ. The λ is the distance a constant phase point on the wave front travels in one period. Since the velocity of the wave in free space is c, the distance traveled in time T is $T \times c$, so that the relation among T, λ, and c is

$$\lambda = T \times c \qquad \text{or} \qquad \lambda = c/f \qquad (T = 1/f) \tag{2–3}$$

EXAMPLE 2–1

Find the wavelength of a light wave in free space with a frequency of

$$0.5 \times 10^{15} \text{ Hz}$$

Solution

$$\lambda = c/f = (300 \times 10^6)/(0.5 \times 10^{15}) = 6 \times 10^{-7}\ \text{m}$$
$$= 600\ \text{nm} = 0.6\ \mu\text{m}$$

The velocity of the constant phase point *p*, (phase velocity) is generally denoted by v_p. It is equal to *c* only in free space. In other materials, v_p is not equal to *c*. The v_p can be expressed as

$$v_p = \lambda \times f \quad \textbf{(2–4)}$$

Equation 2–4 is identical in meaning to Equation 2–3, with v_p replacing *c*. The definitions of λ and v_p assume a monochromatic wave; that is, a wave of a single frequency or color. As you have already noted, light sources emit light that is usually not monochromatic. It consists of a group of frequencies. Most of the discussion, in this chapter, relates to single-frequency waves.

Refractive Index. As was noted earlier, the propagation velocity in a vacuum is $c = 300 \times 10^6$ m/s. When a light wave propagates through material, not a vacuum, the velocity decreases and can be written as

$$v_p = c/n_1 \quad \textbf{(2–5)}$$

or

$$n_1 = c/v_p \quad \textbf{(2–5a)}$$

where n_1 is the **refractive index** (also called the **index of refraction**) of the material. Equation 2–5*a* defines the refractive index as the ratio of light speed in a vacuum to that in the material. (For the electronics student, $n = (e \times \mu)^{1/2}$, where *e* and μ are the permittivity and permeability, respectively.) The refractive index of a vacuum is 1 ($n = 1$) since for a vacuum, $v_p = c$. Since v_p can never exceed *c*, *n* for any material is greater than 1. Table 2–1 lists refractive indices for a variety of materials.

TABLE 2–1
Typical Refractive Indices and Velocities of Light

Material	Refractive Index (*n*)	Velocity of Light ($v_p \times 10^6$ m/s)
Vacuum	1.0	300
Air	1.0003	299.9
Ice	1.3	230.77
Water	1.33	225.56
Ethanol	1.36	220.59
Glass	1.46–1.96	205.48–151
Quartz	1.54	194.8
Diamond	2.42	123.97

It is important to note that the frequency of the wave does not change, which indicates that λ_1 in material with refractive index n_1 is related to λ in a vacuum by

$$\lambda_1 = \lambda/n_1 \tag{2–6}$$

EXAMPLE 2–2

Calculate v_p and λ_1 for a light of frequency

$$f = 0.5 \times 10^{15} \text{ Hz}$$

propagating in glass with refractive index $n_1 = 1.4$.

Solution

$$v_p = c/n_1 = (300 \times 10^6)/1.4 = 214.3 \times 10^6 \text{ m/s}$$
$$\lambda_1 = v_p/f = 214.3 \times 10^6/0.5 \times 10^{15} = 428.6 \text{ nm}$$

An alternate method for finding λ_1 is

$$\lambda = c/f = 300 \times 10^6/0.5 \times 10^{15} = 600 \text{ nm}$$
$$\lambda_1 = \lambda/n_1 = 600/1.4 = 428.6 \text{ nm}$$

The refractive index of any material varies with the wavelength. For example, silicate flint glass has an index of 1.66 at $\lambda = 400$ nm, whereas at $\lambda = 700$ nm, the index becomes 1.62. As you will see later, this relationship affects the performance of various fibers.

In Section 2–2–1, you noted that light energy consists of a group of frequencies, or wavelengths (see Section 3–3), rather than a single frequency. Consequently, the refractive index is related to the velocity of this group (the **group velocity**).

Coherence. Most light sources produce a light wave consisting of a range of frequencies. (The term used to describe this range of frequencies is line width). The incandescent lamp covers a wide frequency range, while the typical laser produces a relatively narrow frequency (or wavelength) range. To the extent that a source produces a single (or narrow range) frequency, the source is coherent. (This type of coherence is called **temporal coherence.** The different source signals may not have unrelated random phases.) Light waves from the coherent source can "interfere" with each other. In other words, they can be combined in space to produce sums and differences of electromagnetic waves, resulting in interference patterns of alternatively high- and low-intensity light spots (see Figure 2–3).

Coherence, or the degree of coherence, can be defined as the ability of a particular wave to produce interference patterns. (This ability requires both

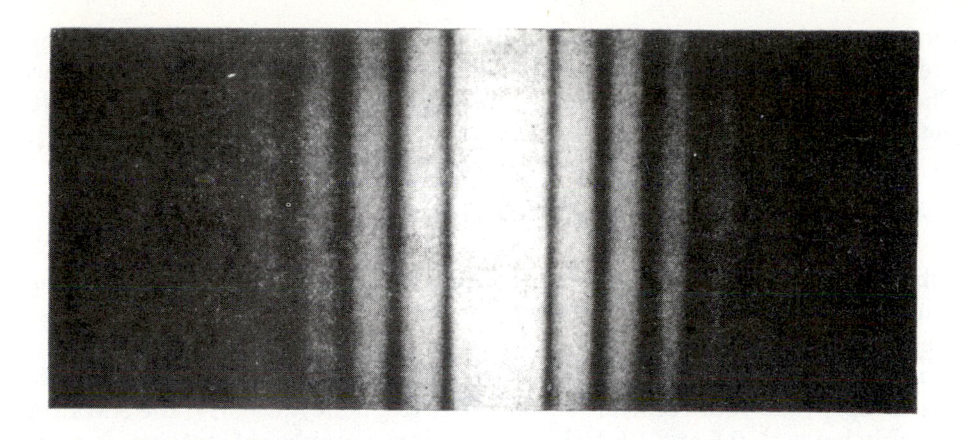

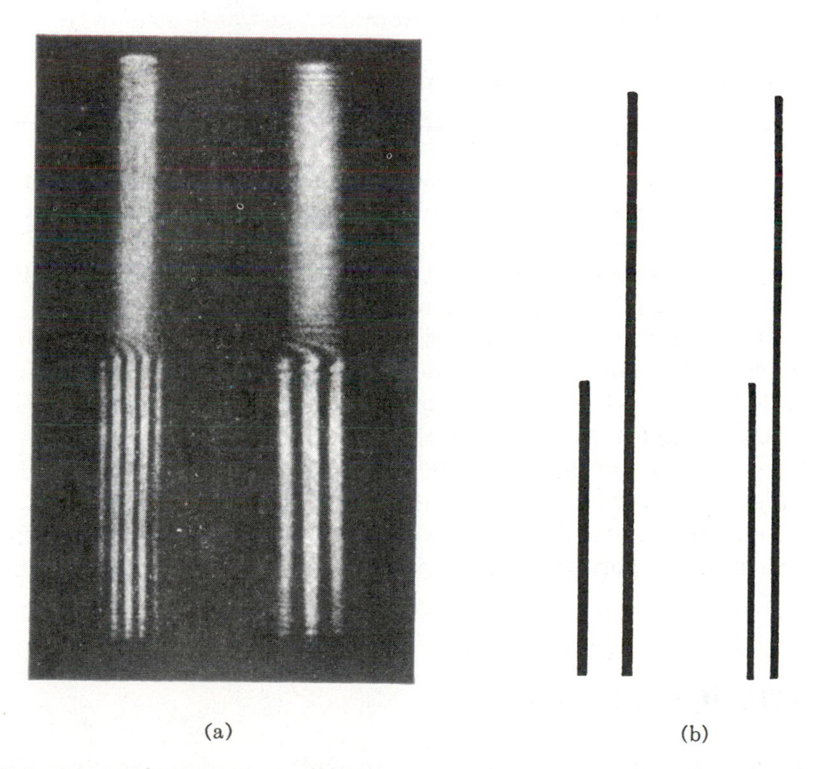

(a) (b)

FIGURE 2–3 Interference patterns.
Source: Sears and Zemansky, *University Physics,* Reading, Mass., Addison-Wesley Publishing Co., 1976. Reprinted with permission.

temporal coherence and **spatial coherence.** The latter refers to the fixed, predictable (nonrandom) phase of the light wave at a particular point in space.) The white-light incandescent lamp cannot produce interference patterns. It is not a coherent light.

2–2–3 Electromagnetic Spectrum

Fiber optic communication systems, as already noted, differ from other communication systems in the frequency of the carrier and in the transmission medium. In AM radio, a carrier frequency of 1 MHz, for example, is amplitude modulated (AM) and then transmitted. To receive the 1-MHz transmission, you simply tune your receiver to the 1-MHz frequency. Clearly, the AM range, roughly between 0.5 and 1.6 MHz, covers a large number of stations. A similar situation exists with optical transmission. You can transmit data on a particular wavelength. When discussing the frequency of light, it is more common to use wavelength than to use frequency. The receiver can then be tuned to that wavelength and receive the transmitted data. It is possible to use different wavelengths to transmit different data just as is done in regular radio transmission. This process is referred to as wavelength multiplexing and will be reexamined in more detail in Section 7–3.

The use of the electromagnetic **spectrum** (i.e., electromagnetic waves of varying frequency [or wavelength]) is controlled by the government through the Federal Communications Commission. However, the use is also a matter of using the right frequency for the right purpose. It is important to note that as the frequency increases (the wavelength decreases), the data rate that can be carried increases; therefore, the short light wavelengths can potentially carry high data rates, particularly in coherent systems.

Because the frequency of light is so high, the size of optical fibers (used in communication) is small, of the order of 10 to approximately 200 μm. All waveguides (conduits that carry electromagnetic waves) have dimensions that are directly related to the wavelengths that they carry. Therefore, since the optical wavelength is of the order of 0.8 to 1.5 μm, the optical waveguide in the fiber is also relatively small (still much larger than the wavelength).

The electromagnetic spectrum is used in a variety of ways. Table 2–2 lists typical uses of the portions of the spectrum, giving both wavelength λ and frequency f.

2–3 OPTICAL RAYS

The behavior of light is sometimes easier to explain by using ray tracings than by using the more detailed description that uses electromagnetic theory. In some cases, photon theory is used. Effects such as **reflection** and **refraction** are much easier to explain using rays. Similarly, the propagation of light in a fiber can be described in terms of rays. Although most light phenomena can be explained using the wave theory, these explanations are beyond the scope of this book. You can view the ray-tracing technique as a simplification of electromagnetic theory. Electromagnetic theory is used when discussing wavelength and frequency.

Another theory of the propagation of light, the photon theory, is used largely in explaining and analyzing light generation and detection. (See Chapters 9 and 10.)

The ray theory applies only when considering waves with a wavelength much smaller than the structure used to guide the wave. The fiber diameter, for example, is many times larger than the optical wavelengths.

2–3–1 Reflection

When a light ray is incident on a reflecting surface, the ray bounces back like a handball when it hits a wall. Typically, a reflecting surface is one that is highly polished, opaque, and coated with special reflective materials.

The **law of reflection** states that the angle of incidence is equal to the angle of reflection. In Figure 2–4, the incident ray is the line $\overline{AO}$, the reflected ray is $\overline{OB}$, and ϕ is perpendicular to the reflecting surface. The incident and reflected angles, θ_1 and θ_2, respectively, are those between the rays and the line perpendicular to the surface:

$$\theta_1 = \theta_2 \tag{2–7}$$

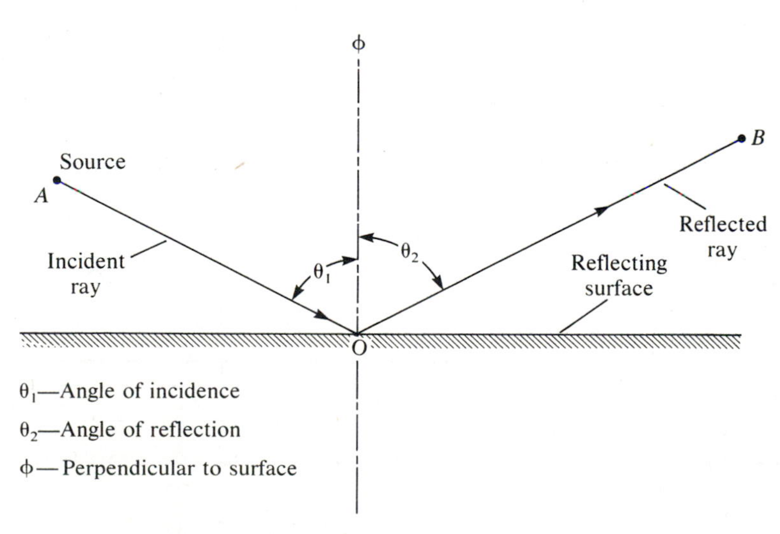

FIGURE 2–4 Incident and reflected rays.

TABLE 2–2
Typical Spectrum Applications

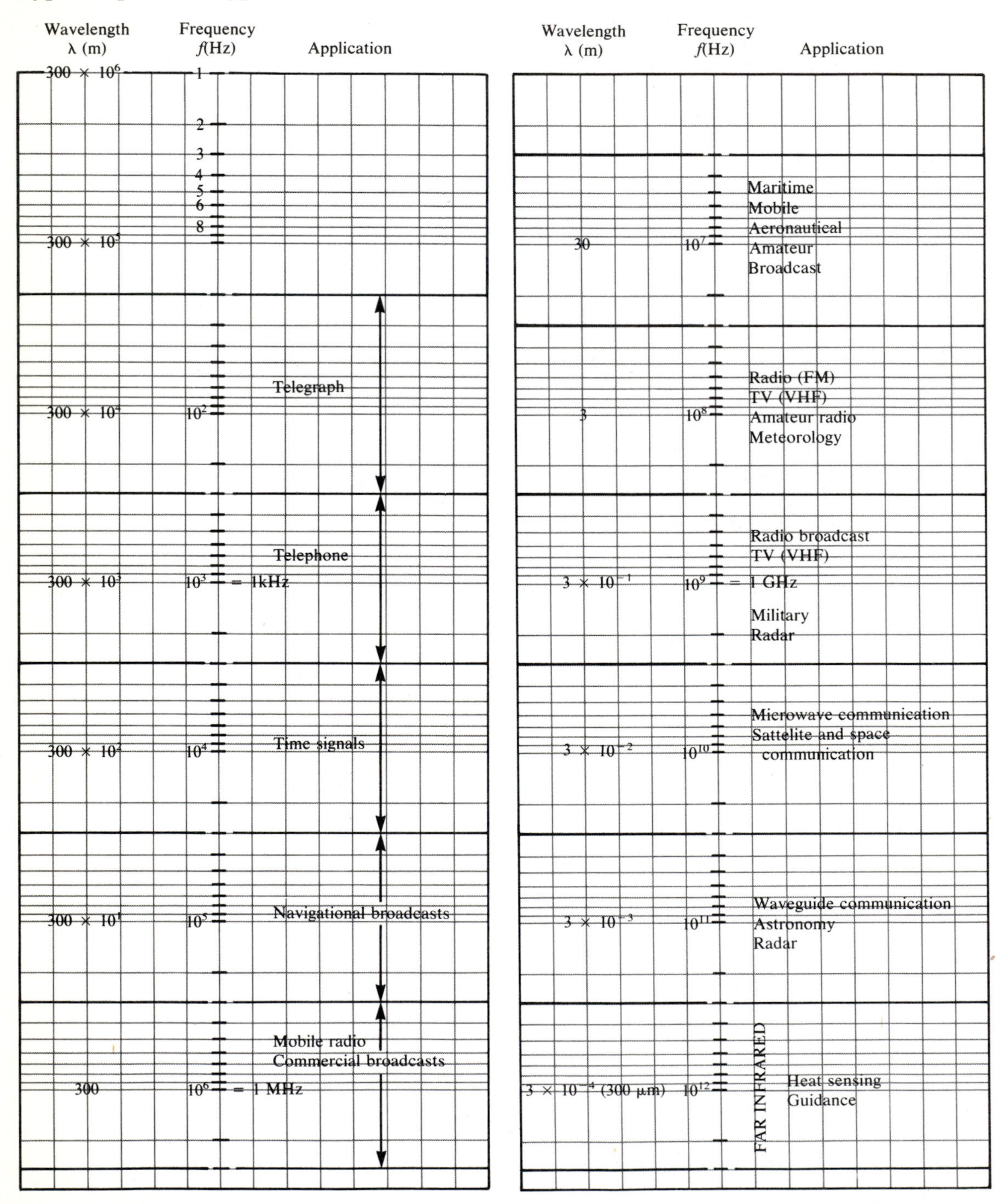

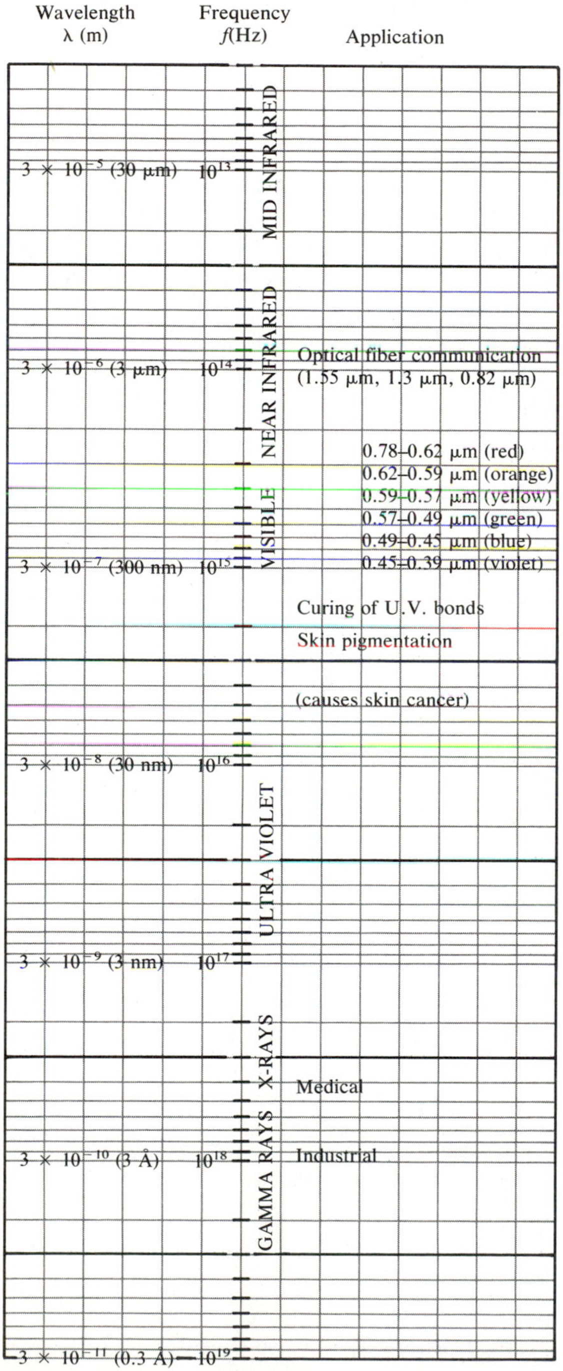

Wavelength λ (m)	Frequency f(Hz)		Application
		MID INFRARED	
3×10^{-5} (30 μm)	10^{13}		
		NEAR INFRARED	Optical fiber communication (1.55 μm, 1.3 μm, 0.82 μm)
3×10^{-6} (3 μm)	10^{14}		
			0.78–0.62 μm (red)
		VISIBLE	0.62–0.59 μm (orange)
			0.59–0.57 μm (yellow)
			0.57–0.49 μm (green)
			0.49–0.45 μm (blue)
3×10^{-7} (300 nm)	10^{15}		0.45–0.39 μm (violet)
			Curing of U.V. bonds
		ULTRA VIOLET	Skin pigmentation (causes skin cancer)
3×10^{-8} (30 nm)	10^{16}		
3×10^{-9} (3 nm)	10^{17}		
		X-RAYS	Medical
3×10^{-10} (3 Å)	10^{18}	GAMMA RAYS	Industrial
3×10^{-11} (0.3 Å)	10^{19}		

kHz = kilohertz = 10^3 Hz
MHz = megahertz = 10^6 Hz
GHz = gegahertz = 10^9 Hz
THz = terahertz = 10^{12} Hz

1 μm = 10^{-6} m
1 nm = 10^{-9} m
1 Å = 10^{-10} m

(The law of reflection and the **law of refraction** can be derived from Fermat's principle of least time; that is, the ray will travel a path requiring the least amount of time.) A direct result of this law is the fact that if θ_1 is 90°, θ_2 is 90° and the reflected ray is in line with the incident ray.

2–3–2 Refraction and Snell's Law

When a ray travels across a boundary between two materials with different refractive indices n_1 and n_2 (Figure 2–5), both refraction and reflection take place. Figure 2–5 illustrates the case where $n_1 > n_2$; that is, where the light travels from high- to low-refractive index materials. (Note that θ_1 equals θ_3 in accordance with the law of reflection.)

The refracted ray (the ray that continues in the same basic direction as the incident ray) is "broken"; that is, the angle θ_2 is not equal to θ_1. The relation between θ_1 and θ_2 is given by Snell's law of refraction. (Willebrord Snell formulated the law in 1621.)

$$n_1 \sin \theta_1 = n_2 \sin \theta_2 \tag{2–8}$$

or

$$\sin \theta_1 / \sin \theta_2 = n_2 / n_1 \tag{2–8a}$$

In accordance with this law, a ray traveling from a high- to a low-index material will move away from the perpendicular. The angle of incidence is smaller than the angle of the refracted ray. The reverse holds for rays traveling from low- to high-index material. The relation between the incident

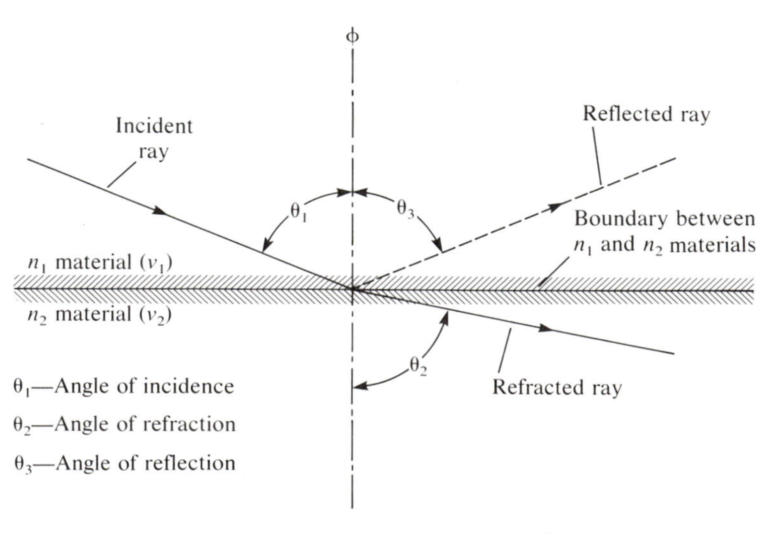

FIGURE 2–5 Incident and refracted rays (where $n_1 > n_2$ so that $\theta_2 > \theta_1$).

and refracted angles can be stated in terms of the propagation velocities in the media:

$$\sin \theta_1 / \sin \theta_2 = v_1 / v_2 \tag{2–9}$$

Note that

$$v_1 = c/n_1 \qquad \text{and} \qquad v_2 = c/n_2$$

where $c = 300 \times 10^6$ m/s (light velocity in free space). Remember that the two materials involved are transparent and allow light propagation.

EXAMPLE 2–3

An experiment is conducted as shown in Figure 2–6(a). The ray reaching the observer at point O forms an angle, $\theta_1 = 30°$, with the perpendicular to the water surface, Figure 2–6(b). Find the distance, x, between the pencil edge and its image.

FIGURE 2–6 Figure for Example 2–3. (a) Pencil and its image. (b) Calculating the position of the image.

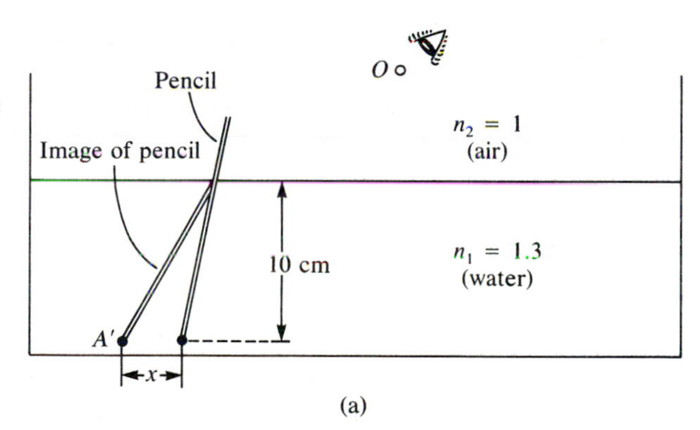

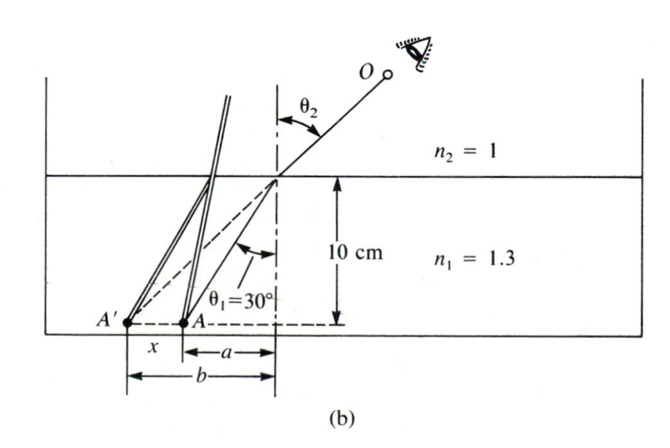

Solution

We find the angle θ_2 using Snell's Law.

$$\frac{\sin \theta_1}{\sin \theta_2} = \frac{n_1}{n_2} = \frac{1}{1.3}$$
$$\sin \theta_2 = 1.3 \times \sin \theta_1 = 1.3 \sin 30 = 0.65$$
$$\theta_2 = \sin^{-1} 0.65 = 40.5°$$

We find the distances a and b.

$$a = 10 \times \tan 30 = 5.77 \text{ cm}$$
$$b = 10 \times \tan 40.5 = 8.54 \text{ cm}$$
$$x = b - a = 2.77 \text{ cm}$$

Example 2–3 and Figure 2–6 show rays traveling from higher to lower refractive index materials, where the refracted beam moves away from the perpendicular. When the light travels from a lower to a higher refractive index, as in Figure 2–7, the refracted beam moves toward the perpendicular.

As was noted before, both reflection and refraction take place. The amount of light reflected depends on the angle of incidence and the refractive indices. For relatively low incident angles, (smaller than about 30°) the reflected light is less than 5% of the incident light. An estimate of the fraction of light energy that is reflected for a 0° incident angle is given by

$$p = [(n_1 - n_2)/(n_1 + n_2)]^2 \tag{2–10}$$

FIGURE 2–7 Refraction (where $n_1 < n_2$ so that $\theta_2 < \theta_1$).

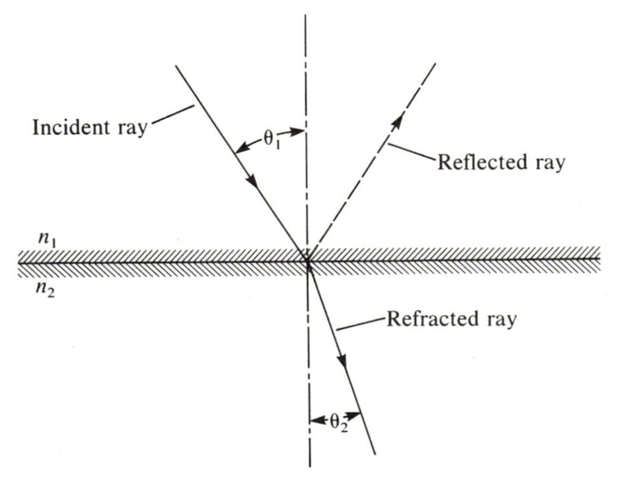

when the light is traveling between materials with refractive indices n_1 and n_2. This type of reflection is called **Fresnel reflection.** As you will see in Section 2–3–3, for larger angles, total internal reflection takes place; that is, all the incident power is reflected.

EXAMPLE 2–4

Find the percent of power lost when light travels between air ($n = 1$) and water ($n = 1.3$). Assume a 0° incident angle.

Solution
Use

$$\begin{aligned} p &= [(n_1 - n_2)/(n_1 + n_2)]^2 \\ &= [(1.3 - 1)/(1.3 + 1)]^2 \\ &= 0.017 \end{aligned}$$

In percent, power loss is

$$\%p = 0.017 \times 100 = 1.7\%$$

2–3–3 Total Internal Reflection

Refer to the situation described in Figure 2–5 and investigate a particular condition, namely, what happens when θ_2, the angle of refraction, equals 90° (Figure 2–8). It is clear that for $\theta_2 = 90°$, the refracted beam is not traveling through the n_2 material. Applying Snell's law of refraction to this case, you get

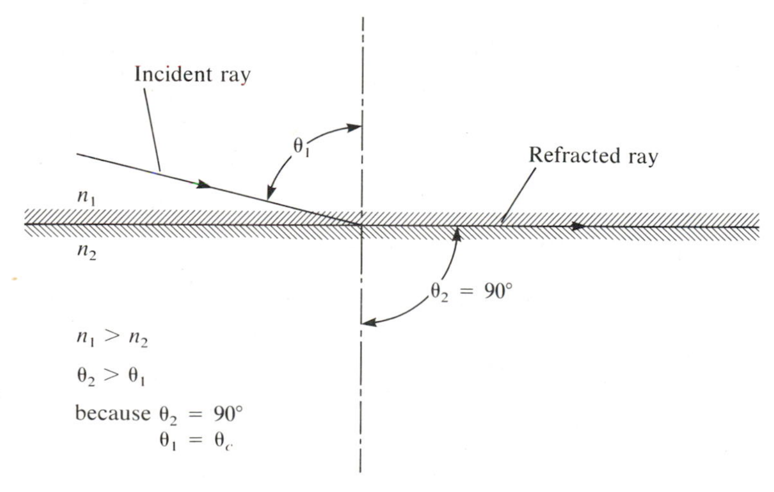

FIGURE 2–8 Refraction at the critical angle.

$$\sin \theta_1/\sin \theta_2 = n_2/n_1$$

Since $\theta_2 = 90°$, you have

$$\sin \theta_1 = n_2/n_1$$

The angle of incidence θ_1 for which $\theta_2 = 90°$ is called the critical angle θ_c:

$$\sin \theta_c = n_2/n_1 \tag{2–11}$$

and

$$\theta_c = \sin^{-1} (n_2/n_1) \tag{2–11a}$$

Since n_2 is less than n_1, the fraction n_2/n_1 is less than 1 and θ_c is less than 90°. You are dealing here with rays that travel between materials of different refractive indices. If a ray is incident on the boundary between n_1 and n_2 materials at the critical angle, the refracted ray will travel along the boundary, never entering the n_2 material.

There are no refracted rays for the case where $\theta_1 \geq \theta_c$. That is, the incident light is reflected when the incident angle is larger than θ_c. This condition is referred to as **total internal reflection,** which can occur only when light travels from high n material to lower n material.

EXAMPLE 2–5

Two layers of glass are placed on top of each other as in Figure 2–9. The light is traveling from $n = 1.45$ to $n = 1.40$. Find the range of angles θ_x for which total internal reflection takes place.

Solution
First find the critical angle θ_c:

$$\theta_c = \sin^{-1}(1.40/1.45) = 74.9°$$

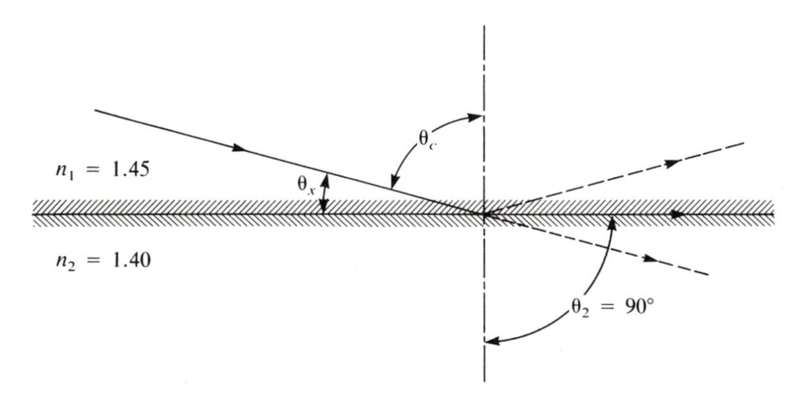

FIGURE 2–9 Figure for Example 2–5. Critical angle of refraction.

Thus, for the critical case $\theta_x = 90 - 74.9 = 15.1°$, and for all angles θ_x less than 15.1°, total internal reflection takes place.

EXAMPLE 2–6

Repeat the preceding problem replacing $n = 1.40$ with $n = 1$ (air). The ray is traveling from glass to air.

Solution

$$\theta_c \sin^{-1}(1/1.45) = 43.6°$$
$$\theta_x \leq 90 - 43.6 \leq 46.4°$$

θ_x ranges from 0° to 43.6°. Here, because there is a larger difference in refractive indices, in comparison with Example 2–4, total internal reflection takes place for larger angles θ_x.

Examples 2–5 and 2–6 show that the light can be restricted to the material with the higher index of refraction if the incident angle is kept above the critical angle. A sandwich of high-index material placed between two slabs of low-index material will allow a beam of light to propagate in the low-index material with relatively little loss. This idea is used in constructing fibers for fiber optic communication (see Chapter 3). Refraction is also used to break down a multicolor, multiwavelength light into separate wavelengths. This separation can be accomplished as shown in Figure 2–10.

The incoming light $\lambda_1 + \lambda_2$ is separated into λ_1 and λ_2 because of the difference in refractive index of the prism for the two wavelengths. Here, n for λ_1 is larger than n for λ_2, so λ_1 undergoes a more pronounced refraction at both prism surfaces.

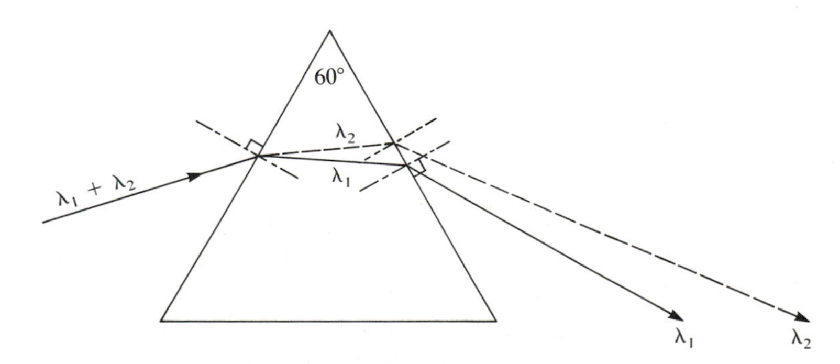

FIGURE 2–10 Refractions of λ_1 and λ_2 in a 60° prism.

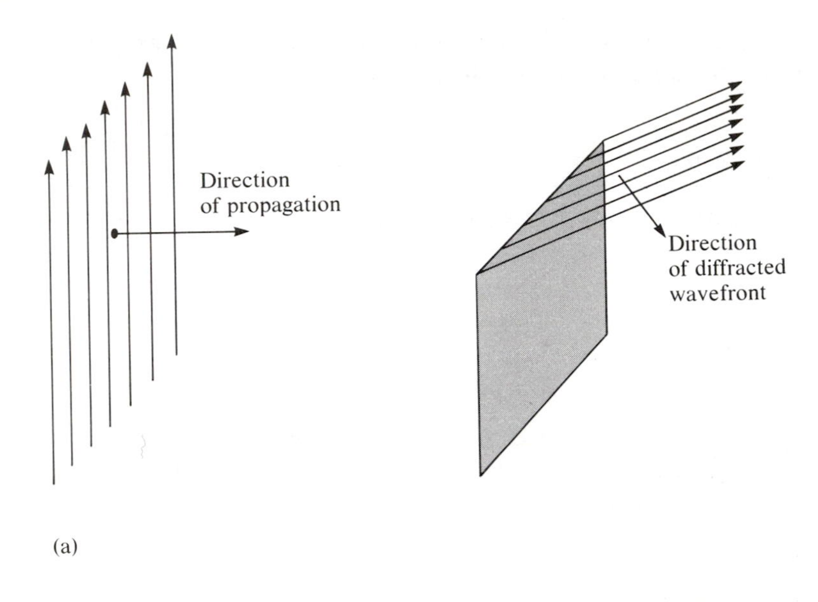

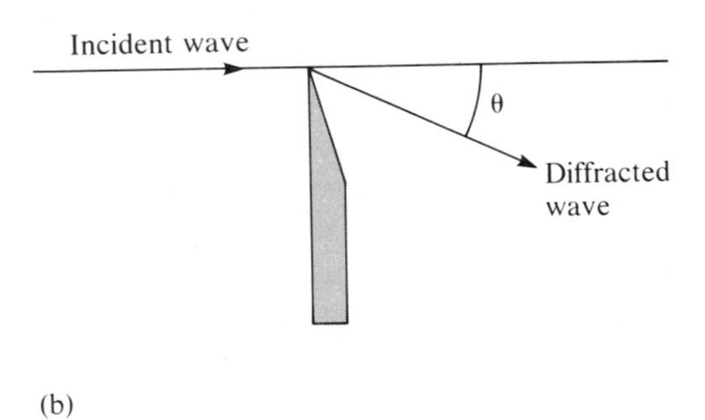

FIGURE 2–11 Diffraction. (a) Perspective view. (b) Propagation vectors.

2–3–4 Diffraction

Diffraction is another process that causes light rays to bend, (that is, to deviate from a straight line). It is often used in the same way that refraction is used. The diffraction phenomenon is used in a number of optical components such as lasers and optical filters. Diffraction occurs when a light wavefront is blocked by a sharp object. Figure 2–11 illustrates an obstructed wavefront and the diffracted wave. The angle of diffraction θ depends on the wavelength λ of the incident light. The result of this relationship between θ and λ is that the diffracted light is dispersed much like light refracted through a prism, giving rise to the rainbow effect (Figure 2–12).

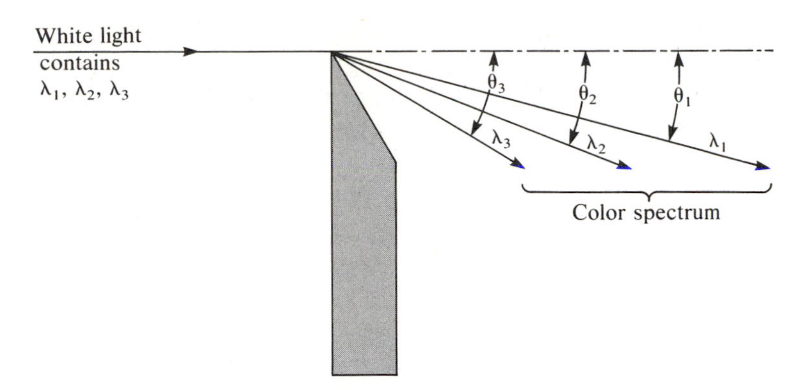

FIGURE 2–12 Diffraction by wavelength.

Remember that diffraction and refraction are two distinctly different phenomena.

Diffraction also takes place when a light passes through a narrow slit or a small hole (relative to the wavelength) with well-defined edges. This process is in contrast with the straight line ray propagation, which applies when the conduit is much larger than λ. The diffraction through slits is used in constructing **diffraction gratings** to obtain more efficient diffraction. There are two types of diffraction gratings, transmissive and reflective.

The transmissive diffraction grating is constructed by scribing thin lines in a dark film coating that covers a transparent material. This process yields a plate with alternating opaque and transparent slits. A simple version of a diffraction grating is shown in Figure 2–13. The distance between the slits S is constant over the whole plate. S is of the order of magnitude of λ. Light passes through the slits and is diffracted.

To visualize a reflective diffraction grating, pretend that the slits in Figure 2–13 are actually reflecting surfaces. The cumulative effect of the reflection from all the reflective strips results in a diffracted wavefront that travels in the general direction back to the light source.

FIGURE 2–13 Diffraction grating.

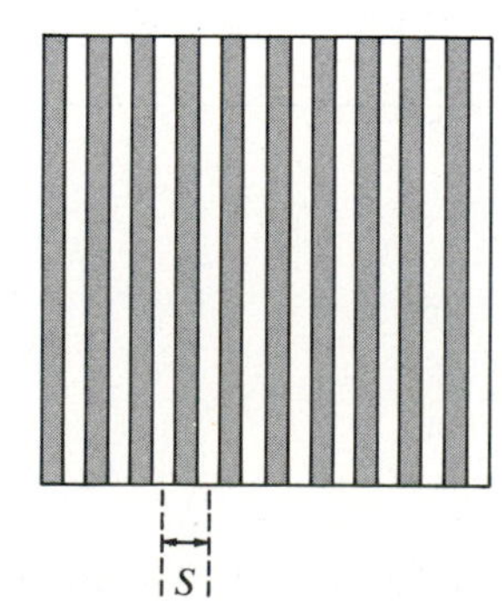

The law governing the transmissive diffraction grating is demonstrated in Figure 2–14. A monochromatic light (single wavelength λ) is incident on the grating at angle θ_i. Many diffracted waves are produced. The one proceeding in the direction of the incident light is called the zero-order wave. Other waves of different orders (such as first order or second order) are diffracted at distinctly different angles. Next, the laws governing only the first-order wave are covered.

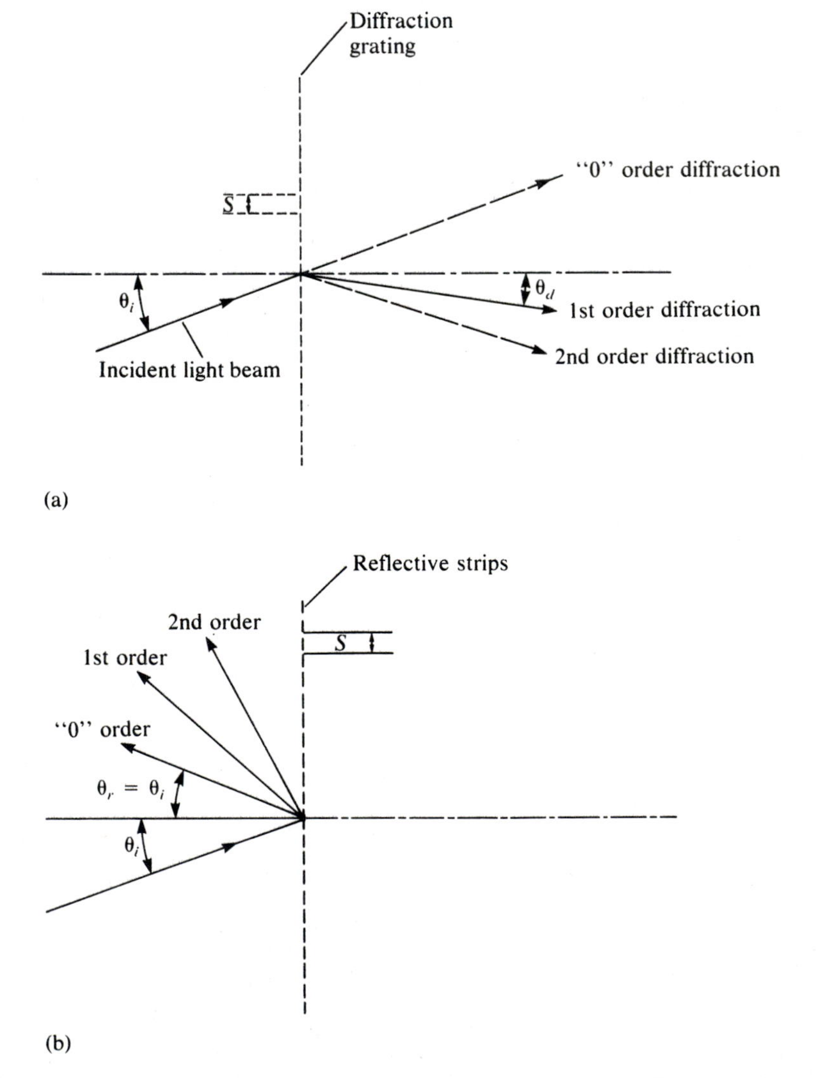

FIGURE 2–14 Light diffraction by a diffraction grating. (a) Transmissive diffraction grating. (b) Reflective diffraction grating.

The relationship among the wavelength λ, the grating spacing (or distance between slits) S, the angle of incidence θ_i, and the angle of diffraction θ_d is given in Equation 2–12:

$$m \times (\lambda/S) = \sin \theta_i + \sin \theta_d \qquad \textbf{(2–12)}$$

where m can be ± 1, ± 2, and so forth.

With $m = 1$, Equation 2–12 can be rewritten

$$\sin \theta_d = \lambda/S - \sin \theta_i \qquad \textbf{(2–12a)}$$

The angle of diffraction (first order) can be found from

$$\theta_d = \sin^{-1}(\lambda/S - \sin \theta_i) \qquad \textbf{(2–12b)}$$

EXAMPLE 2–7

In Figure 2–14, $\theta_i = 30°$ and $S = 1.3\ \mu m$.

1. Find the diffraction angle (first order) for $\lambda = 0.82\ \mu m$.
2. Find the diffraction angle for $\lambda = 1.5\ \mu m$.

Solution

1. $\theta_d = \sin^{-1}[(0.82/1.3) - \sin 30]$
   $= \sin^{-1} 0.131 = 7.5°$
2. $\sin^{-1}[(1.5/1.3 - \sin 30] = \sin^{-1} 0.654 = 40.8°$

The two rays are illustrated in Figure 2–15.

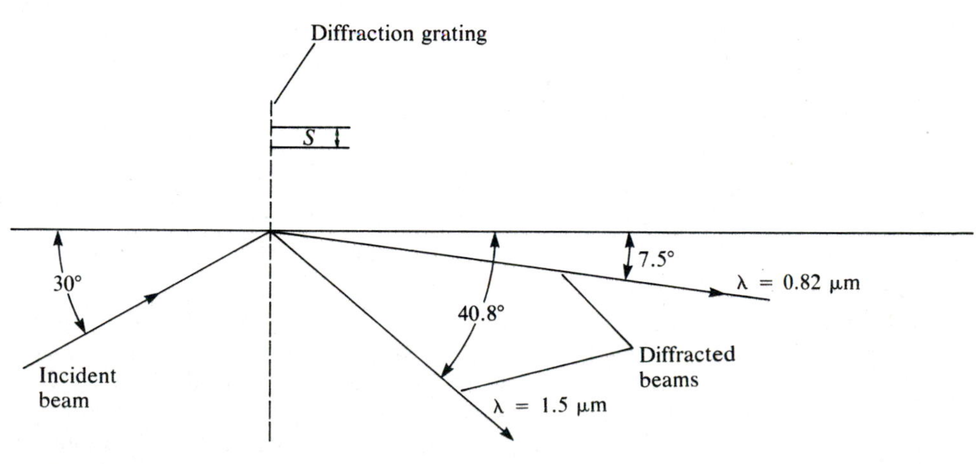

FIGURE 2–15 Solution for Example 2–7.

2–3–5 Other Optical Phenomena

Optical Dispersion. The refractive index varies with the wavelength λ. Consequently, the angle of refraction will vary with λ, leading to the rainbow display. The different colors, which have different wavelengths, have different angles of refraction as they pass through the clouds (the atmosphere) and are separated. This relation between λ and n, the refractive index, has a strong effect of the propagation of light pulses (digital information) in the optical fiber.

Absorption and Scattering. As the light travels through a medium, its energy decreases because of losses in the material. These losses are due to a number of effects, two being **absorption** and **scattering.** Both are dependent on the wavelength.

The first loss mechanism is absorption. There are two basic causes of absorption in glass. One is the interaction of the light waves with the molecular structure of the material, referred to as intrinsic absorption. This type of loss becomes significant for wavelengths below about 0.75 μm and above about 1.7 μm (for germanium-oxide-silicon-oxide glass).

The second cause of absorption is impurities in the glass. The impurities could be various metals (such as chromium (Cr), copper (Cu), or vanadium (V) or hydroxyl (OH ion.) This absorption is highly wavelength dependent. For example, the worst absorption caused by copper occurs at $\lambda = 0.850$ μm and by chromium at $\lambda = 0.625$ μm. In recent years, however, metallic impurities have been reduced to less than one part in 10^{10} ($10^{-8}\%$), practically eliminating this loss.

The absorption characteristics of the OH ion behave like a resonance circuit with sharp peaks at $\lambda = 1.38$ μm (highest absorption) and secondary peaks at $\lambda = 1.24$ μm and $\lambda = 0.95$ μm. The OH ions in the fibers have been reduced in recent years by sophisticated manufacturing processes. Nevertheless, the contribution of the OH ions to total fiber loss is still significant. This impurity results in low-loss windows (ranges of wavelengths between the high-loss OH peaks) for which the losses in the fiber are relatively low.

The second loss mechanism is scattering. Light energy that hits obstructions scatters in all directions, and most of it does not continue in the original propagation direction. This scattered light energy is lost. The obstructions are typically variations in the density of the material and the refractive index.

When the obstructions are small, less than λ, the scattering is called **Rayleigh scattering.** Losses from Rayleigh scattering are related to $1/\lambda^4$; that is, as λ increases, the loss decreases as the fourth power of λ. When the obstructions are larger than λ, the scattering effect is called **Mie scattering.** Because of improved manufacturing processes, this loss has been minimized to the point that it is relatively insignificant.

In later chapters, you will examine the choice of wavelength for optical fiber communications. One criterion for selection will be the fiber losses just discussed.

SUMMARY AND GLOSSARY

Chapter 2 discussed fundamental laws relating to the propagation of light, such as the relation of the speed of light to the index of refraction and the laws of reflection, refraction, and diffraction. You must understand and memorize these relations and laws. You must also understand the causes of light loss and their relation to the wavelength.

The terms listed below should serve as a review, and you should be familiar with all of them.

Absorption. A loss process in which impurities take up (absorb) some of the propagating light energy.

Diffraction. The process that causes an electromagnetic wave to bend as it passes by a sharp edge (for example, the edge of a razor blade) obstructing the wave.

Diffraction grating. A set of opaque, or reflective slits (about the size of λ), spaced a constant distance S. (This grating produces an efficient diffraction effect.)

Electromagnetic waves. A wave that propagates by the interchange of energy between electric and magnetic fields.

Fresnel reflection. The reflection from the boundary between materials with differing refractive index, where refraction rather than reflection is the dominant effect.

Group velocity. The velocity of a group of wavelengths. (Most light consists of a range of wavelengths rather than a single wavelength.)

Index of refraction. See "refractive index."

Law of reflection. The relation between the angles of an incident light ray and the reflected light ray. (They are equal.)

Law of refraction. The law relating the angle of incidence to the angle of refraction in a refracted light beam.

Mie scattering. Scattering loss caused by imperfections and obstructions larger than the wavelength.

Phase velocity. The velocity of a constant phase point on a propagating wave.

Photon. The packet of light energy, analogous to the basic atomic parts (electron, proton, etc.).

Photon theory. The theory that views light as being made up of packets of energy, largely used in explaining the generation and detection of light.

Planck's constant. The constant h that relates the energy of the photon to its wavelength ($h = 6.626 \times 10^{-34}$ J-s).

Polarization. The direction of the electric field of a wave determines its polarization.

Quantum theory of light. See "photon theory."

Rayleigh scattering. Scattering caused by small obstructions approximately the size of λ.

Ray tracing. A method of analyzing the propagation of light waves by considering only their straight line direction of propagation.

Reflection. A light beam incident on an opaque polished surface returned in the general direction of the incident light (reflection from a mirror).

Refraction. A light beam passing from one material to another that bends as it enters the second material.

Refractive index. The ratio of the speed of light in a vacuum to the speed of light in a material n, where n is greater than or equal to 1.)

Scattering. The process that causes light to be dispersed in all directions. (See "Mie scattering" and "Rayleigh scattering.")

Spatial coherence. An electromagnetic wave with a fixed and nonrandom phase at a point in space.

Spectrum. Range of frequencies of an electromagnetic wave. (The overall spectrum of electromagnetic waves runs from very low frequencies through radio frequencies, very high frequencies, light, X-rays, etc. [see Table 2–2].)

Speed of light. The speed of light in a vacuum: $c = 300 \times 10^6$ m/s.

Temporal coherence. An electromagnetic wave of a narrow range of frequencies. Ideally, a single-frequency (single-wavelength) wave.

Total internal reflection. When a ray incident on the interface between a high and low refractive index material is totally reflected rather than refracted. The angle of incidence is larger than the critical angle.

Transverse electromagnetic field. A wave structure with mutually perpendicular electric field, magnetic field, and direction of propagation.

Wavelength. The distance a constant phase point on the wave travels in one period T of the sinusoidal wave.

FORMULAS

$$E_p = h \times f \qquad h = 6.626 \times 10^{-34} \text{ J-s} \tag{2–1}$$

Energy of a photon in terms of its frequency.

$$c = 300 \times 10^6 \text{ m/s} \tag{2–2}$$

Speed of light in a vacuum.

$$\begin{aligned} \lambda &= T \times c \\ \lambda &= c/f \end{aligned} \tag{2–3}$$

Relation among wavelength, the sine wave period, and frequency.

$$v_p = \lambda \times f \tag{2–4}$$

Phase velocity related to frequency and wavelength.

$$v_p = c/n_1 \tag{2–5}$$

Speed of light in a material n_1.

$$n_1 = c/v_p \tag{2–5a}$$

Definition of n_1.

$$\lambda_1 = \lambda/n_1 \tag{2–6}$$

Change in λ *due to* n_1.

$$\theta_1 = \theta_2 \tag{2–7}$$

Equality of the angle of incidence to the angle of reflection.

$$n_1 \sin \theta_1 = n_2 \sin \theta_2 \tag{2–8}$$

Relation between the refractive indices and the angle of incidence and the angle of refraction (Snell's law).

$$\sin \theta_1/\sin \theta_2 = n_2/n_1 \tag{2–8a}$$

Variation of Equation 2–8 showing the ratio of the refractive indices.

$$\sin \theta_1/\sin \theta_2 = v_1/v_2 \tag{2–9}$$

Relation of speed of light to the angles of incidence and refraction.

$$p = [(n_1 - n_2)/(n_1 + n_2)]^2 \quad \textbf{(2–10)}$$

Fresnel reflection, power loss for normal incident light.

$$\sin \theta_c = n_2/n_1 \quad \textbf{(2–11)}$$

Critical angle.

$$\theta_c = \sin^{-1} (n_2/n_1) \quad \textbf{(2–11a)}$$

Critical angle.

$$m \times (\lambda/S) = \sin \theta_i + \sin \theta_d \quad \textbf{(2–12)}$$

Relation among the diffraction grating spacing S, the wavelength λ, the incident angle θ_i, and the angle of diffraction (first order) θ_d.

$$\sin \theta_d = \lambda/S - \sin \theta_i \quad \textbf{(2–12a)}$$

Variation of Equation 2–12 for first order with $m = 1$.

QUESTIONS

1. Light is an electromagnetic wave. List some other electromagnetic waves and their frequencies, wavelengths, and uses.
2. In simple terms, explain what a photon is.
3. What phenomenon is most frequently explained by photon theory?
4. What does the term *coherence* mean?
5. The three wavelengths $\lambda = 0.82\ \mu m$, $1.3\ \mu m$, and $1.5\ \mu m$ are the most popular in fiber optic communication. To what spectrum do they belong?
6. If the frequency of a wave has doubled (in free space), what do you expect to happen to the wavelength?
7. Is the velocity of light greater in the clouds (refractive index of 1.3) than in free space? Explain.
8. What is the difference between refraction and reflection?
9. What is the difference between refraction and diffraction?
10. Explain the difference between absorption loss and scattering loss.
11. One cause of absorption is impurities in glass. What impurity is responsible for most of the absorption loss in modern fiber? Is this loss wavelength sensitive? Explain.
12. What is Rayleigh scattering? For a wavelength change from $\lambda = 0.5\ \mu m$ to $\lambda = 1.2\ \mu m$, by what factor do you expect the Rayleigh scattering to change? (Does it increase or decrease?)

PROBLEMS

1. Calculate the wavelength λ in meters, in free space, of the following:
   a. Typical sound wave electrical signal, frequency of 3 kHz
   b. The WINS radio carrier, frequency of 1.010 MHz
   c. Channel 2 television, frequency of 57 MHz
   d. VHF television Channel 36, frequency of 605 MHz
   e. A far infrared signal, frequency of 1,000 GHz
   f. An infrared signal, frequency of 0.35×10^{15} Hz
   g. Yellow light, frequency of 0.5×10^{15} Hz
2. Find the velocity of light in the following materials:
   a. Fused silica, $n = 1.46$
   b. Polystyrene, $n = 1.6$
   c. Sapphire, $n = 1.8$
   d. Silicon, $n = 3.5$
3. An electron loses an amount of energy equal to 3.2×10^{-19} J (2 eV), which is converted into radiation. What is the frequency of the radiation?
4. An infrared beam, $\lambda = 0.82\ \mu$m, in free space, passes through the following materials. Find its wavelength inside each.
   a. Ice
   b. Water
   c. Quartz
   d. Diamond
5. A light beam is launched into a fiber made of quartz. The length of the fiber is 500 m. How long will it take the light to travel through the fiber? Assume that the light travels in a perfectly straight line along the fiber center.
6. How long will it take a light beam to reach the bottom of a lake 500 m deep?
7. A beam is traveling to a satellite that is 30 km high. Assume that the first 5 km consists of heavily polluted air with $n = 1.01$ and that the rest is a vacuum. Find the time required to make a round trip. (The beam is reflected by the satellite.)
8. A beam incident on point O is reflected and passes through point A (Figure 2–16). Find the angle of incidence.

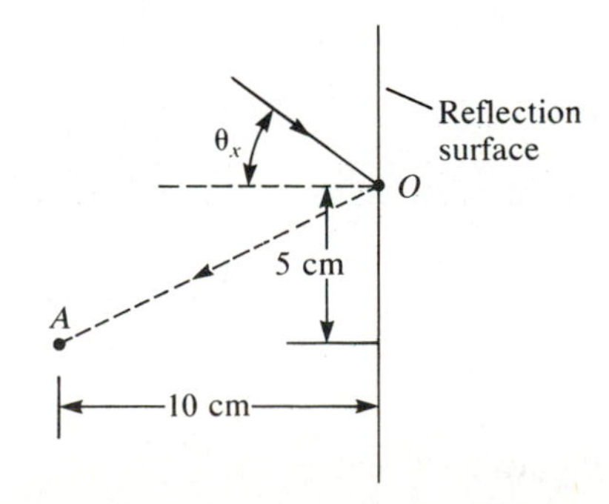

FIGURE 2–16 Figure for Problem 8.

9. Find the angle θ so that the reflected beam would travel perpendicular to the incident beam (Figure 2–17).

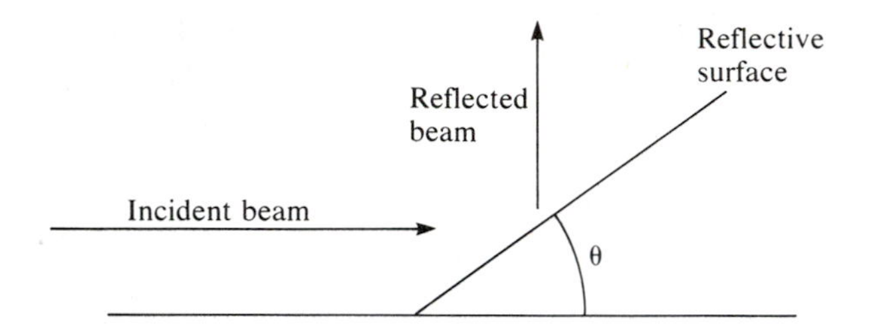

FIGURE 2–17 Figure for Problem 9.

10. Find the angle θ_x in the diagram shown (Figure 2–18).

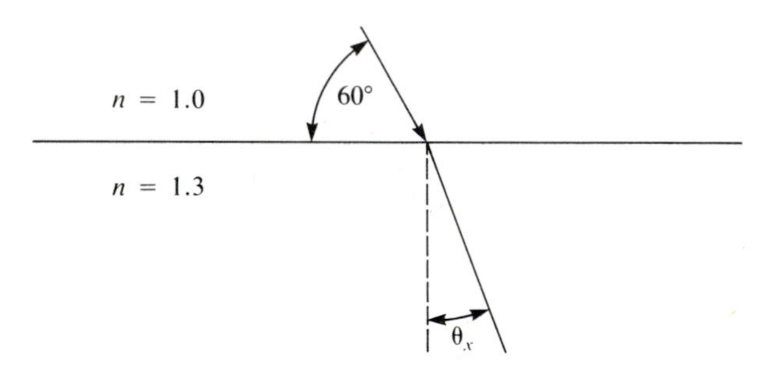

FIGURE 2–18 Figure for Problem 10.

11. A beam travels from water ($n = 1.3$) to air ($n = 1.0$) at angle $\theta = 20°$. What was the angle of incidence?
12. A beam is passing through a slab of glass, as shown in Figure 2–19, and is shifted to the left as it exits the slab. Find the distance x. (Hint: $x = a - b$)

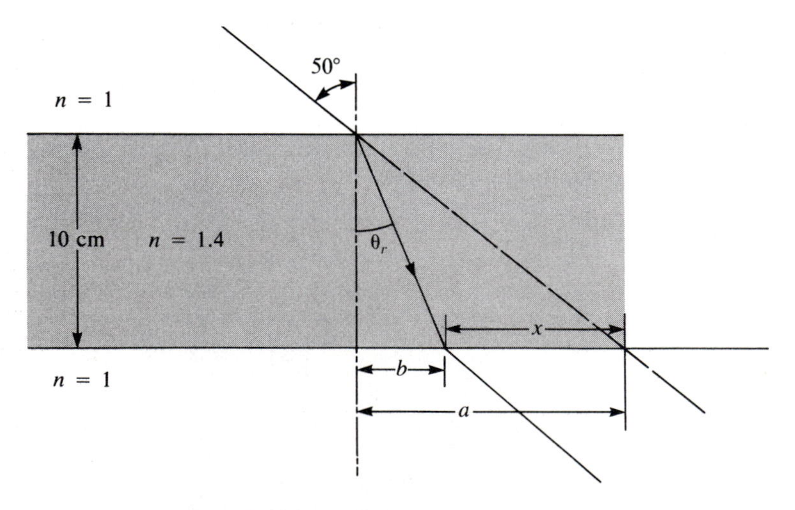

FIGURE 2–19 Figure for Problem 12.

13. Two beams are incident on a glass rod, as shown in Figure 2–20.
  a. Find the time it takes beam A to travel through the rod.
  b. Repeat step a for beam B.
  c. Find the difference between the two intervals.

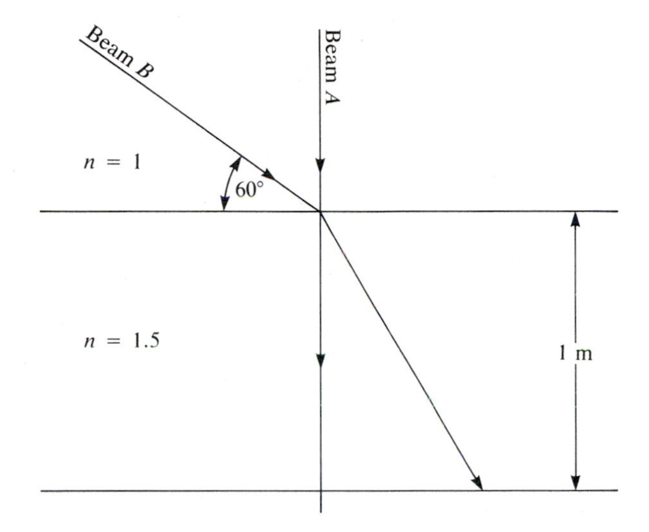

FIGURE 2–20 Figure for Problem 13.

14. A beam that contains two wavelengths, $\lambda_1 = 0.82\ \mu m$ and $\lambda_2 = 1.3\ \mu m$, enters a glass bar at 30° and is reflected by a mirror at the bottom of the bar (Figure 2–21). Since the refractive index of the glass is different for λ_1 and λ_2, the two wavelengths follow different paths, as shown in the figure. If for λ_1, $n = 1.4$, and for λ_2, $n = 1.46$, find the distances a_1 and a_2 and the interval $\Delta a = a_1 - a_2$. The Δa is the dispersion caused by the glass bar as it separates the light source into its wavelengths.

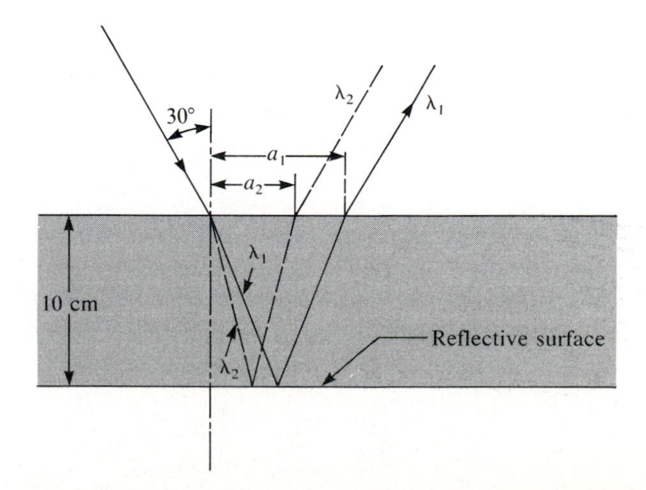

FIGURE 2–21 Figure for Problem 14.

15. A transmissive diffraction grating with a distance between slits of $S = 0.8$ μm is used as shown in Figure 2–22. A ray of light with $\lambda = 0.5$ μm is incident at an angle $\theta = 20°$. Find the angle θ_x of the first-order diffracted wave.

FIGURE 2–22 Figure for Problem 15.

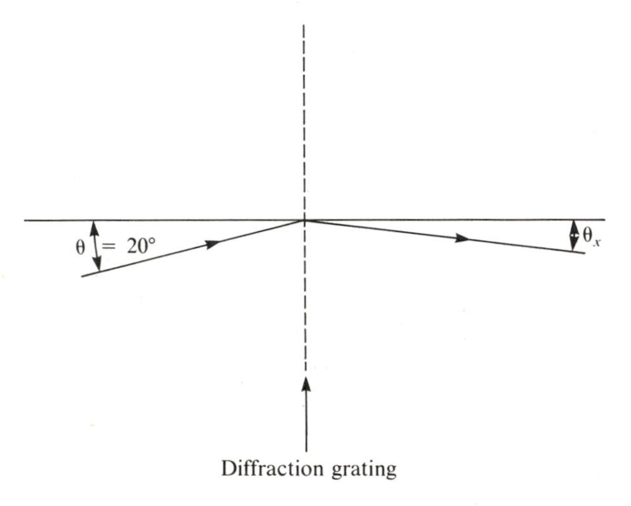

16. A diffraction grating is often used to separate wavelengths. If, in Figure 2–23, the light source consists of three separate wavelengths—$\lambda_1 = 0.6$ μm, $\lambda_2 = 0.7$ μm, and $\lambda_3 = 0.8$ μm—find the angle of diffraction (first order) of the three waves.

FIGURE 2–23 Figure for Problem 16.

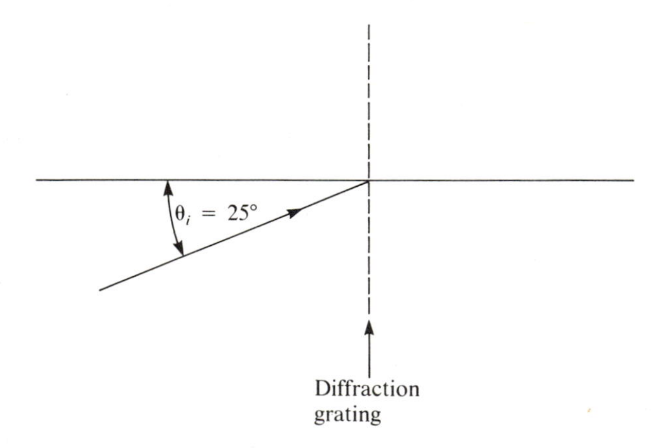

17. Find the percentage of Fresnel reflection when a ray passes between materials with $n = 1.0$ to $n = 1.4$. Assume 0° incident angle.

3
Principles of Fiber Optics

CHAPTER OBJECTIVES

This chapter discusses light propagation in a step-index fiber. Total internal reflection, which is the basic underlying principle, and some details of light propagation in fiber are presented, including the modes of propagation, the angle of propagation, and the acceptance angle. An important term, numerical aperture, which is connected to the relative refractive index difference between core and cladding, is discussed. You will be able to calculate these parameters numerically and be introduced to the relations among the numerical aperture, the line width, and the data rate of the fiber.

3–1 INTRODUCTION

Figure 3–1 illustrates a **step-index fiber.** In this fiber the refractive index changes in step fashion, from the center of the fiber, the core, to the outer shell, the cladding. It is high in the core and lower in the cladding. The light in the fiber propagates by bouncing back and forth from the core-cladding interface.

To simplify the discussion of propagation, you will use ray-tracing techniques. That is, you will follow a sample ray through the fiber. Mostly, you will assume that the sample ray passes through the center of the fiber. Such rays are called **meridional rays.** Section 3–2–3 briefly covers nonmeridional, or **skew rays.**

The ray propagating in the fiber must be launched into the fiber at one end. The conditions necessary to inject such rays efficiently depend on the fiber structure, as well as on the characteristics of the light source. Note that in communication applications the power introduced into the fiber is typically 10–100 μW with a light-emitting diode (LED) source and approximately 1 mW with a laser source.

3–2 LIGHT PROPAGATION

3–2–1 Total Internal Reflection

A typical step-index fiber is shown in Figure 3–1. Two rays are shown in Figure 3–1. One (the solid line) is injected at a lower angle than is the other (the dashed line). Follow the dashed ray first (the dashed line).

At interface *A*, between air and the core, refraction takes place, and the ray continues at a smaller angle, closer to the center line; that is, $\theta_{L2} > \theta_1$. The ray then gets to the core-cladding interface at point *B*. Again, refraction takes place and the ray bends and continues in the cladding. Finally, the ray bends again, as it exits the fiber at the cladding-air interface, at point *C*. However, this time the ray leaves the fiber. This ray is not confined and does not propagate through the fiber.

Now, follow the second ray (the solid line). Again, refraction takes place at point *A*. At point *B′*, the core-cladding interface, total internal reflection occurs. This ray is confined to the fiber core. For convenience, assume that the angle of incidence at the core-cladding interface is the critical angle and call it α_c (a specific case of θ_c in Equation 2–11 for a fiber, where the index for the cladding is n_2 and for the core is n_1). From Equation 2–11*a*, $\alpha_c = \sin^{-1}(n_2/n_1)$. An incident ray with an angle larger than α_c will propagate in the fiber.

The critical ray (the solid line) in Figure 3–1 makes an angle θ_c with the fiber center. Rays with propagation angles larger than θ_c will not propagate. Note that $\theta_1 > \theta_c$, and that θ_1 ray exits the fiber and is not confined to the fiber.

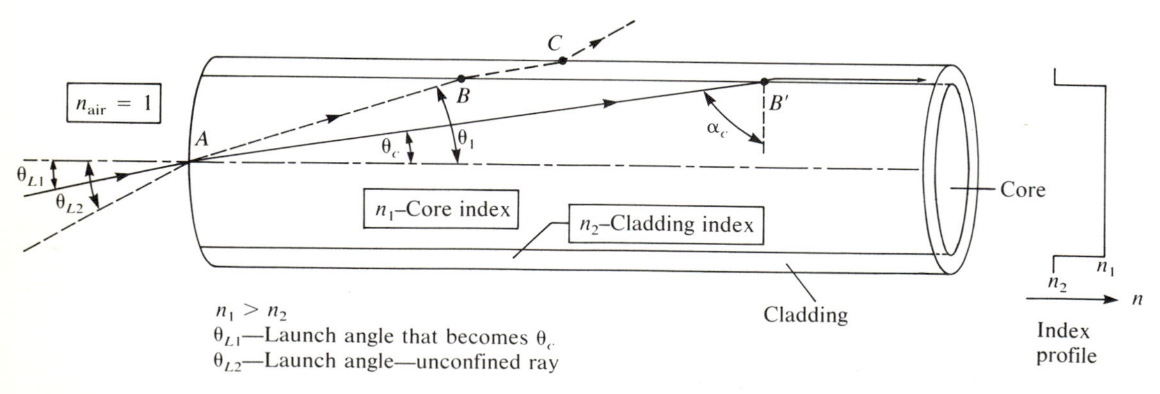

FIGURE 3–1 Light propagation in a step-index fiber.

The angle θ_c is called the **critical propagation angle** (θ_c is not the same here as θ_c in Chapter 2). From the geometry,

$$\sin \alpha_c = \cos \theta_c = n_2/n_1 \quad \textbf{(3–1)}$$

It is important to note that **total internal reflection** can occur only when light travels from high index to low index media.

EXAMPLE 3–1

Find θ_c in a fiber with $n_{\text{core}} = 1.43$ and $n_{\text{clad}} = 1.40$.

Solution

$$\cos \theta_c = n_2/n_1 = 1.40/1.43$$
$$\theta_c = 11.7^\circ$$

The ray propagating at the critical angle, $\theta_c = 11.7^\circ$ in Example 3–1, is incident at the core-cladding interface at $\alpha_c = 90 - 11.7 = 78.3^\circ$. The values given in Example 3–1 are typical for step-index fibers used in communications (single-mode fibers have shallower critical angles). θ_c of about 12° is very shallow and requires special care to make sure the light enters and is confined to the fiber. The angles θ_{L1} and θ_{L2} in Figure 3–1 are not equal to θ_1 and θ_c (see Section 3–2–4). θ_{L1} and θ_{L2} are the angles of incidence of the light entering the fiber end, and θ_1 and θ_c are the corresponding angles of light propagation inside the fiber.

3–2–2 Mode Propagation

All rays with angles less than θ_c will propagate in the fiber. On the basis of electromagnetic theory, these rays propagate at distinct angles. If the critical ray is propagating at $\theta_2 = 11.7^\circ$, as in the Example 3–1, other rays will propagate at distinct angles below 11.7°. Figure 3–2 shows three distinct rays, propagating at θ_1, θ_2, and θ_3. These rays are referred to as modes of propagation. The total number of modes propagating in the fiber increases as θ_c increases. The θ_c depends on n_2/n_1, as does the number of modes. It turns out

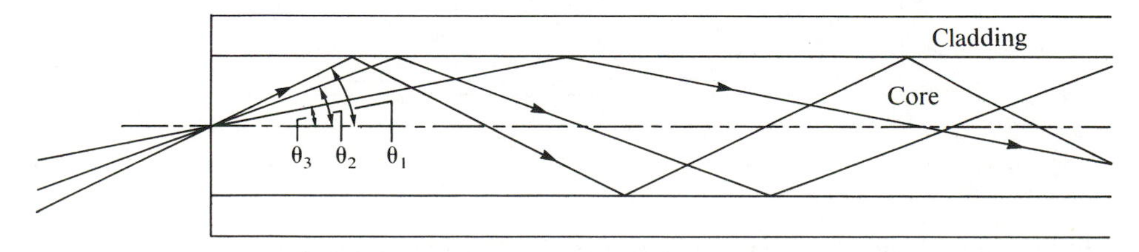

FIGURE 3–2 Three propagation modes: θ_1, θ_2, and θ_3.

that the total number of modes increases as the **relative refractive index difference** $(n_2 - n_1)/n_1$ increases.

It is common to distinguish between high-order modes, those with propagation angles close to the critical angle θ_c, and low-order modes, those with propagation angles much lower than the critical angle. The high-order modes tend to send light energy into the cladding. This energy is ultimately lost, particularly at fiber bends.

Mode Conversion (Mode Coupling). Whether the light energy propagates mostly in high-order modes, low-order modes, or a particular mix of modes depends on launch conditions (the angle of incidence of the rays entering the fiber end) and on the extent that **mode coupling** (the transfer of light energy from one mode to another) takes place. If the light source to fiber connection causes a large part of the light energy to be coupled to the fiber at relatively large angles, high-order modes will be set up (Figure 3–3(a)). This tends to cause losses, particularly at fiber bends. Figure 3–3(b) shows a light source that couples the light at shallow angles, and thus low-order modes are set up and energy loss is reduced. In Figure 3–3(c), the light source is misaligned and tends to set up higher and leaky modes. It is most efficient to avoid the situations shown in Figure 3–3(a) and 3–3(c). The **mode distribution** (the relative amount of energy carried by each mode) initially set up in the fiber is substantially altered by mode coupling (or mode conversion).

The mode distribution after about 1 or 2 km of fiber reaches what is called a **steady-state mode distribution.** This means that the distribution of light energy among the modes is relatively constant from there on. Each mode is

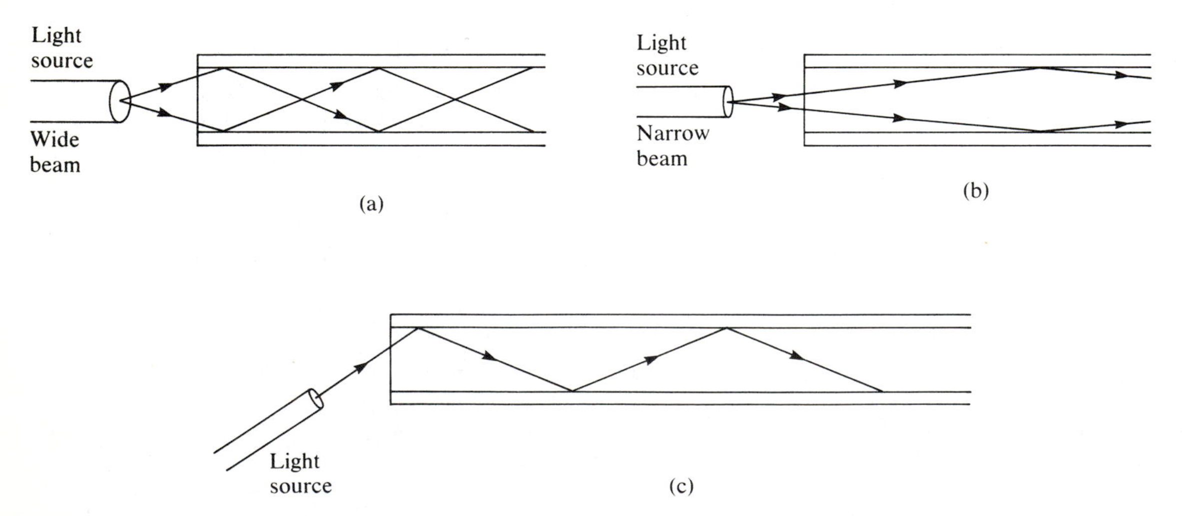

FIGURE 3–3 Production of high- and low-order modes. (a) High-order modes. (b) Low-order modes. (c) Off axis launch: high-order modes.

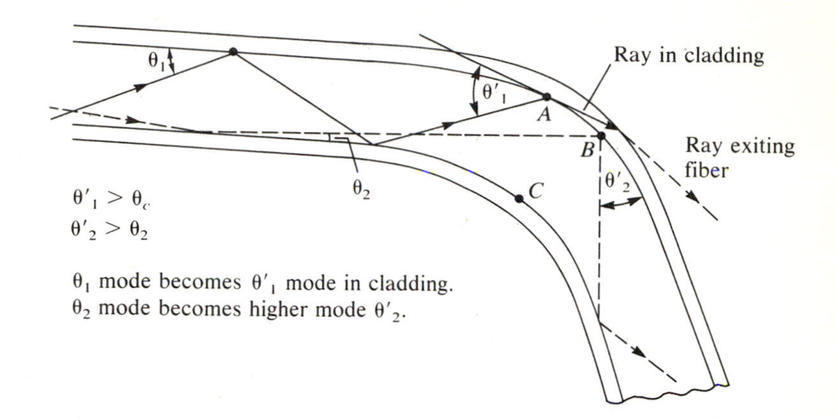

FIGURE 3–4 Effects of macrobend.

carrying its fair share of light. (Although mode coupling continues throughout the fiber, mode distribution remains relatively unchanged.)

Mode coupling (the conversion of one mode to another) is caused by fiber bends, large and small, macrobends and microbends, respectively. Figure 3–4 shows what happens to two rays as they pass through a macrobend in the fiber. When the ray with angle θ_1 reaches point A, its propagation angle becomes larger than θ_1. In Figure 3–4, this angle is assumed to be larger than θ_c (the critical angle), and the ray exits the fiber. The θ_1 mode has been converted to a leaky mode (very high order mode) and thus lost. The mode propagating at angle θ_2 is converted to a higher-order mode θ_2' due to the bend. Here, $\theta_2' > \theta_2$. Note that mode conversion to both higher-order and lower-order modes usually takes place. (To prove this point, see what happens to a ray incident at point C. The incident and reflected angles must be the same.) Figure 3–5 shows mode conversion caused by a small indentation in the fiber, a microbend. Conversions to high-order and low-order modes are shown.

It is sometimes desirable to set up the steady-state mode distribution over a short fiber length by deliberately introducing minute bends in the fiber. The fiber is pressed between two blocks covered with fine sandpaper (a **mode mixing block**). The sandpaper introduces indentations in the fiber and causes

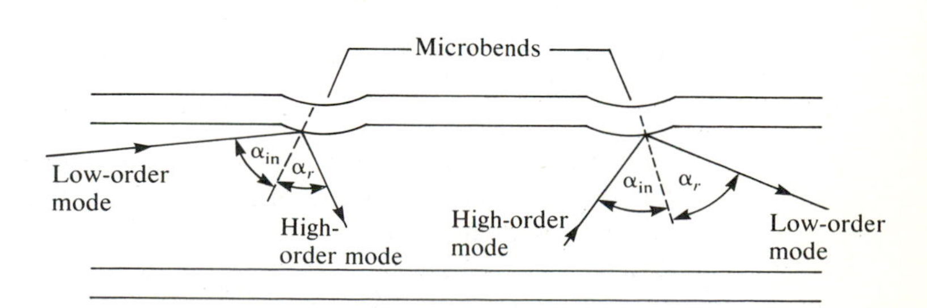

FIGURE 3–5 Effects of microbends.

increased mode mixing. This method also causes increased loss because some of the modes become leaky.

A common question regarding light propagation in the fiber is whether a ray ever travels directly along the fiber, parallel to the fiber axis. The answer is that such a mode ($\theta = 0°$) would very quickly be converted to higher-order modes because of fiber bends.

3–2–3 Skew Waves

Thus far, all rays have been assumed to be meridional, passing through the fiber center. In reality, a large number of rays travel through the fiber without going through the fiber's center line. Skew waves (also called skew rays) represent a significant part of the total light transmission. Fortunately, the analysis of meridional rays gives a close approximation of what actually takes place so that it is not necessary to include the complex analysis of skew waves. Skew waves are a result of the way the light is injected into the fiber, and it is nearly impossible and also unnecessary to avoid them.

3–2–4 Acceptance Angle and Numerical Aperture

The propagation angle must be equal to or less than the critical angle. This means that the light entering the fiber must be shallow enough to maintain this condition. Figure 3–6 traces two entering rays that become the critical rays in the fiber. If you follow the solid line ray, there is refraction at point A so that θ_a does not equal θ_c. The refractive indices involved are those of air, $n = 1$, and the core n_1. Only rays that enter the fiber edge within the angle $2\theta_a$ will be accepted by the fiber. The angle $2\theta_a$ is the **acceptance angle.** In three dimensions, it is an **acceptance cone,** limited by the angle $2\theta_a$.

It is useful to relate the angle θ_a to the refractive indices of the fiber. By Snell's law, at point A (Figure 3–6),

$$\sin \theta_a / \sin \theta_c = n_1 / n_{\text{air}} = n_1$$

and

$$\sin \theta_a = n_1 \times \sin \theta_c \tag{3–2}$$

The term $\sin \theta_a$ is called the **numerical aperture** (N.A.) and

$$\text{N.A.} = \sin \theta_a = n_1 \sin \theta_c \tag{3–3}$$

To obtain N.A. in terms of the refractive indices n_1 and n_2, where n_1 is the core index (n_{core}) and n_2 is the cladding index (n_{clad}), use Equations 3–1 and 3–3 and the trigonometric identity

$$\cos^2 \theta = 1 - \sin^2 \theta$$

you get

$$\text{N.A.} = (n_1{}^2 - n_2{}^2)^{1/2} \tag{3–4}$$

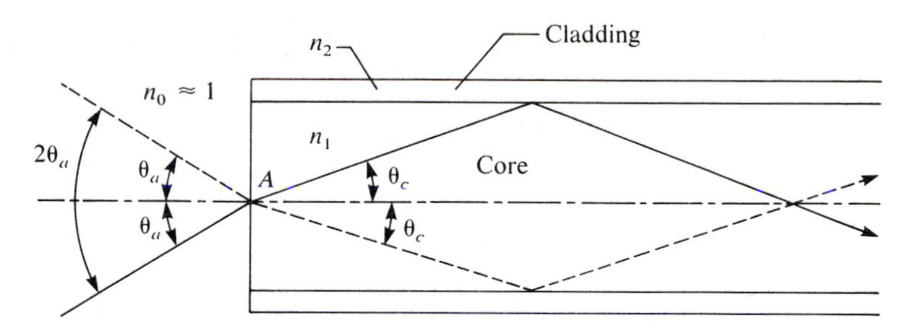

FIGURE 3–6 Acceptance angle.

The "half acceptance" angle θ_a is given by

$$\begin{aligned} \theta_a &= \sin^{-1}(\text{N.A.}) \\ &= \sin^{-1}(n_1{}^2 - n_2{}^2)^{1/2} \end{aligned} \tag{3–5}$$

You can express the N.A. in terms of the **relative refractive index difference** Δ, which is defined as

$$\begin{aligned} \Delta &= (n_1{}^2 - n_2{}^2)/(2 \times n_1{}^2) \\ &= (\text{N.A.})^2/(2 \times n_1{}^2) \end{aligned} \tag{3–6}$$

From Equations 3–4 and 3–6,

$$\begin{aligned} (\text{N.A.})^2 &= n_1{}^2 - n_2{}^2 = 2 \times n_1{}^2 \times \Delta \\ \text{N.A.} &= n_1 \times (2 \times \Delta)^{1/2} \end{aligned} \tag{3–7}$$

EXAMPLE 3–2

A fiber has the following characteristics: $n_1 = 1.35$ (core index) and $\Delta = 2\%$ (or, as a decimal ratio, 0.02). Find the N.A. and the acceptance angle.

Solution

$$\begin{aligned} \text{N.A.} &= n_1 \times (2 \times \Delta)^{1/2} = 1.35 \times (2 \times 0.02)^{1/2} \\ &= 0.27 \end{aligned}$$

$$\theta_a = \sin^{-1}\text{N.A.} = \sin^{-1} 0.27 = 15.66°$$

$$\text{Acceptance angle} = 2 \times \theta_a = 31.33°$$

EXAMPLE 3–3

A step-index fiber has $n_{\text{core}} = 1.44$ and $n_{\text{clad}} = 1.40$. Find (1) the N.A., (2) Δ, and (3) the acceptance angle.

Solution

1. From Equation 3–4,

$$\text{N.A.} = [(1.44)^2 - (1.40)^2)]^{1/2} = 0.337$$

2. $\Delta = (1.44^2 - 1.40^2)/(2 \times 1.44^2)$ (approximate method)
$= 0.027$
$= 2.7\%$

3. $\theta_a = \sin^{-1} 0.337$ (Equation 3–5)
$= 19.7°$

Acceptance angle $= 2 \times \theta_a = 39.4°$

It is often convenient to simplify the expression for Δ. An approximation for Δ is obtained by rewriting

$$\Delta = (n_1^2 - n_2^2)/(2 \times n_1^2)$$
$$= [(n_1 + n_2^1) \times (n_1 - n_2)]/(2 \times n_1^2).$$

When n_1 is approximately equal to n_2,

$$\Delta = [2 \times n_1 \times (n_1 - n_2)]/(2 \times n_1^2)$$
$$= (n_1 - n_2)/n_1 \qquad \textbf{(3–8)}$$

Remember that the N.A. represents the acceptance angle. A large N.A. represents a large acceptance angle and vice versa. A large N.A. also implies large Δ, a large difference in refractive index. As you will see in Section 5–1, a large N.A. produces a large number of modes and presents some serious performance problems. Typically, Δ is of the order of 0.01–0.03 (1–3%).

3–3 LINE WIDTH

The actual sources of light used in fiber optics produce a light that has a band of frequencies. Typically, they are not **monochromatic.** That is, they are not single-frequency sources. This band corresponds to a range of wavelengths. The **line width** of light energy and of a light source is the width in wavelengths between the two points where the light energy drops off to one half its maximum power. In Figure 3–7, the power is maximum at $\lambda = 820$ nm and drops to half its maximum at $\lambda = 810$ nm and $\lambda = 830$ nm. The line width is, therefore, $830 - 810 = 20$ nm. We will use the notation $\Delta\lambda$ to denote line width. $\Delta\lambda$ corresponds to a **bandwidth** Δf. Δf can be expressed for narrow line widths as

$$\Delta f = (\Delta\lambda/\lambda_0) \times f_0 \qquad \textbf{(3–9)}$$

where $\Delta f = f_2 - f_1$, f_1 and f_2 are the half-power frequencies of the light source, and Δf is the bandwidth of the light source.

λ_0 = center wavelength
f_0 = center frequency

That is, the bandwidth is the product of the **relative line width** $\Delta\lambda/\lambda_0$ and the center frequency. (This can be derived directly, by realizing that $\Delta f/f = \Delta\lambda/\lambda$. The relative bandwidth and relative line width are equal.)

FIGURE 3–7 Line width.

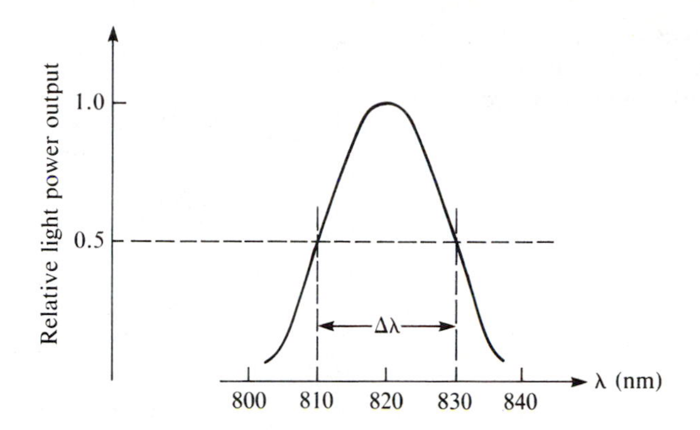

The line width of the source has serious effects on the overall performance of the fiber optic transmission system. A large line width yields a larger number of modes for the same N.A. and lowers the maximum data rate of the system.

EXAMPLE 3–4

A typical LED emits light at $\lambda_0 = 0.82\ \mu m$ with $\Delta\lambda = 40$ nm. Find (1) the relative line width in percent and (2) Δf.

Solution

1. $\Delta\lambda/\lambda_0 = (40 \times 10^{-9})/(820 \times 10^{-9})$
$= 0.0488$

In percent,

$$0.0488 \times 100 = 4.88\%$$

2. $f_0 = c/\lambda_0$
$= 300 \times 10^6/(0.82 \times 10^{-6})$
$= 0.3658 \times 10^{15}$

f_0 is the center frequency, corresponding to $\lambda_0 = 0.82\ \mu m$.

$$\Delta f = 40/820 \times 0.3685 \times 10^{15}$$
$$= 17.84 \times 10^{12} \text{ Hz}$$

Note that in this example, an enormous bandwidth Δf is involved.

3–4 PROPAGATION VELOCITIES

The refractive index n of most materials varies with the wavelength involved. This makes the speed of light v depend on wavelength. (Remember the relation $v = c/n$, in a vacuum, where $n = 1$, $v = c$.) For silicon glass, in the range of wavelengths used in fiber optics, the variations in n and v with wavelength

FIGURE 3–8 *n* and *v* versus λ (for silicon glass). (Arrows indicate vertical scale to be used.)

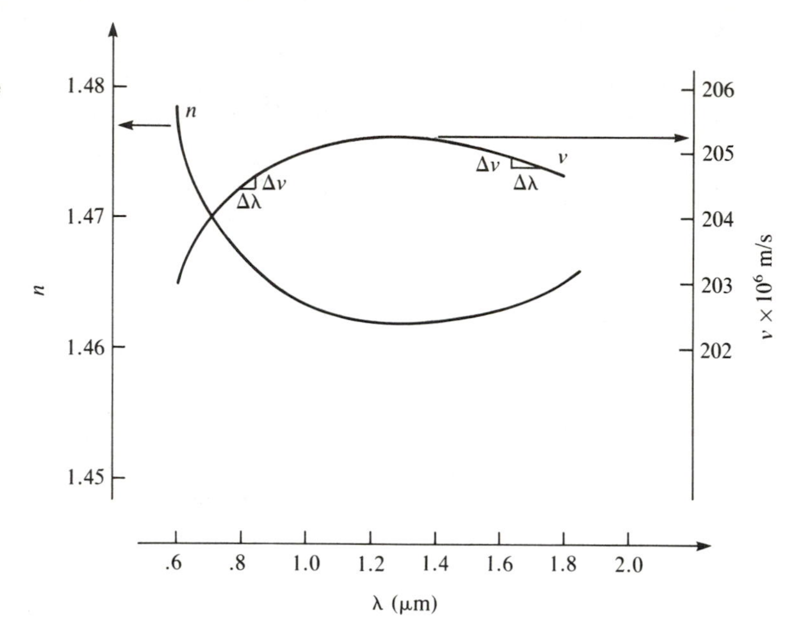

change direction. For wavelengths from about 0.6 to 1.3 μm, *n* decreases with increasing wavelength, indicating a negative slope ($\Delta n/\Delta\lambda$ is negative). For the same wavelengths, *v* increases with wavelength, indicating a positive slope ($\Delta v/\Delta\lambda$ is positive). For 1.3–1.7 μm, *n* increases with increasing wavelength while *v* decreases, a reversal in the slopes of *n* and *v*.

Figure 3–8 shows this relationship among *n, v,* and increasing wavelength λ. At about 1.3 μm, the slope of both *n* and *v* is approximately 0. This means that there are no variations in *n* or *v* as the wavelength changes (for small Δλ). Because changes in *v* with wavelength greatly reduce the data rate the fiber can carry, it is advantageous to operate at about 1.3 μm. That is, it is desirable to use sources that operate at close to 1.3 μm. If the line width of the source is very narrow, say, 1 or 2 nm, and you operate near 1.3 μm, you can expect $\Delta v/\Delta\lambda$ to be near 0 and the bandwidth (or data rate that the fiber can carry) to be extremely high. These systems can operate at frequencies in excess of 2 GHz. (See the discussion of dispersion in Section 4–3.)

SUMMARY AND GLOSSARY

The terms defined here reflect what you have learned in this chapter. These terms are used throughout this text and in industry. You should recognize them and understand what they mean. Use this glossary to review the material you have just read.

Acceptance angle. The range of angles within which an injected light beam will enter the fiber. (This angle is related to the numerical aperture.)

Acceptance cone. The acceptance angle in three dimensions. (The cone is formed by rotating the acceptance angle, with the fiber center line as the axis.)

Bandwidth. Frequency range corresponding to the line width $\Delta\lambda$. (See "line width" and Equation 3–9.)

Critical propagation angle. Rays with propagation angles larger than the critical angle are not confined to the fiber. (They leave the fiber.)

Line width. The range of wavelengths between the two points of half-power emission of a light source. (See "relative line width.")

Meridional ray. A ray that passes through the fiber center line.

Mode conversion. The transfer of light energy from one mode to another.

Mode coupling. See "mode conversion."

Mode distribution. The amount of energy carried by each mode in an optical fiber, usually given in relative terms.

Mode mixing block. A device designed to cause mode conversion in the fiber.

Mode propagation. The propagation of light energy in an optical fiber takes place at distinct angles of propagation called modes of propagation, or simply modes.

Monochromatic. Of single color or single frequency. (A light source which has a very narrow line width is monochromatic.)

Numerical aperture. The sine of one-half the acceptance angle.

Propagation angle. The angle a beam inside a fiber makes with the fiber axis. (See "critical propagation angle.")

Relative refractive index difference. Approximately the ratio of the refractive index difference over the core index. (See Equations 3–6 and 3–8.)

Relative line width. The ratio of the line width to the center wavelength of the source, $\Delta\lambda/\lambda$.

Skew ray. A ray that propagates in the fiber without crossing the fiber center line. (Skew ray is the same as skew wave.)

Steady-state mode distribution. After a certain length of fiber, the power carried by each mode does not change any more. The mode distribution is said to be in the steady-state distribution.

Step-index fiber. A fiber made of a core and cladding with two refractive indices, n_{core} and n_{clad}.

Total internal reflection. Rays traveling at shallow angles (below the critical propagation angle) from a high-index material to a low-index material undergo total internal reflection and do not cross into the low-index material. This behavior is the same as that of a reflected ray.

FORMULAS

In the following formulas, n_1, core index and n_2, cladding index.

$$\sin \alpha_c = \cos \theta_c = n_2/n_1 \qquad (3\text{–}1)$$

The relation between the refractive indices and the critical angles.

$$\sin \theta_a = n_1 \times \sin \theta_c \qquad (3\text{–}2)$$

Definition of half acceptance angle θ_a.

$$\text{N.A.} = \sin \theta_a = n_1 \sin \theta_c \qquad (3\text{–}3)$$

Definition of N.A.

$$\text{N.A.} = (n_1^2 - n_2^2)^{1/2} \qquad (3\text{–}4)$$

Relation between N.A. and the refractive indices.

$$\begin{aligned} \theta_a &= \sin^{-1}(\text{N.A.}) \\ &= \sin^{-1}(n_1^2 - n_2^2)^{1/2} \end{aligned} \qquad (3\text{–}5)$$

Half acceptance angle.

$$\begin{aligned} \Delta &= (n_1^2 - n_2^2)/(2 \times n_1^2) \\ &= (\text{N.A.})^2/(2 \times n_1^2) \end{aligned} \qquad (3\text{–}6)$$

Definition of the relative refractive difference Δ.

$$\text{N.A.} = n_1 \times (2 \times \Delta)^{1/2} \qquad (3\text{–}7)$$

Relation between Δ *and N.A.*

$$\Delta = (n_1 - n_2)/n_1 \qquad (3\text{–}8)$$

Approximation for Δ.

$$\Delta f = (\Delta\lambda/\lambda_0) \times f_0 \qquad (3\text{–}9)$$

Relation between the frequency bandwidth Δf *and the line width* $\Delta\lambda$.

QUESTIONS

1. What is the difference between meridional and the skew waves in a fiber?
2. Light beams are injected into the fiber at angles larger than the acceptance angle. Will they propagate in the fiber? Explain.
3. Would light be confined in an optical fiber if the total internal reflection phenomenon did not exist? Explain.
4. What are propagation modes?
5. What is mode coupling? Give causes for mode mixing.
6. The simple flashlight sends beams at a relatively wide angle. Would it be easier to couple it to a fiber with a low acceptance angle or a high acceptance angle? Explain.
7. How are the following related to the fiber data rate?
   a. Propagation modes
   b. Acceptance angle
   c. Line width
8. What is the steady-state mode distribution? Does mode coupling stop after steady-state mode distribution is reached?
9. Are low-order modes or high-order modes more likely to become leaky modes and to leave the fiber?
10. How is the relative refractive index difference related to propagation modes?
11. Will light propagate in a glass tubing (air inside with a glass shell)? Explain.

PROBLEMS

1. A step-index fiber has $n_{core} = 1.41$ and $n_{clad} = 1.37$. Find the range of propagation angles (all angles below the critical angle).
2. For the fiber in Problem 1, find
   a. The N.A.
   b. The acceptance angle
3. In a step-index fiber, the relative refractive index difference is 2% and $n_{clad} = 1.40$. Find
   a. The n_{core}
   b. The critical propagation angle
   c. The N.A.
   d. The acceptance angle

4. Find Δ for the step-index fibers listed here, using the exact and approximate formulas. Compare the resulting values.

	$n_{core}(n_1)$	$n_{clad}(n_2)$
a.	1.42	1.415
b.	1.38	1.36
c.	1.68	1.34

5. An optical fiber is being designed. It must have an acceptance angle of 75°. Its cladding index is 1.38. Find its core index.
6. A fiber is made of a core with an index of 1.40 and no cladding (air cladding). Find
   a. The N.A.
   b. The acceptance angle
7. For the fiber in Problem 6, what is the highest-order mode (critical propagation angle)?
8. A light source has the response in Figure 3–9. Find its line width.

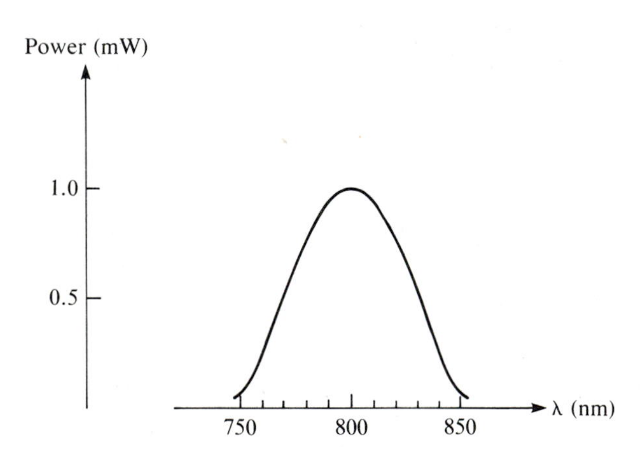

FIGURE 3–9 Figure for Problem 8.

9. For the source of Problem 8, find Δf.
10. A source has a 2% line width at a center wavelength of 1.3 μm. Find Δf.
11. Repeat Problem 10 for a source with a 10% line width.
12. A source with a center wavelength of 0.8 μm and a line width of 80 nm ($\pm$40 nm) is used with a fiber made of silicon glass (Figure 3–8). Find
    a. The Δn involved
    b. The Δv involved

4
Fiber Characteristics

CHAPTER OBJECTIVES

After studying this chapter, you will be able to identify the basic causes of energy loss in fiber and fiber dispersion and the effect that the loss and dispersion have on fiber performance. The discussion on the choice of wavelength is designed to allow you to analyze the relations among loss, dispersion, and wavelength and to highlight the reasons some wavelengths are preferred over others.

4–1 INTRODUCTION

The characteristics of optical fibers depend both on the specific material composition and the physical shape and size. Things such as glass composition, diameter of the fiber, and the way the index of refraction varies within the fiber directly affect the fiber performance. Imperfections and concentricity affect losses in the fiber, as do small bends (microbends) in the fiber. In general, when discussing fiber characteristics, this text will concentrate on fiber losses and the data rate that the fiber can handle.

4–2 FIBER LOSSES

There are three fundamental causes of light loss in fiber.

1. **Material loss.** Absorption by material. This includes absorption due to the light interacting with the molecular structure of the material, as well as loss because of material impurities.
2. **Light scattering.** The light scattered by the molecules of the material by structural imperfections and impurities. The scattered light does not propagate down the fiber; it is lost.
3. **Waveguide and bend losses.** Losses caused by imperfections and deformations of the fiber structure. The term waveguide, or **optical waveguide,** is often used in the place of optical fiber.

All these losses are wavelength dependent. By carefully choosing the operating wavelength, the losses can be minimized. Losses introduced by connecting fibers to each other or to other devices will be discussed in Sections 8–6 and 11–5.

4–2–1 Definitions

Let us first briefly review the basic definitions of power losses. In Figure 4–1, P_{in} is the power into the fiber and P_{out} is the power out of the fiber. Fiber loss is defined as

$$\text{Loss} = P_{out}/P_{in} \tag{4–1}$$

Using logarithmic units, decibels, loss is defined as

$$\text{Loss}|_{dB} = 10 \times \log(P_{out}/P_{in}) \tag{4–2}$$

Because you should expect the loss to increase with fiber length, the text usually refers to decibels per kilometer (loss per kilometer of fiber).

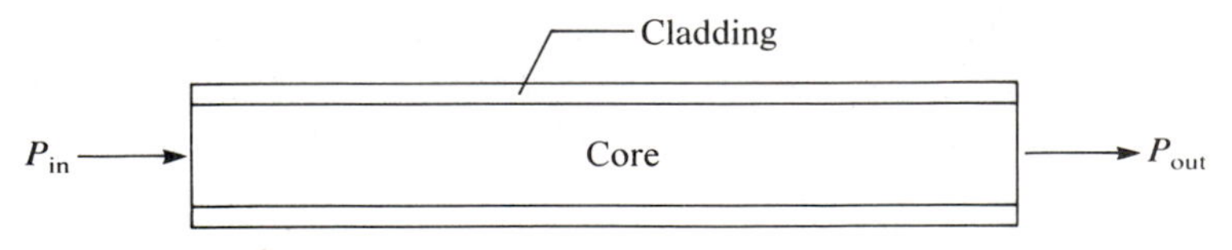

FIGURE 4–1 Step-index fiber with power in P_{in} and power out P_{out}.

EXAMPLE 4–1

A fiber of 100-m length has $P_{in} = 10\ \mu W$ and $P_{out} = 9\ \mu W$. Find the loss in dB/km.

Solution
The $(P_{out}/P_{in})|_{dB} = 10 \times \log(9/10) = -0.458$ dB. Because this loss is for 100 m = 0.1 km, for 1 km you have

$$-0.457 \times (1/0.1) = -4.58 \text{ dB/km}$$

a loss per km of 4.58 dB. The negative sign means loss.

EXAMPLE 4–2

In calculating absorption losses, it was found that 3% of the power input to a 10-m fiber was absorbed (lost) by **OH ions.** (OH [hydroxyl] is an ion with one hydrogen atom [H] and one oxygen atom [O]). Find the loss in dB/km.

Solution
For the 10-m (0.01-km) fiber,

$$P_{out} = (1 - 0.03) \times P_{in} = 0.97 \times P_{in}$$

Three percent, or 0.03, of P_{in} was absorbed. Hence, (1 − 0.03) of P_{in} was transmitted.

$$\begin{aligned}(P_{out}/P_{in})|_{dB} &= 10 \times \log(P_{out}/P_{in}) \\ &= 10 \times \log[(0.97 \times P_{in})/P_{in}] \\ &= -0.132 \text{ dB}\end{aligned}$$

For 1 km, this becomes

$$(-0.132) \times (1/0.01) = -13.2 \text{ dB/km}$$

a loss of 13.2 dB/km.

The minus signs in Examples 4–1 and 4–2 indicate that P_{out} is smaller than P_{in}. One often says, simply, "a loss of 13.2 dB/km," where the word **loss** implies the negative sign.

EXAMPLE 4–3

A communications system uses a 10-km fiber, which has a 2.5-dB/km loss characteristic. Find the output power if the input power is 400 μW.

Solution
The total loss for the 10-km fiber is $10 \times (-2.5) = -25$ dB. The minus sign must be used because P_{out} is smaller than P_{in}.

$$10 \times \log(P_{out}/P_{in}) = -25$$
$$\log(P_{out}/P_{in}) = -2.5$$
$$P_{out}/P_{in} = 0.00316 \qquad \text{(antilog of } -2.5\text{)}$$
$$P_{out} = (400\ \mu W) \times (0.00316) = 1.264\ \mu W$$

4–2–2 Material Losses

The loss due to the atomic structure of the material itself is relatively small. For example, in a germanium-silicon glass, this loss is less than 0.1 dB/km, with λ between 0.8 and 1.6 μm. It is near zero for λ of about 1.3 μm.

Losses due to impurities can be reduced by better manufacturing processes. In improved fibers, metal impurities are practically negligible. The largest loss is caused by OH ions. These cannot be sufficiently reduced. The OH impurity causes loss for particular wavelength bands. The worst loss, about 4 dB/km, occurs near $\lambda = 1.4\ \mu$m, for an impurity concentration of 1 ppm. Figure 4–2 gives the OH absorption peaks for a concentration somewhat larger than 1 ppm. Note the three loss peaks that occur at wavelengths of 0.93, 1.25, and 1.40 μm.

4–2–3 Scattering

When light is scattered by an obstruction, the result is power loss. The term "obstruction" refers to density variations in the material that result in changes in the refractive index. When the index variations (that is, the obstructions) are molecular in size, the power loss is due to Rayleigh scattering. These small obstructions, which are inherent in the manufacturing process and cannot be eliminated, behave like point sources, scattering light in all directions. Rayleigh scattering loss greatly depends on the wavelength. It varies as $1/\lambda^4$, the longer the wavelength, the lower the loss. At 1.8 μm, for example, the Rayleigh losses for a typical glass fiber are about 0.1 dB/km. Figure 4–2 shows some of the effects of Rayleigh scattering losses.

Figure 4–3 illustrates scattering by large obstructions, macrobends, or fiber deformation. This type of loss is minimized by improved manufacturing techniques and is now nearly negligible.

4–2–4 Waveguide and Microbend Losses

Structural variations in the fiber, or fiber deformation, cause radiation of light away from the fiber. Figure 4–4 illustrates radiation caused by a change in diameter. Here, the angle α_1 at the deformation is smaller than the critical angle, so that the ray leaves the fiber (solid line). With the absence of the

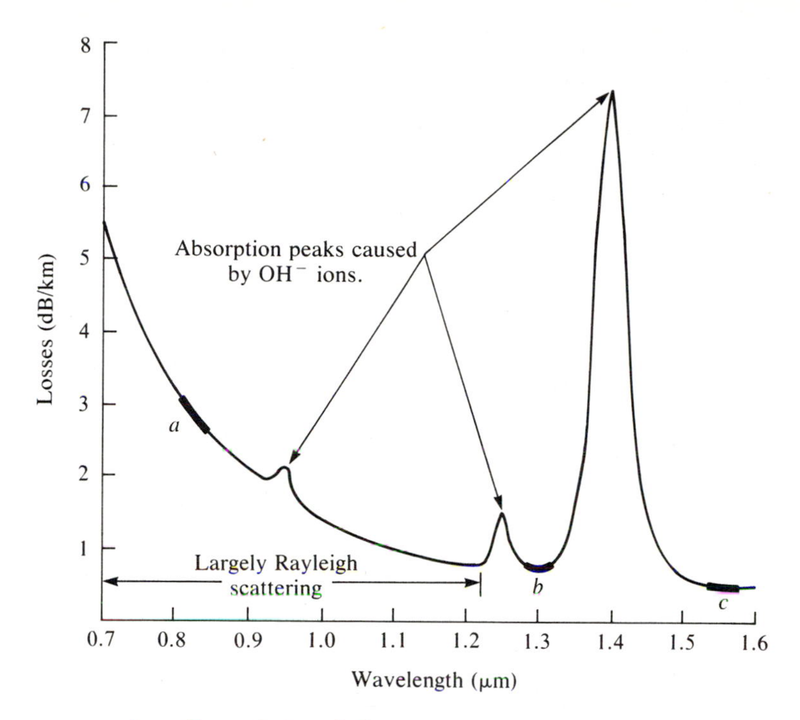

FIGURE 4–2 Fiber losses over the 0.7- to 1.6-μm spectrum.

FIGURE 4–3 Scattering by large obstructions.

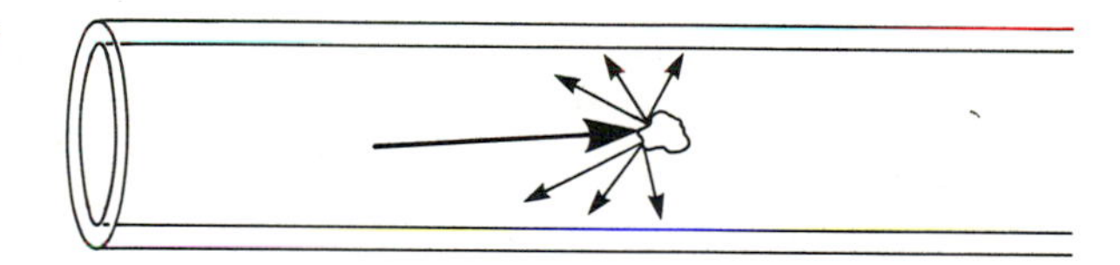

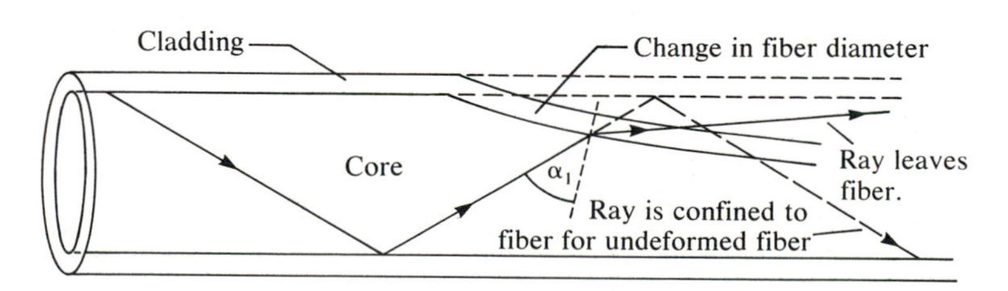

FIGURE 4–4 Radiation loss caused by diameter changes.

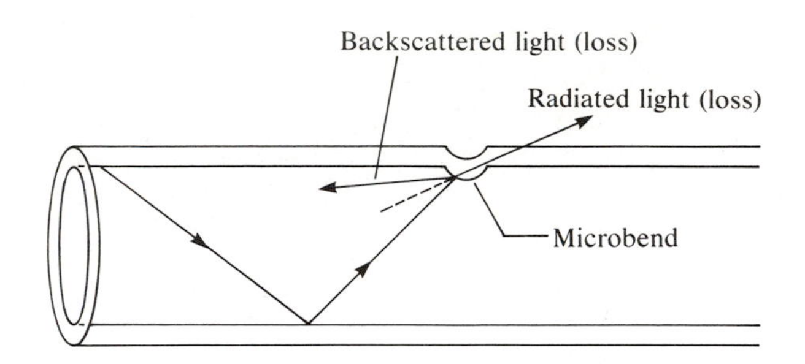

FIGURE 4–5 Microbend loss and backscatter.

deformation, the ray would be confined to the fiber, as illustrated by the dashed line.

Microbends, very minute disturbances in core size, also cause radiation of light. Figure 4–5 demonstrates this effect. Any variation in refractive index along the fiber will cause a similar energy loss. In most of these instances, the loss results from scattering of light.

4–3 DISPERSION

The term **dispersion** is used to describe pulse broadening effect by fibers. Figure 4–6 shows that the pulse that appears at the output of the fiber is wider than the input pulse. As the signal, which is a pulse of light, travels along the fiber, it becomes wider because of various propagation phenomena. You will see how different aspects of wave propagation affect dispersion. Dispersion can be defined as the output light pulse width produced by an idealized input pulse

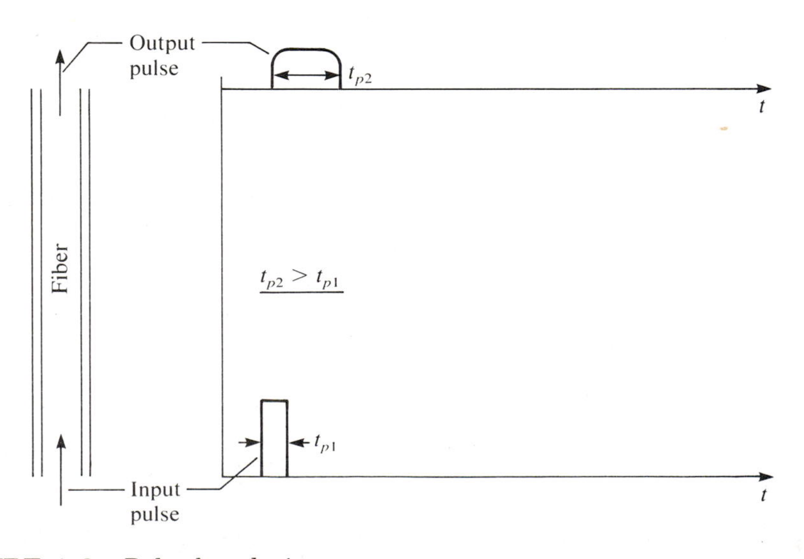

FIGURE 4–6 Pulse broadening.

of near-zero width. In other words, if you assume that the input pulse width is zero, the pulse width at the output is totally a result of fiber dispersion. In practical terms, you would have an input pulse of width t_{p1} and an output pulse t_{p2}, with t_{p2} larger than t_{p1}. The dispersion Δt can be defined by

$$\Delta t = (t_{p2}{}^2 - t_{p1}{}^2)^{1/2} \tag{4–3}$$

Note that dispersion is measured in units of time, typically, nanoseconds, (10^{-9} s) or picoseconds (10^{-12} s).

The total dispersion of a fiber depends on its length. A longer fiber causes more pulse broadening; it has larger dispersion. As a result, when specifying dispersion of a particular fiber, the manufacturer gives the dispersion per unit length, typically nanoseconds per kilometer (ns/km) or picoseconds per kilometer (ps/km). This permits the user to calculate the expected total dispersion for a given length of fiber.

$$\Delta t = L \times (\text{dispersion/km}) \tag{4–4}$$

Here, Δt is the fiber dispersion, L is the fiber length in km, and dispersion/km is given by the manufacturer.

We divide dispersion into two general categories, intermodal and intramodal.

4–3–1 Intermodal Dispersion

This type of dispersion, **intermodal dispersion,** results from the fact that the wave propagates in modes. It is a dispersion between the modes, caused by the difference in propagation time for the different modes. To understand this type of dispersion, consider two extreme modes propagating along the fiber axis (illustrated in Figure 4–7(a)): the critical mode propagating at angle θ_c and another mode where the propagation angle is zero (zero mode). A pulse of light launched into the fiber will propagate along the fiber in both modes.

For mode zero, travel time will be minimal and can be expressed by

$$\begin{aligned} t_{d0} &= L/(c/n_1) \\ &= L\,(n_1/c) \qquad \text{(minimum propagation delay)} \end{aligned} \tag{4–5}$$

where L is fiber length, n_1 is the core refractive index, and c/n_1 is the speed of the light in the fiber. For the ray traveling at angle θ_c, the delay will be maximal and can be expressed by

$$\begin{aligned} t_{dc} &= (L/\cos\theta_c)/(c/n_1) \\ &= (L \times n_1)/(\cos\theta_c \times c) \qquad \text{(maximum propagation delay)} \end{aligned} \tag{4–6}$$

The difference between the shortest delay t_{d0} and the largest delay t_{dc} is the time during which pulse energy will arrive at the fiber output. (Assume a zero pulse width.) It represents the pulse width at the output. Using Equations 3–1,

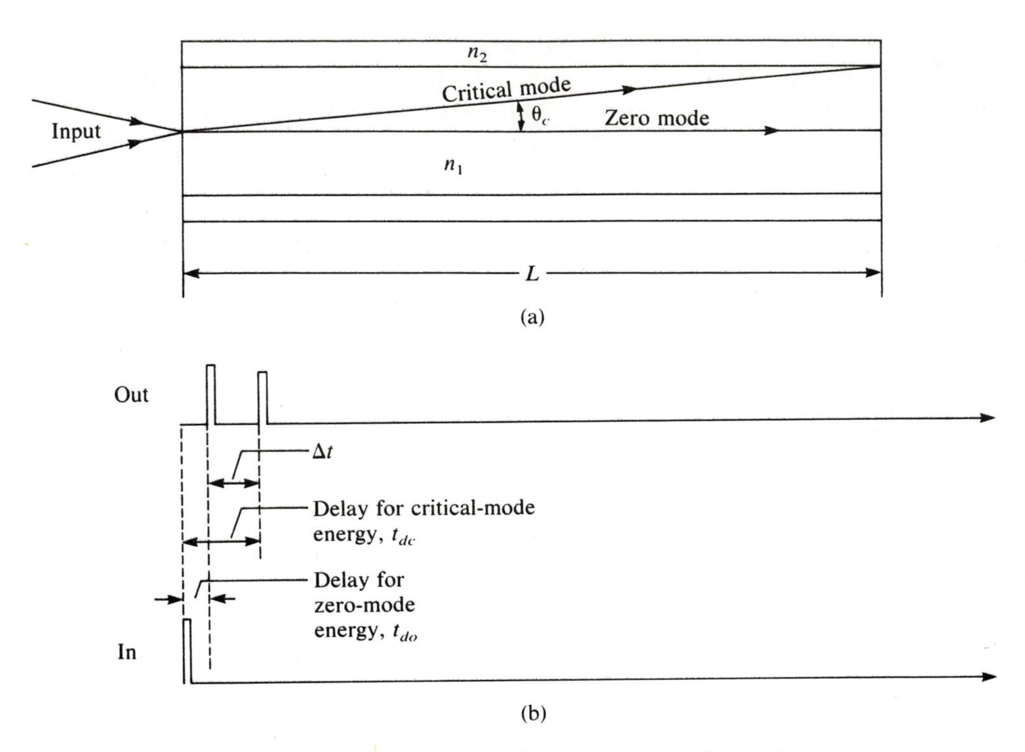

FIGURE 4–7 Intermodal dispersion. (a) Ray tracing. (b) Time delay.

4–5, and 4–6, you get an expression for this pulse width Δt:

$$\begin{aligned}\Delta t &= t_{dc} - t_{d0} \\ &= [(L \times n_1)/c] \times [(n_1 - n_2)/n_2]\end{aligned} \tag{4–7}$$

where n_1 and n_2 are the refractive indices of the core and cladding, respectively.

Figure 4–7(b) illustrates t_{dc}, t_{d0}, and Δt. Because $\Delta = (n_1 - n_2)/n_1$, the relative refractive difference (for $\Delta << 1$, the denominator can be n_1 or n_2), you get

$$\Delta t = (L \times n_1/c) \times \Delta \tag{4–8}$$

Do not confuse Δ, the relative refractive difference, with Δt, the pulse width. Using Equation 3–6, you can express Δt in terms of the numerical aperture (N.A.) as

$$\Delta t = [L \times (\text{N.A.})^2]/(2 \times n_1 \times c) \tag{4–9}$$

or

$$\Delta t/L = (\text{N.A.})^2/(2 \times n_1 \times c) \tag{4–10}$$

Δt is the amount of pulse broadening, and $\Delta t/L$ is the broadening per unit length of fiber. The next example will illustrate the problems introduced by pulse broadening.

EXAMPLE 4–4

A train of light pulses is transmitted through a 400-m fiber with $n_{core} = 1.4$ and $n_{clad} = 1.36$. Sketch the output pulses for (1) a pulse rate of 10×10^6 pulses per second (10 Mb/s) and (2) a pulse rate of 20×10^6 pulses per second (20 Mb/s). Also, find the dispersion per kilometer. For each assume that the input pulse is of near-zero width.

Solution

You compute the N.A. by

$$\text{N.A.} = (n_1 - n_2)/n_1 = (1.6 - 1.35)/1.6 = 0.1567$$

$$\Delta = (n_1{}^2 - n_2{}^2)/(2 \times n_1{}^2)$$

$$= (1.4^2 - 1.36^2)/(2 \times 1.4^2) = 0.02816$$

The Δt for the 400 m length is

$$\Delta t = (L \times n_1/c) \times \Delta$$

$$= [(400 \times 1.4)/(300 \times 10^6)] \times 0.02816 = 52.6 \text{ ns}$$

(Units used are meters and seconds.)

1. The output pulse train is shown in Figure 4–8(a). Note the relatively large separation between the pulses.

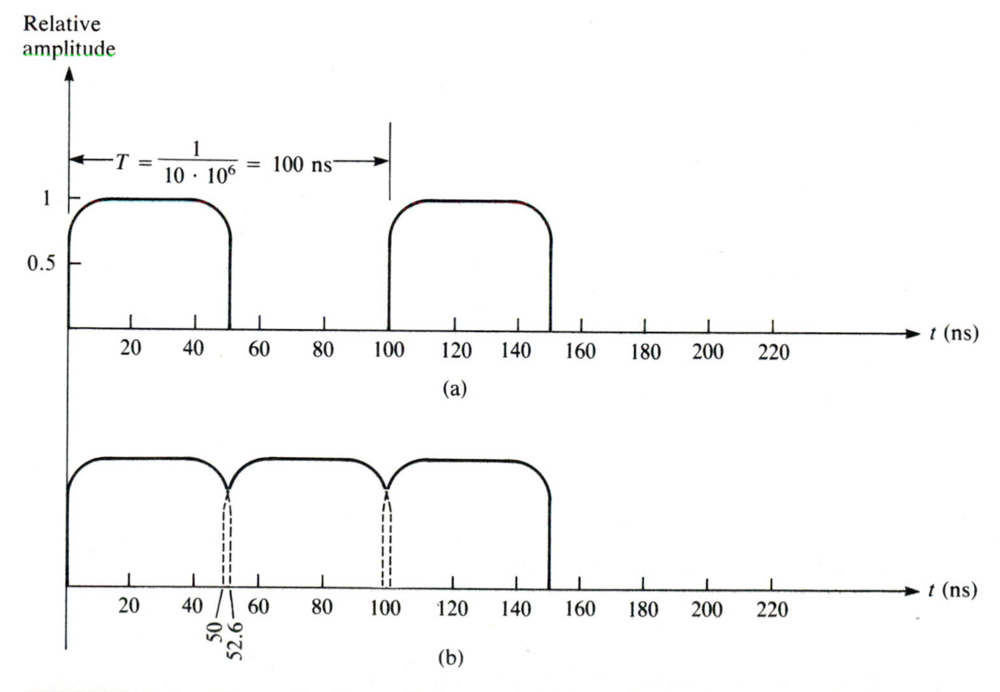

FIGURE 4–8 Figure for Example 4–4. (a) 10-Mb/s transmission. (b) 20-Mb/s transmission.

2. The pulse train is shown in Figure 4–8(b). Here, the pulses just overlap.

Finally, the dispersion/km of fiber is 52.6/0.4 = 131.4 ns/km.

Note that for 10 Mb/s, the individual pulses are clearly recognizable, while for 20 Mb/s they are merged and nearly indistinguishable. The consequence is that the 400-m fiber cannot be used at the 20-Mb/s data rate without special reshaping of the pulses after approximately 200 m.

The intermodal dispersion is directly related to the N.A. A large N.A. results in a large dispersion. Similarly, a large acceptance angle, although making it easy to inject power into a fiber, also causes an increase in dispersion, resulting in a reduction in the data rate.

4–3–2 Intramodal Dispersion

Intramodal dispersion, sometimes called **chromatic dispersion,** is a direct result of the fact that the light in the fiber consists of a group of frequencies. This dispersion is related to the line width $\Delta\lambda$. This dispersion is often given in terms of picoseconds per kilometer per nanometer of line width. (Increased line width results in increased dispersion.)

Earlier, you learned that the light velocity in the fiber varies with λ (see Section 3–3). If the line width is 820 to 850 nm (30 nm), the 820-nm part of the pulse energy is slower than the 850-nm part, producing a pulse broadening effect. For a source with λ above about 1.3 μm, the situation is reversed. If you have a line width of 1320 to 1350 nm, the λ = 1320-nm part is faster than the λ = 1350-nm part, again producing pulse broadening. For a small $\Delta\lambda$, near λ = 1300 nm, the shorter and longer wavelengths travel at about the same velocity. This leads to minimal intramodal dispersion at λ = 1300 nm (1.3 μm). When considering the total dispersion from different causes, use the approximation

$$\Delta t_{tot} = [(\Delta t_1)^2 + (\Delta t_2)^2 + \ldots]^{1/2} \qquad \textbf{(4–11)}$$

EXAMPLE 4–5

A fiber has the following specifications:

Dispersion	
Intermodal	5 ns/km
Intramodal	100 ps/(km × nm)
Line width	40 nm
Fiber length	5 km

1. Find the total intermodal dispersion.
2. Find the total intramodal dispersion.
3. Find the total dispersion.

Solution

1. Intermodal dispersion = 5 (ns/km) × 5 km = 25 ns.
2. For each nanometer of line width, there are 100 ps/km. For $\Delta\lambda = 40$ nm, you estimate the intramodal dispersion to be

$$40 \times 100 \text{ ps/km} = 4 \text{ ns/km}$$

   For 5 km, total intramodal dispersion = 4 ns/km × 5 km = 20 ns.
3. An estimate of total dispersion is obtained from Equation 4–11:

$$\Delta t_{tot} = (25^2 + 20^2)^{1/2}$$
$$= (25^2 + 20^2)^{1/2} = 32 \text{ ns}$$

4–4 CHOICE OF WAVELENGTH

In the earlier stages of the development of fiber optic technology, the wavelength used most often was 820 nm largely because of the availability of both sources and detectors in this range. For better performance, the choice of wavelength must be based on minimizing loss and minimizing dispersion.

In terms of loss, there are three low-**loss windows,** where losses from the various causes are relatively low. Table 4–1 gives some of the relevant data for high-quality glass fibers. On the basis of these data, the range 1550 to 1610 is most suitable. From the point of view of dispersion, the low intramodal dispersion wavelength of about 1300 nm is most suitable. In recent years, sources and detectors for both of these ranges have been developed. The 1550 to 1610-nm range is particularly promising in terms of losses. To minimize dispersion in this range, special fibers have been developed. These fibers are constructed so that the minimum dispersion is shifted to about 1550 nm. These are called **dispersion-shifted fibers.** These improvements are especially important in fibers in which the intermodal dispersion is nonexistent **(single-mode fibers).** Note that intermodal dispersion is usually much larger than intramodal dispersion.

TABLE 4–1
Low-Loss Windows

λ (nm)	Approximate Loss per Kilometer (dB)
820–880	2.2
1200–1320	0.6
1550–1610	0.2

SUMMARY AND GLOSSARY

Study the glossary terms. They represent the material covered in this chapter. Make sure you know how to calculate fiber loss and fiber dispersion, both intermodal and intramodal. You should understand what the causes of loss and dispersion are and their relation to system performance.

Backscatter. Light scattered back in the direction of the light source, typically caused by an obstruction, microbend, or macrobend.

Bend losses. Losses caused by imperfection and deformations of the fiber structure.

Chromatic dispersion. See intramodal dispersion. (This dispersion is called chromatic [color dependent] because it depends on line width [which is related to color].)

Dispersion. Pulse broadening effect. (The effect that causes the output pulse to be wider than the input pulse. See "chromatic dispersion," "dispersion-shifted fiber," "intermodal dispersion," and "intramodal dispersion.")

Dispersion-shifted fiber. A fiber designed to have its minimum dispersion at $\lambda = 1.5\ \mu m$ (in contrast to the standard fiber for which the minimum dispersion occurs at about $\lambda = 1.3\ \mu m$).

Intermodal dispersion. Dispersion caused by the delay between different modes. Typically, you consider the extremes: the delay between the shortest path (zero mode) and the longest path (the critical mode).

Intramodal dispersion. Dispersion that is independent of modes, related to the line width of the source, and caused by variations in the refractive index as a function of wavelength.

Light scattering. Reflection of light in many directions, typically caused by an obstruction.

Loss. The ratio of output power to input power expressed in decibels.

Loss windows. Three specific ranges of wavelengths for which fiber loss has a relative minimum (three low-loss windows).

Material loss. Loss caused by the material and its impurities rather than by structural defects in the fiber.

OH ion. An impurity in the fiber, typically caused by water condensate present during the manufacturing process.

Optical waveguide. Another name for an optical fiber.

Single-mode fiber. A fiber designed to allow light propagation in one mode only, the lowest-order mode.

Waveguide. See "optical waveguide."

FORMULAS

$$\text{Loss} = P_{out}/P_{in} \tag{4–1}$$

Definition of system power loss.

$$\text{Loss}|_{dB} = 10 \times \log(P_{out}/P_{in}) \tag{4–2}$$

System power loss in decibels.

$$\Delta t = (t_{p2}{}^2 - t_{p1}{}^2)^{1/2} \tag{4–3}$$

The dispersion pulse broadening Δt in terms of the input and output pulse widths.

$$\Delta t = L \times (\text{dispersion/kilometer}) \tag{4–4}$$

The Δt related to fiber length and the dispersion per kilometer of the fiber.

$$t_{d0} = L\,(n_1/c) \tag{4–5}$$

The minimum propagation time (in the zero-order mode).

$$t_{dc} = (L \times n_1)/(\cos\theta_c \times c) \tag{4–6}$$

Maximum propagation time. Propagation time for the critical mode.

$$\Delta t = [(L \times n_1/c)] \times [(n_1 - n_2)/n_2] \tag{4–7}$$

Intermodal dispersion Δt in terms of the refractive indices. (Approximation when $n_1 \approx n_2$.)

$$\Delta t = (L \times n_1/c) \times \Delta \tag{4–8}$$

Δt in terms of n_1 and Δ.

$$\Delta t = [L \times (\text{N.A.})^2]/(2 \times n_1 \times c) \tag{4–9}$$

Δt in terms of N.A.

$$\Delta t/L = (\text{N.A.})^2/(2 \times n_1 \times c) \tag{4–10}$$

Dispersion per unit distance in terms of N.A. and n_1.

$$\Delta t_{tot} = [(\Delta t_1)^2 + (\Delta t_2)^2 + \ldots]^{1/2} \tag{4–11}$$

Total dispersion as a function of various dispersion effects.

QUESTIONS

1. What are the two basic mechanisms of light loss? Discuss them.
2. What are material losses? Discuss the two types of material loss.
3. What is light scattering? Give two types.
4. What are waveguides and bend losses? Discuss them.
5. Can the numerical loss in a fiber (expressed in decibels) be positive? Explain.
6. Which of the losses listed below are usually given in losses per unit distance? Which are not?
   a. Material
   b. Scattering
   c. Absorption
   d. Waveguide
   e. Microbend
   f. Macrobend
7. What is dispersion?
8. What is the difference between intermodal and intramodal dispersion? Discuss the difference.
9. How is total dispersion related to the data rate of the fiber?
10. What are low-loss windows? Give the wavelengths for the three low-loss windows.
11. Why is it desirable to operate at a wavelength of about 1.3 μm?
12. Why is it advantageous to operate at $\lambda = 1.5$ μm?
13. What are dispersion-shifted fibers?

PROBLEMS

1. Find the loss, in decibels, for a fiber with input power of 200 μW and output power of 10 μW.
2. Find the output power for a 5-km fiber with a 4-dB/km loss and with $P_{in} = 500$ μW.
3. A fiber, 8 km in length, has an input power of 300 μW and an output power of 10 μW. What is the loss per kilometer?

4. Two fibers, one 3 km long with 4-dB/km loss and the other 9 km long with 1.2-dB/km loss, are spliced to form a 12-km fiber. Ignore losses due to splicing.
   a. What is the average dB/km loss?
   b. For an input of 260 μW, what is the power out?
5. Given a fiber 5 km long, with $n_{\text{core}} = 1.44$ and $n_{\text{clad}} = 1.4$, find the intermodal dispersion.
6. Find the intermodal dispersion per kilometer for a fiber with $\Delta = 2\%$ and $n_1 = 1.5$.
7. A fiber has a specified N.A. of 0.22 and n_{core} of 1.4. Find the intermodal dispersion for a 500-m length.
8. A system introduces pulse broadening of 50 ns. The input pulse is 20 ns wide. Find the output pulse width.
9. A 5-km fiber with intermodal dispersion of 5 ns/km (and no other dispersion effects) is used with an input pulse of 20 ns. What is the output pulse width?
10. The dispersion of a 5-km fiber is measured using a 10-ns input pulse and is found to have an output pulse width of 15 ns. Find
    a. The total dispersion (for the 5-km fiber)
    b. The dispersion per kilometer
11. A single-mode fiber has an intramodal dispersion of 10 ps/(km $\times$ nm) at a given wavelength. (Single-mode fibers have no intermodal dispersion.) The fiber is used with a laser with a line width of 4 nm. Find
    a. The dispersion for a 10-km length of fiber
    b. The output pulse width if the input pulse to the 10-km fiber is 2 ns wide
12. A single-mode fiber 2 km long has a dispersion of 2 ps/(km $\times$ nm). Find
    a. The dispersion if the line width of the source is 2 nm
    b. The dispersion if the line width of the source is 15 nm
13. A 10-km single-mode fiber was used in conjunction with a laser with a line width of 3 nm. The following measurements were taken: input pulse width = 40 ps and output pulse width = 60 ps. Estimate the dispersion per kilometer per nanometer of the fiber.
14. A multimode fiber has intermodal dispersion of 5 ns/km and intramodal dispersion of 50 ps/(km $\times$ nm). It is used with a light-emitting diode (LED) with a line width of 40 nm. Find the total dispersion per kilometer.
15. A 100-km length of the fiber in Problem 13 is used with the LED in Problem 14 and an input pulse width of 1 ns. Calculate the output pulse width.

5
Optical Fibers

CHAPTER OBJECTIVES

In this chapter, you will become familiar with a variety of different fibers and some of their important characteristics. Included are the step-index and graded-index fibers, as well as a number of different structures of single-mode fibers. The fiber bundle is also mentioned, as are several plastic and plastic-clad fibers. One of the basic differences among different types of fibers is their refractive index profile. The profiles discussed here are the step-index, graded-index with parabolic profile, triangular, and the *W* profiles.

5–1 INTRODUCTION

The simplest and one of the earliest fiber structures is the multimode step-index fiber. Two other structures are the graded-index fiber and the single-mode step-index fiber. These three structures are illustrated in Figure 5–1. Figure 5–1 also shows the **refractive index profile** and some typical dimensions.

For overall performance in terms of data rate and attenuation, the single-mode fiber has the best characteristics (highest data rate and least attenuation), while the multimode fiber has the worst. This does not mean that one should always use the single-mode fiber. It is the most costly and the most difficult to work with because of its small size. (The core is only 5–10 μm in diameter.)

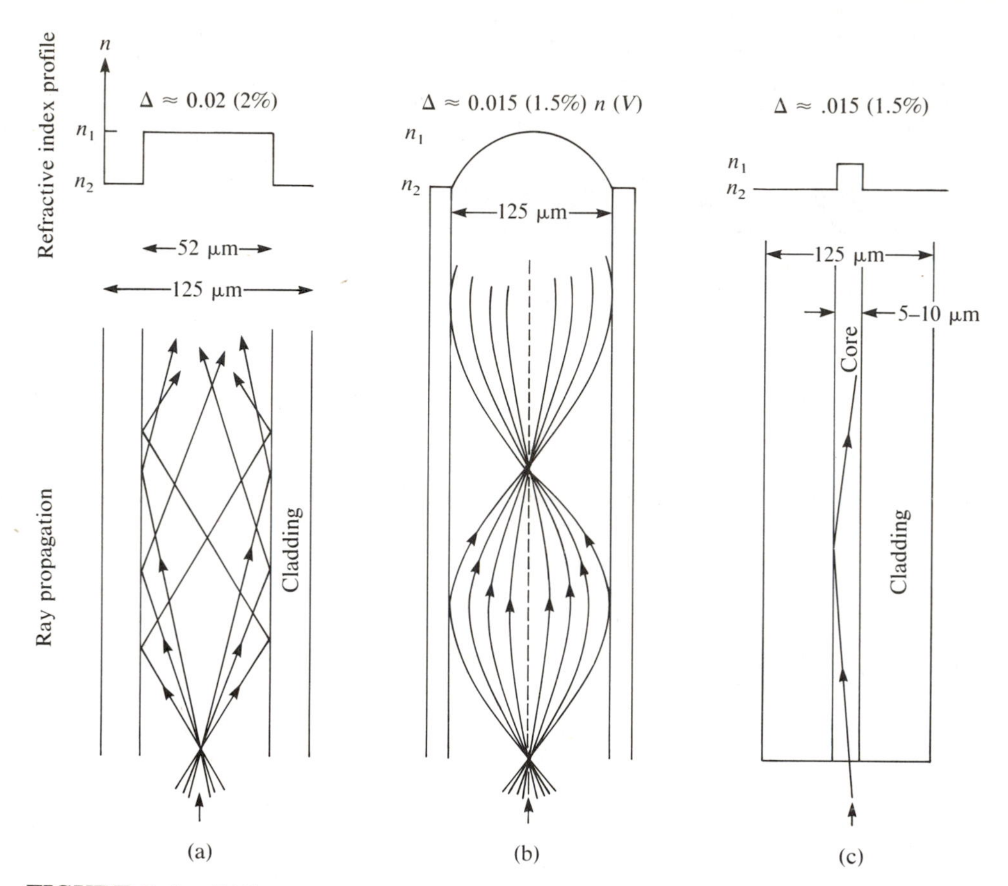

FIGURE 5–1 Different types of fibers. (Dimensions are typical for communication fibers.) (a) Step-index. (b) Graded-index. (c) Single mode.

This chapter discusses the characteristics of the three types of fiber and how these relate to their performance in a communication application.

5–2 STEP-INDEX: MULTIMODE FIBERS

Some of the basic parameters of the step-index fiber were introduced in Section 3–1 as part of the general discussion of light propagation in fibers. From Figures 5–1(a) and 5–1(c), it may appear that there is no significant difference between the two step-index fibers other than size. Indeed, this is the only fundamental difference. The consequences of this size difference are substantial.

In the multimode step-index fiber illustrated in Figure 5–1(a), light propagates in many modes. The total number of modes M_N increases with increase in the numerical aperture (N.A.). For a large number of modes, M_N can be approximated by

$$M_N = V^2/2 \qquad \textbf{(5–1)}$$

The normalized frequency V is a relation among the fiber size, the refractive indices, and the wavelength. (To explain the meaning of V, it is necessary to employ sophisticated electromagnetic theory, which is beyond the scope of this text. V is used here only as a way of simplifying the formula for M_N.) V is the **normalized frequency,** or simply the ***V* number,** and is given by

$$V = [(2 \times \pi \times a)/\lambda] \times \text{N.A.} \qquad \textbf{(5–2)}$$

$$V = [(2 \times \pi \times a)/\lambda] \times n_1 \times (2 \times \Delta)^{1/2} \qquad \textbf{(5–2a)}$$

where a is the fiber core radius, λ is the operating wavelength, n_1 the core refractive index, and Δ the relative refractive index difference.

Because more modes imply more dispersion, we calculate M_N for two step-index fibers and compare their dispersions.

EXAMPLE 5–1

Calculate the number of modes and the intermodal dispersion per kilometer for the following step-index fibers, at $\lambda = 820$ nm:

1. $n_1 = 1.41$, $n_2 = 1.4$ and core diameter = 50 μm. (With these refractive indices, $\Delta = 0.0071$, a rather low value. This type of operation is referred to as **weakly guided waves.**)
2. $n_1 = 1.45$, $n_2 = 1.4$ and core diameter = 50 μm

Solution

1. $\text{N.A.} = (1.41^2 - 1.4^2)^{1/2} = 0.1676$

$$V = [(2 \times \pi \times 25 \times 10^{-6})/(0.82 \times 10^{-6})] \times 0.1676$$
$$= 32.1$$
$$M_N = 32.1^2/2 = 515$$

M_N is always an integer, obtained by dropping the fractional part.

Dispersion for 1 km (see Equation 4–9),

$$\Delta t = (10^3 \times 0.1676^2)/(2 \times 1.41 \times 300 \times 10^6)$$
$$= 33 \text{ ns/km}$$

2. $\text{N.A.} = (1.45^2 - 1.4^2)^{1/2} = 0.3775$

$$V = [(2 \times \pi \times 25 \times 10^{-6})/(0.82 \times 10^{-6})] \times 0.3775$$
$$= 72.3$$
$$M_N = 72.3^2/2 = 2612$$

Dispersion for 1 km:

$$\Delta_t = (10^3 \times 0.3775^2)/(2 \times 1.45 \times 300 \times 10^6)$$
$$= 164 \text{ ns/km}$$

Example 5–1 demonstrates that (1) larger N.A. means more modes, and (2) more modes (larger N.A.) also means higher dispersion. Note that a larger diameter yields more modes and higher dispersion. It is useful to stress that higher dispersion means lower data rate, which leads to a less efficient transmission. Because there are a large number of modes, it is clear that the total dispersion is mostly intermodal. (The intramodal dispersion is nearly negligible in comparison with the intermodal dispersion.)

You may be tempted to argue for a great reduction in N.A. to reduce dispersion. This is not advisable for two reasons. First, injecting light into fiber with low N.A. becomes difficult. Remember that lower N.A. means lower acceptance angle, which requires the entering light to have a very shallow angle. (See Chapter 3). Second, leakage of energy is more likely, and hence losses increase.

The core diameter of the typical multimode communication fiber varies between 50 μm and about 200 μm, with the cladding thickness typically equal to the core radius.

Figure 5–2 gives some data relating to typical multimode step-index fibers. The multimode step-index fiber is relatively easy to manufacture. It is less costly than the graded-index or the single-mode fibers.

5–3 GRADED-INDEX FIBERS

A **graded-index** fiber is drawn in Figure 5–3. Here, the refractive index n in the core varies as you move away from the center. At the fiber center, you have n_1; at the cladding you have n_2; and in between, you have $n(r)$, where n is a

Material Core/ Cladding	Core Diameter (μm)	Cladding Diameter (μm)	Loss (dB/km @ 850 nm)	BW (MHz × km)	N.A.
Silica/silica	50	125	2.4	800	0.2
Silica/silica	100	140	4.5	300	0.29
Silica/silica	200	300	8	20	0.2
Silica/silica	1000	1250	6	—	0.4
Silica/plastic	250	320	8	20	0.3
Silica/plastic	200	600	30 (@ 820 nm)	90	0.38
Plastic/plastic	368	400	270 (@ 790 nm)	50	0.42

FIGURE 5–2 Typical step-index multimode fibers.

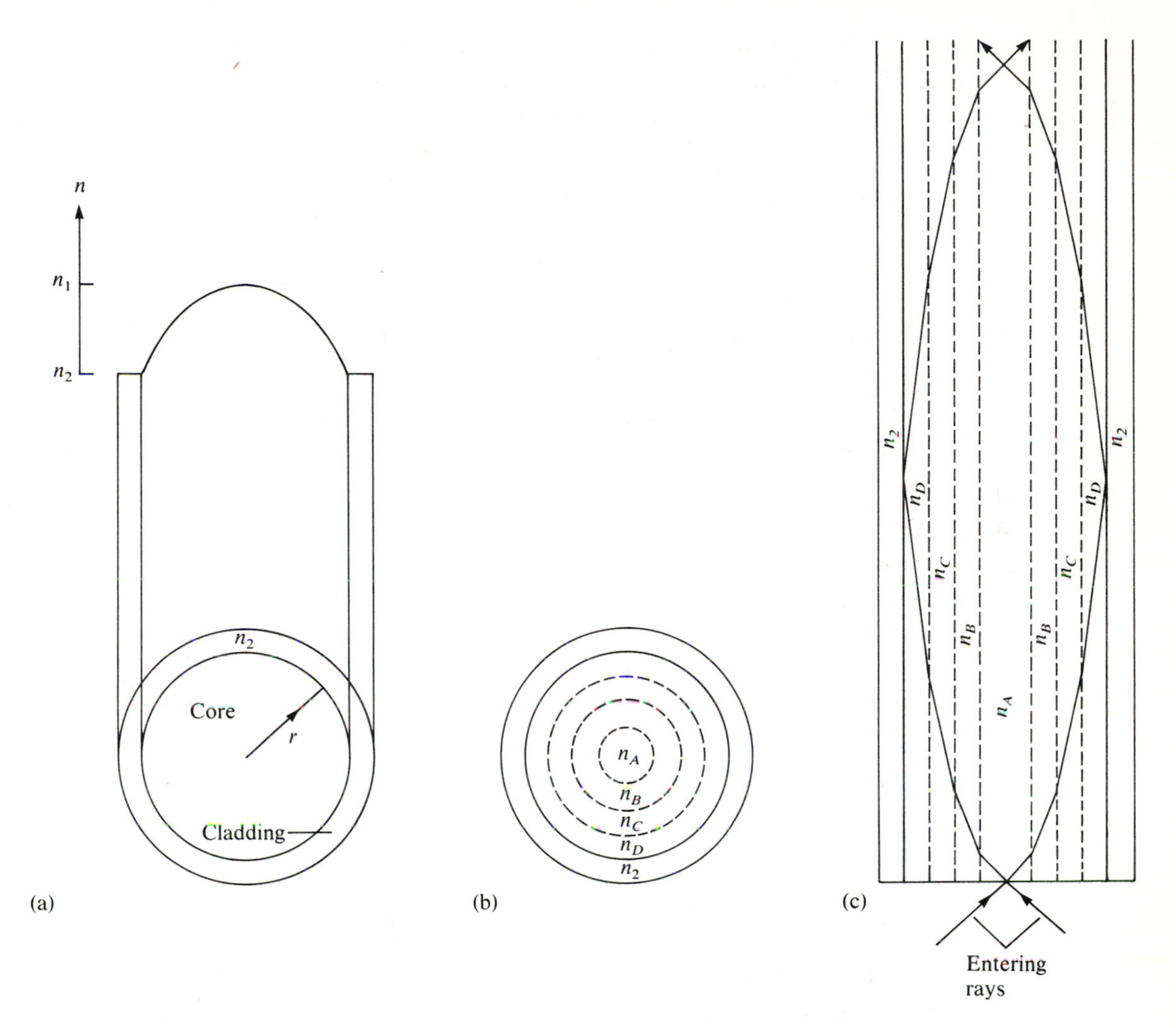

FIGURE 5–3 Graded-index fiber. (a) Index profile. (b) Stepwise index profile. (c) Ray tracing in stepwise index profile.

function of the particular radius. (See Figure 5–3(a).) Figure 5–3(b) simulates the change in n in a stepwise manner. Each dashed circle represents a different refractive index, decreasing as you move away from the fiber center. A ray incident on these boundaries between $n_A - n_B$, $n_B - n_C$, etc., is refracted. Eventually, at n_2, the ray is turned around and totally reflected. This continuous refraction yields the ray tracings shown in Figure 5–3(c).

The mathematical description of $n(r)$ is given by

$$n(r) = n_1 \times \{1 - [2 \times \Delta \times (r/a)^2]\}^{1/2} \tag{5–3}$$

a is the total radius of the core. (The power of r/a here is 2, yielding a parabolic profile. In general, the power is α (e.g., $[r/a]^{\alpha}$), where α is carefully selected to obtain the best fiber characteristics.) Δ depends only on the refractive indices at the center of the core (n_1) and the outer edge (n_2).

To appreciate the meaning of Equation 5–3, first evaluate $n(r)$ at the two end points, namely, $r = 0$ and $r = a$. For $r = 0$, you get $n(0) = n_1$, the refractive index at the core center. For $r = a$, you have

$$n(a) = n_1 \times (1 - 2 \times \Delta)^{1/2} \qquad \textbf{(5–4)}$$

The value $n(a)$ is identical to n_2 (or n_{clad}) in a step-index fiber; in other words, $n(a) = n_2$. For r between 0 and a, the refractive index varies parabolically.

EXAMPLE 5–2

A graded-index fiber has $n_1 = 1.5$ and $n_2 = 1.45$. Find the index at a point where $r = a/2$, midpoint on the radius.

Solution
First find Δ.

$$\begin{aligned}\Delta &= (n_1^2 - n_2^2)/(2 \times n_1^2) \\ &= (1.5^2 - 1.45^2)/(2 \times 1.45^2) \\ &= 0.1475/4.205 \\ &= 0.035 \qquad (3.5\%)\end{aligned}$$

Then, by substituting $a/2$ for r in Equation 5–3, you get

$$\begin{aligned}n_{(r=a/2)} &= 1.5 \times [1 - 2 \times 0.035 \times (1/2)^2]^{1/2} \\ &= 1.487\end{aligned}$$

Note that at $r = a/2$, the refractive index lies between n_1 and n_2.

The N.A. of a graded-index fiber varies radially along r because n keeps varying. For your purposes, however, the N.A. of a graded-index fiber will be computed exactly as that for a step-index fiber, namely, N.A. $= (n_1^2 - n_2^2)^{1/2}$. (This results in a relatively small error.)

For Example 5–2,

$$\text{N.A.} = (1.5^2 - 1.45^2)^{1/2} = 0.384$$

If you calculate the N.A. for $n_1 = 1.5$ (center of core index) and $n_2 = 1.487$ (the index at $r = a/2$), you get

$$\text{N.A.} = (1.5^2 - 1.487^2)^{1/2} = 0.197$$

This is quite different from the N.A. calculated previously for the complete fiber.

The number of modes in a graded-index (parabolic) fiber is about half that in a similar step-index fiber.

$$M_N = V^2/4 \qquad \textbf{(5–5)}$$

The lower number of modes in the graded-index fiber results in lower dispersion than is found in the step-index fiber.

For the graded-index fiber, the dispersion is approximately

$$\Delta t = (L \times n_1 \times \Delta^2)/(8 \times c) \quad \textbf{(5–6)}$$

EXAMPLE 5–3

Calculate (1) the number of modes and (2) the intermodal dispersion per kilometer of a parabolic graded-index fiber with the same size n_1 and n_2 as those in Example 5–1 part 1:

$$n_1 = 1.41,\ n_2 = 1.4,\ \text{and diameter} = 50\ \mu\text{m}$$

Solution

You find V, as in Example 5–1 part 1.

1. $V = 32.1$
   $M_N = V^2/4 = 257$
2. $\Delta = (n_1 - n_2)/n_1 = 0.0071$

Dispersion for 1 km, using Equation 5–6,

$$\Delta t = [10^3 \times 1.41 \times (0.0071)^2]/(8 \times 300 \times 10^6)$$
$$= 29.6\ \text{ps/km}$$

(This is not the total dispersion expected because intramodal dispersion was not considered.)

When comparing the dispersion in Example 5–1 part 1, 33 ns/km, to the dispersion in Example 5–3, 29.6 ps/km, we note more than a 1000 to 1 improvement.

The size of the graded-index fiber is about the same as the step-index fiber. The manufacture of graded-index fiber is more complex. It is more difficult to control the refractive index well enough to produce accurately the variations needed for the desired index profile.

The effective acceptance angle of the graded-index fiber is somewhat less than that of an equivalent step-index fiber. This makes coupling fiber to the light source more difficult.

5–4 STEP-INDEX: SINGLE-MODE FIBERS

As its name implies, the light energy in a **single-mode fiber** is concentrated in one mode only. This is accomplished by reducing Δ and/or the core diameter to a point where the V is less than 2.4. In other words, the fiber is designed to have a V number between 0 and 2.4. This relatively small value means that the fiber radius and Δ, the relative refractive index difference, must be small.

EXAMPLE 5–4

1. Find the largest Δ for a single-mode fiber with $n_1 = 1.4$, $a = 5\mu$m, and $\lambda = 1.5\ \mu$m.
2. Find the N.A.

Solution

1. $V = (2 \times \pi/\lambda) \times a \times n_1 \times (2 \times \Delta)^{1/2}$

To support single-mode operation, $V = 2.4$.

$$2.4 = [2 \times \pi/(1.5 \times 10^{-6})] \times 5 \times 10^{-6} \times (2 \times \Delta)^{1/2}$$
$$\Delta = (2.4^2/29.307^2)/2 = 0.00335 = 0.335\%$$

2. $\text{N.A.} = 1.4 \times (2 \times 0.00335)^{1/2} = 0.115$

The common value of Δ for multimode fibers is about 2%, and the N.A. is about 0.2. As Example 5–4 shows, both Δ and N.A. are very small for single-mode fibers. The low N.A. means a low acceptance angle. In Example 5–4, the acceptance angle θ_{ac} is

$$\theta_{ac} = 2 \times \sin^{-1} \text{N.A.} = 13.16°$$

This means that the incident ray must be nearly perpendicular to the fiber edge. Figure 5–4 compares the acceptance angle for N.A. = 0.115 and N.A. = 0.2 (acceptance angles of 13.16° and 23.1°, respectively).

No intermodal dispersion exists in single-mode fibers because only one mode exists. With careful choice of material, dimensions, and λ, the total dispersion can be made extremely small, less than 0.1 ps/(km × nm), making

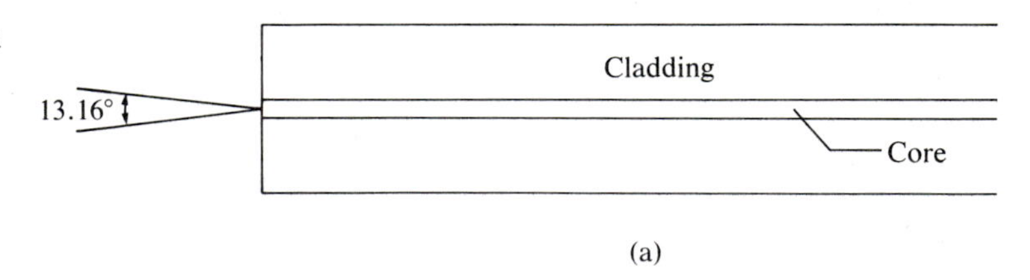

FIGURE 5–4 Low and high numerical aperture (N.A.) (a) Step-index, single-mode fiber, N.A. = 0.115. (b) Step-index multimode fiber, N.A. = 0.2.

this fiber suitable for use with high data rates. Single-mode fibers have been used experimentally at data rates of 1000 GHz.

In single-mode fiber, unlike the case in multimode fibers, part of the light propagates in the cladding. The cladding must then have a low loss and be relatively thick. Typically, for a core diameter of 10 μm, the cladding diameter is about 120 μm. The overall fiber size is about the same as that of other fibers. Handling and manufacturing is more difficult. Presently, the single-mode fiber is the lowest loss and the highest data rate fiber; it is also the most costly. It is, however, considered cost effective when the cost is related to the data rate.

Single-mode fibers are becoming more popular (and are likely to become less expensive) in many other specialized applications.

5–5 OTHER FIBERS

Most new fiber development is aimed at improving the fibers for communication applications, usually involving reduction of loss and dispersion and improvement in source and detector coupling.

5–5–1 Fiber Bundles

Before discussing some of these new fibers, an older fiber version is described. One of the earlier fibers consisted of a **fiber bundle,** packed together as illustrated in Figure 5–5. Each fiber in the bundle has an outer diameter of the order of 120 μm, and the bundle may be as large as 0.25–0.50 in. in diameter. The efficiency of transmission is low because a good portion of the bundle is occupied by cladding and open space between the fibers. In addition, some of the individual fibers break during manufacturing and are effectively opaque. The size of the fiber makes it easy to couple it to even simple incandescent lights. This type of bundle is the one predominantly used in illumination applications.

FIGURE 5–5 Fiber bundle (simplified).

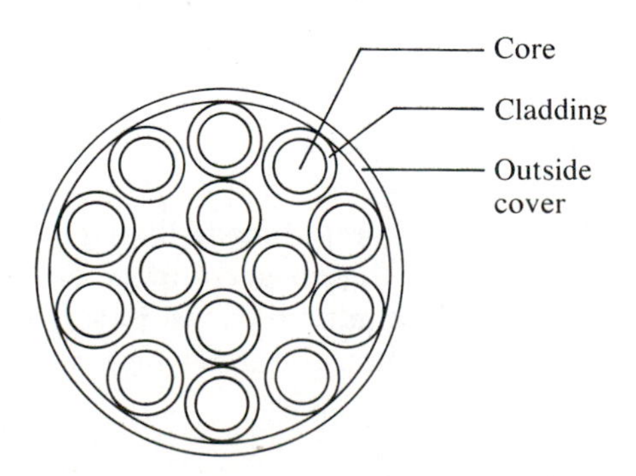

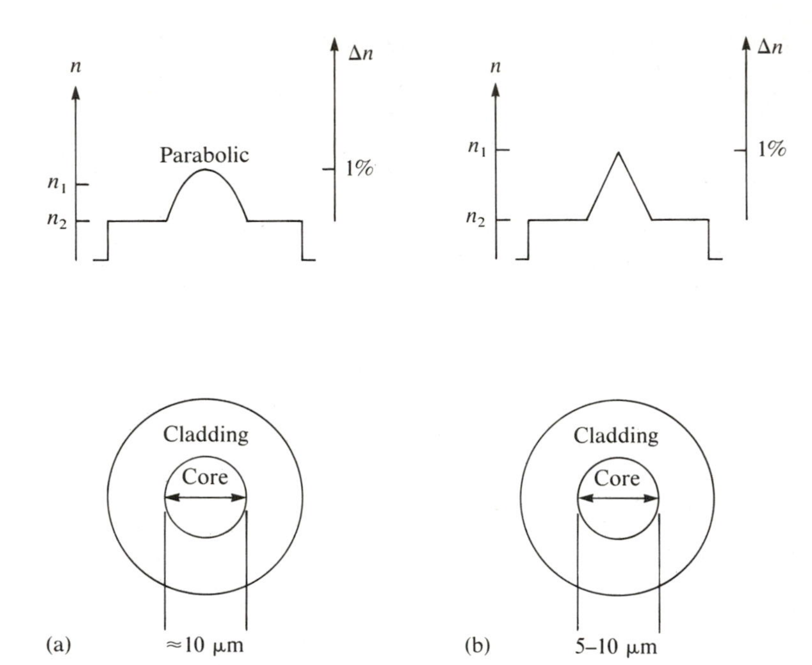

FIGURE 5–6 Single-mode fibers. (a) With parabolic index profile. (b) With triangular index profile.

5–5–2 Single-Mode with Parabolic and Triangular Profiles

Two variations of the single-mode fiber are the graded-index fiber with **parabolic refractive index profile** and a fiber with a **triangular refractive index profile.** These are shown in Figure 5–6. The basic single-mode characteristics are still maintained. The V numbers for these fibers may be larger than 2.4, so that the diameters may also be larger. For the parabolic index profile, the core diameter may be increased by 1.4, and for the triangular fiber, by 1.7. The larger core size makes handling, connecting, and coupling the fiber easier and results in less loss.

5–5–3 The *W* Fibers

Another variation of the single-mode fiber is the ***W* refractive index profile** fiber (sometimes called **depressed-index cladding fiber**) shown in Figure 5–7(a). The cladding is made up of two parts, with the outer section having a refractive index larger than the inner one, $n_3 > n_2$. This fiber has three main advantages:

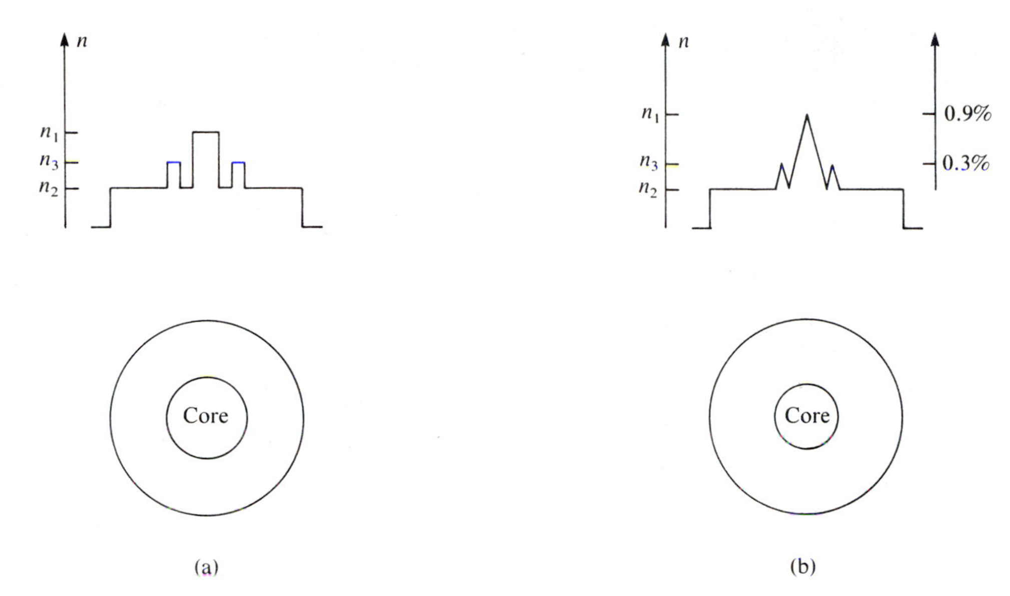

FIGURE 5–7 *W*-index profile fibers. (a) Step-index-*W* profile. (b) Triangular *W* profile—single-mode. (Corning SMF/DS fiber).

1. Its diameter may be larger than the equivalent step-index fiber.
2. Its bending loss tends to be smaller.
3. By manipulating the index profile, that is, the values of n_1, n_2, n_3, and their diameters, it is possible to shift the minimum wavelength at which dispersion occurs and then construct a fiber where the minimum dispersion coincides with the minimum loss. Recall that, in the standard step-index fiber, minimum loss occurs at about $\lambda = 1.5\ \mu m$, while minimum dispersion occurs at $\lambda = 1.3\ \mu m$. Figure 5–7(b) gives the profile of such a **dispersion-shifted fiber,** the single-mode dispersion–shifted fiber, made by Corning Glass, Inc.

5–6 PLASTIC FIBERS

When discussing plastic fibers, you must distinguish between plastic-clad and all-plastic fibers. In the former, only the cladding is plastic, while in the latter, the core and cladding are plastic. The plastic-clad fiber has characteristics similar to those of the all-glass fiber; however, the plastic-clad fiber is much more sensitive to abrasive damage.

The all-plastic fiber is stronger mechanically but has poorer transmission characteristics. Its loss is much higher, and its bandwidth (BW) is lower than glass fiber. It is rarely used in commercial communication systems. Because

TABLE 5–1
Typical Fiber Specifications

Fiber Type	Manu-facturer	Model	Refrac-tive Profile	Core Diameter (μm)	Cladding Diameter (μm)	n_{core}	n_{clad}	N.A.[1]	Loss[2] (dB/km)	BW[3] (MHz-km)
All-silicon step-index	Quartz Products Corp.	QSF 133/200 ASW		133	200			0.2	5	20
Silica core, plastic-clad step-index	DuPont	Pifax S-120 Type 30		200	600	1.46	1.409	0.38	30	0.85
All-silicon graded-index	Valtec/ Phillips Corp.	MG05		50	125			0.2	5	400
All-plastic step-index	Dupont	Pifax PIR140 Type B		368	400	1.48	1.419	0.42	270 @ 790 nm	0.58
All-silicon, single-mode step-index	Corning Glass Inc.	SMF-28		8.3	125	1.471 @ 1300 nm	$\Delta = 0.36\%$	0.12	1.8 @ 850 nm 0.19 @ 1550 nm	>2000 @ 1300 nm (565 Mb/s-40 km)
Silicon, single-mode, *W* profile dispersion-shifted	Corning Glass Inc.	SMF-D5		≈ 16	125	1.476	$\Delta_{max} = 0.9\%$ $\Delta_{ring} = 0.3\%$	—	2.06 @ 850 nm 0.22 @ 1550 nm	>4000 @ 1550 nm (565 Mb/s-80 km)

[1]N.A. = numerical aperture.
[2]Loss at 820 + 0850 nm unless otherwise specified.
[3]BW = bandwidth.

it is easier to handle and does not break as easily, it is often used for short-distance, low BW applications. It is used in closed circuit television, demonstration systems, and other similar applications.

Some typical data relating to a number of different fibers are shown in Table 5–1. Table 5–1 represents only a fraction of the fibers available today and an even smaller fraction of those that will be available tomorrow.

SUMMARY AND GLOSSARY

Use the following list of terms to review the material in this chapter.

Depressed-index cladding fiber. See "*W* refractive index profile."

Dispersion-shifted fiber. A fiber designed to have its minimum intramodal dispersion at about 1.5 μm, coinciding with the minimum loss window.

Fiber bundle. A fiber cable made up of a large number of individual optical fibers, each with its own cladding, in a bundle.

Graded-index. See "parabolic refractive index profile."

Normalized frequency. The V number, where V is a function of fiber radius and the numerical aperture, and the wavelength of operation.

Parabolic refractive index profile. The refractive index changes as the square of the ratio r/a, highest for $r = 0$ and lowest for $r = a$. r is the distance of an arbitrary point in the core from the center; a is the core radius. See Figure 5–6(a).

Refractive index profile. The way the refractive index varies as the distance from the fiber center varies. (See "parabolic refractive index profile," "triangular refractive index profile," and "*W* refractive index profile.")

Single-mode fiber. A fiber in which only one propagation mode exists.

Triangular refractive index profile. The refractive index changes linearly as the point moves from the center, $r = 0$, to the core edge, $r = a$. (See Figure 5–6(b).)

***V* number.** See "normalized frequency."

Weakly guided wave. When the difference between core index and cladding index is very small, the light wave is said to be "weakly guided" in the fiber.

***W* refractive index profile.** The cladding has two values of n: low close to the core and at the cladding outer edge and higher in between. (See Figure 5–7(a).)

FORMULAS

$$M_N = V^2/2 \tag{5–1}$$

An approximation for the number of modes in a step-index fiber.

$$V = [(2 \times \pi \times a)/\lambda] \times \text{N.A.} \tag{5–2}$$

The value of V in Equation 5–1.

$$V = [(2 \times \pi \times a)/\lambda] \times n_1 \times (2 \times \Delta)^{1/2} \tag{5–2a}$$

V expressed in terms of Δ instead of N.A.

$$n(r) = n_1 \times \{1 - [2 \times \Delta \times (r/a)^2]\}^{1/2} \tag{5–3}$$

The equation giving the refractive index at any r for the parabolic graded-index fiber.

$$n(a) = n_1 \times (1 - 2 \times \Delta)^{1/2} \tag{5–4}$$

The refractive index at the point where r = a, the core edge, for a graded-index fiber.

$$M_N = V^2/4 \tag{5–5}$$

The number of modes in a parabolic graded-index fiber (V as in Equation 5–2).

$$\Delta t = (L \times n_1 \times \Delta^2)/(8 \times c) \tag{5–6}$$

Dispersion for graded-index fiber.

QUESTIONS

1. To what does the term "multimode" refer?
2. Why is it desirable to reduce the number of modes propagating in a fiber?
3. How is the number of modes in a fiber related to
   a. The N.A.?
   b. The relative refractive index difference Δ?
   c. The radius of the fiber core?
4. How is the data rate of the fiber (its capacity) related to the number of modes in the fiber? Explain.
5. In what way is the graded-index fiber better than the multimode step-index fiber?

6. How would you qualitatively compare the dispersion in multimode step-index and graded-index fibers?
7. What kind of dispersion is dominant in
   a. Multimode step-index fiber?
   b. Multimode graded-index fiber?
   c. Single-mode fiber?
8. What parameters determine whether a step-index fiber is multimode versus single mode?
9. Why is the cladding in the single-mode fiber so thick relative to its core? (The cladding thickness is 10 times the core radius for single-mode fibers and approximately equal to the core radius for multimode fibers.)
10. What is one of the major difficulties in using the single-mode fiber?
11. What is a typical application of a fiber bundle?
12. What is the major advantage of the graded-index and triangular profile, single-mode fibers over the step-index single-mode fibers?
13. What are the advantages of the *W* profile single-mode fiber over the step-index single-mode fiber?
14. What is a dispersion-shifted fiber? Is it always a single-mode fiber? Why?
15. Why is intramodal dispersion largely ignored in multimode fibers and not in single-mode fibers?
16. What are the advantages and disadvantages of the all-plastic fiber?

PROBLEMS

1. Calculate the number of modes for the following fibers. (All have a core diameter of 60 μm and operate with $\lambda = 0.8$ μm.)
   a. $n_1 = 1.58$ and $n_2 = 1.40$
   b. $n_1 = 1.45$ and $\Delta = 1.5\%$
   c. $n_1 = 1.42$ and $n_2 = 1.40$
2. Calculate the intermodal dispersion per kilometer for the fibers in Problem 1.
3. Assume the fibers in Problem 1 are graded-index fibers with parabolic profile. Find the number of modes in each case.
4. Find the intermodal dispersion for the fibers in Problem 3.
5. A graded-index parabolic profile fiber has $n_1 = 1.46$, $n_2 = 1.42$, and a core radius a of 25 μm (50-μm diameter). Find the refractive index at a point where $r = 20$ μm.
6. A graded-index fiber (parabolic) has $n_1 = 1.52$, $\Delta = 2\%$, and $a = 30$ μm. Find
   a. The N.A. of the fiber
   b. The N.A. for a section of the core between $r = 0$ and $r = 20$ μm
   c. The N.A. for the section of the core between $r = 20$ μm and $r = 30$ μm
7. A fiber has $n_1 = 1.4$ and $n_2 = 1.39$ and operates at $\lambda = 1.3$ μm. Find the largest core diameter for which the fiber will be single mode.

8. If a fiber has a parabolic profile (graded-index) with $n_1 = 1.4$ and $\Delta = 0.3\%$ and is operated at $\lambda = 1.5\ \mu m$, what is the largest core radius for which this fiber is single mode?
9. The diameter of a step-index fiber core is 10 μm. With $n_1 = 1.4$ and $\lambda = 1.5$ μm, what is the smallest possible n_2 to ensure single-mode operation? (Calculate to four significant digits.)
10. Repeat Problem 9 for a core diameter of 4 μm.
11. Repeat Problem 9 for $\lambda = 0.82\ \mu m$.

6
Principles of Fiber Optic Communication

CHAPTER OBJECTIVES

This chapter is an introduction to communication techniques with an emphasis on fiber optic digital communication. You will learn how analog data is transmitted digitally; how data rate, bandwidth, and dispersion are related. This chapter will also enable you to calculate electrical and optical bandwidth as well as understand the significance of the data **format** used.

6–1 INTRODUCTION

The fiber optic technology has been developed largely for use in communication. Figure 6–1 illustrates a block diagram of a typical communication system. Two basic disciplines are involved: fiber optic technology and communication technology. The fiber optic part involves the light source and detector, with the associated drive and receive circuits and the fiber itself. Connectors and a splice have been included as practical considerations. The communication portion relates to the processing of data, modulation, and demodulation.

In the fiber optic communication discussed here, information is carried as **intensity modulated** light. The intensity of the light represents the input voltage or current. A sine wave input produces a light with an intensity that varies sinusoidally. (This is in contrast to coherent transmission, which is the method most used with radio waves.)

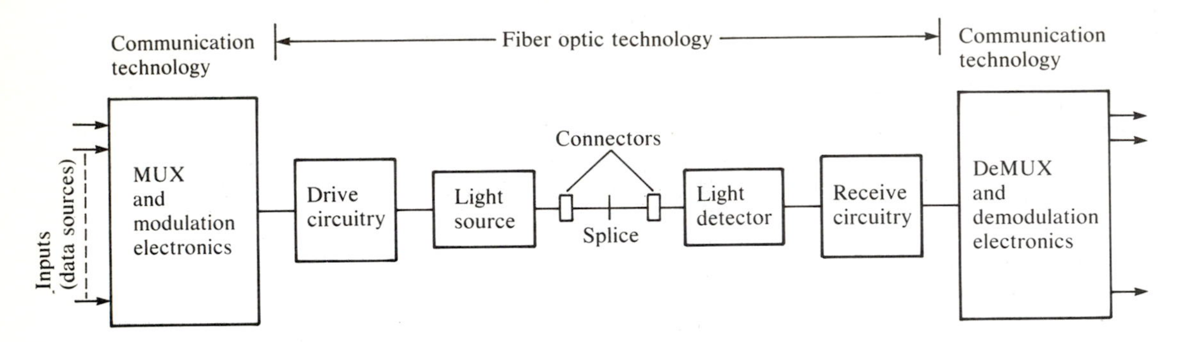

FIGURE 6–1 Block diagram of communication system.

6–2 ANALOG AND DIGITAL SYSTEMS

Physical quantities may be represented in analog or digital form. The **analog** representation is continuous. The signal used in the representation is continuous and can have any numerical value between zero and some maximum value.

In contrast, the signal used in **digital** representation consists of two levels called high and low or "1" and "0," respectively. To represent symbols or numerical values, use the **binary code,** a two-level code or number system. As an example, the symbol "e" may be represented by the **ASCII code** (a standard 7-bit binary code) 1100101 and the number 10 by its binary value representation, 1010. These codes often appear as a train of voltage (or current) pulses, as shown in Figure 6–2.

Analog quantities may be represented with limited resolution in digital form. For example, a 5-V voltage may be represented by the digital value 110010 (50 in decimal). Note that the digital representation is discrete (can be done only in steps) rather than continuous. The representation of analog quantities in digital fashion may be done electronically by analog-to-digital converters **(ADC).** Similarly, digital numerical values may be converted to analog signals by use of digital-to-analog converters **(DAC).**

Fiber optic communication, particularly in voice-telephone applications, is largely digital. Such a system includes both ADCs, and DACs, as shown in Figure 6–3(a). Here the transmitted light intensity is either high (on) or low (off) because only two signal levels are recognized.

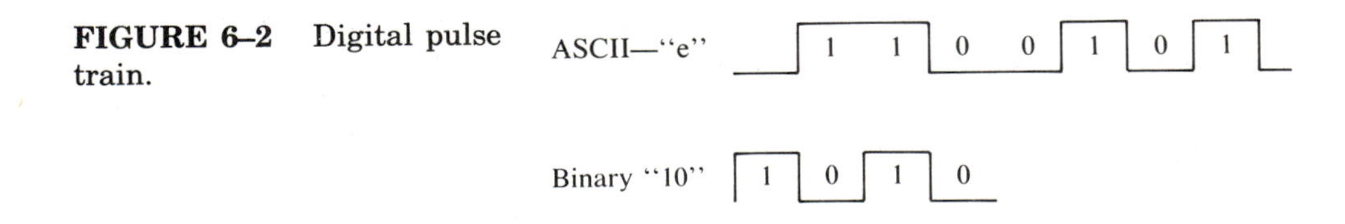

FIGURE 6–2 Digital pulse train.

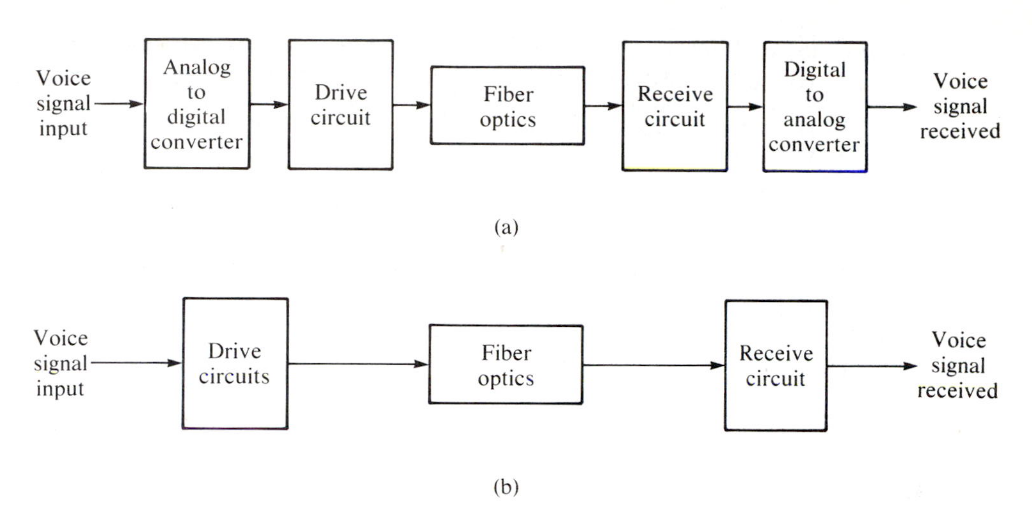

FIGURE 6–3 Voice transmission. (a) Block diagram of digital system. (b) Block diagram of direct analog system.

In direct analog transmission, Figure 6–3(b), the light intensity directly represents the voice signal amplitude. As the instantaneous signal level changes, the light intensity changes.

In converting an analog signal to a digital one, **sampling** is used. You are actually converting samples of the signal that closely approximate the actual signal. You must, of course, have enough such samples to represent the changing signal. Because of this sampling and analog-to-digital conversion, the rate (or bandwidth) of the analog signal is much lower than that of the digital pulse train used to represent it. Typically, eight pulses are used to represent each sample, and about four samples are required for each analog cycle. (The minimum number of samples required is two, which can be used only in perfect, noiseless transmission. This number is based on the Nyquist **sampling theorem** [1924].) With four samples per cycle, a 4-kHz analog signal (4000 cycles per second) requires $8 \times 4 \times 4000 = 128{,}000$ pulses per second. The number of samples per second (cycles per second times samples per cycle) is the **sample rate.** In spite of the relatively large increase in data that is required, digital transmission is the most dominant.

6–2–1 Analog Transmission

In analog transmission, the signal to be transmitted is used through an amplifier-driver circuit to control the current through a light source. In this way, the light intensity is directly related to the input signal. At the receiving end, the light is converted to current or voltage by a photodetector. The expectation is that the output waveform is an accurate copy of the input waveform, with or without gain. To obtain a faithful signal reproduction, the system, consisting of transmitting section, fiber, and receiver portion, must

1. Be highly linear,
2. Have a time response (or BW) compatible with the input signal, and
3. Have low internal noise.

To obtain system **linearity,** look for the individual components to be linear. In terms of the light source, this means that the relation between driving current and light intensity must be linear. An idealization of such a relationship is shown in Figure 6–4(a). Figure 6–4(b) shows two typical plots for two light sources, a light-emitting diode (LED) and a laser. (See Chapter 8.)

FIGURE 6–4 Typical LED and laser characteristics. (a) Linear relationship: light power versus current. (b) Laser and LED characteristics.

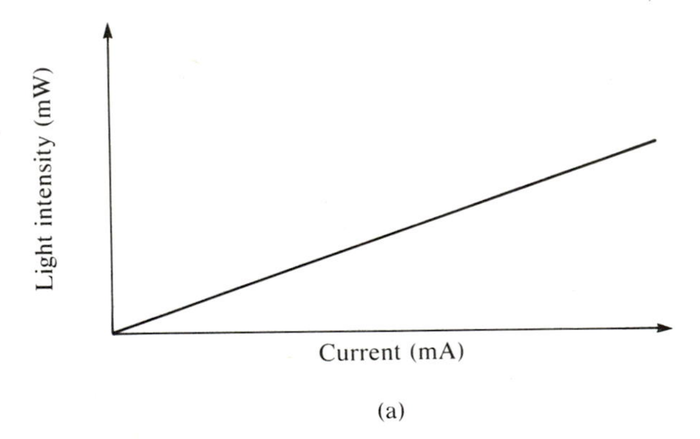

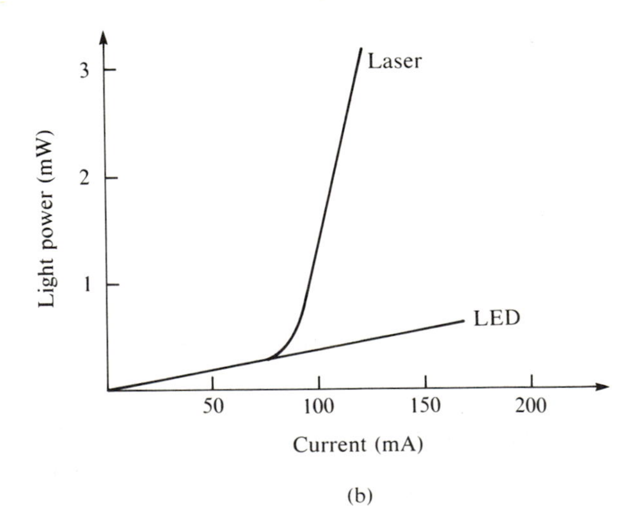

The LED has a response that is nearly linear, making it suitable for analog transmission. Its power output is relatively low, and the power coupled into the fiber is even lower. This makes the transmission susceptible to noise. The laser source has a sharp nonlinearity, similar to the "knee" (the bend in the curve) in a forward biased diode. Its power output is substantially higher. It is possible to operate the laser in the linear region below the knee, with no real advantage. To operate on the steeper, higher-power linear portion beyond the knee requires special stabilizing circuitry. It often involves complex temperature stabilization. Because of these difficulties, the laser is rarely considered for direct analog transmission. The fiber is passive and may be considered perfectly linear.

The detectors, like the light sources, must also be linear. Again, practical devices are rarely perfectly linear. It is not uncommon, however, to obtain a transmission with distortion caused by linearity well below 1%.

System **response time** (or BW) must be equal to or better than that of the input signal. BW refers to the highest frequency the system can handle with minimum signal distortion. If, for example, the input signal has a frequency content as high as 10 kHz, the system must be capable of handling this frequency. Its BW must be 10 kHz or higher. Analog fiber optic systems are available with BW in the 10- to 20-MHz range.

One of the greatest difficulties in analog systems is noise interference. The sources of electrical noise are induced **external noise** and internally generated noise. The external noise is usually reduced by proper shielding. (Here noise induced into the electronic circuitry rather than into the optical fiber is discussed.) The **internal noise,** which is inherent in the operation of both the light source and the detector sections, is hard to overcome. For a signal to be faithfully reproduced, the signal power must be larger than the noise power. In more technical language, the **signal-to-noise ratio** (S/N) must be large. With the low power of LEDs and the inherent noise involved, analog transmission can only be used over relatively short distances. Long stretches of fiber introduce larger signal attenuation, which reduces the S/N. This severe distance restriction is one of the reasons that direct analog transmission is not used in long-distance communication systems. It is popular in closed circuit television transmission and some voice communication systems, over relatively short distances.

6–2–2 Digital Transmission

Digital transmission consists of sending and receiving light pulses. Two levels are recognized: high and low or "on" and "off." Some of the parameters of the pulse waveforms are defined here (see Figure 6–5). Also defined here are some general terms relating to digital data transmission.

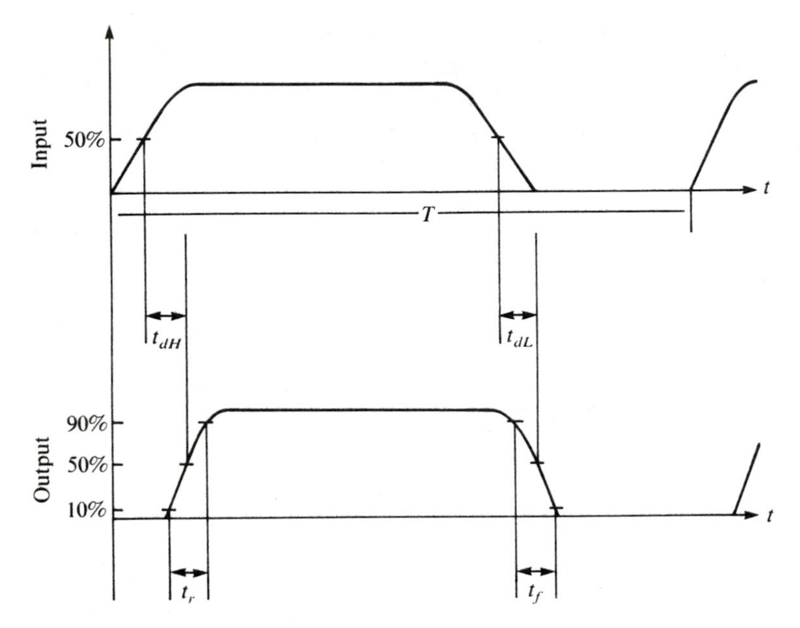

FIGURE 6–5 Pulse parameters.

- ***Rise time*** t_r. The time it takes for the signal (voltage, current, light intensity, etc.) to go from 10 to 90% of its maximum value (amplitude).
- ***Fall time*** t_f. The time required to go from 90 to 10% of the amplitude.
- *Delay time* t_d. The time delay between 50% of amplitude at the input and the corresponding 50% of amplitude at the output.
- ***Period*** *T*. The time it takes to complete one repeating cycle.
- *Pulse repetition rate*. The number of **pulses per second.** Also called the pulse frequency.
- *Bit rate*. The number of bits per second. A bit is represented by either a "1" (high) or a "0" (low).
- ***Baud*** *rate*. The number of transitions (or signaling events) per second. It changes with the bit pattern and coding used.

For digital transmission, linearity has no significance. You should only be concerned with whether the light is high (on) or low (off). Noise immunity is much higher for digital transmission than for analog transmission because the detecting circuit needs only to recognize two levels that are usually well separated. Note that noise as it affects time jitter (variations in the time) is a more important problem. (See Chapter 10.)

The system does not have to be fast enough or to have a short response time to be able to reproduce the rise and fall times. A distortion in t_r and/or t_f may be acceptable as long as the high and low levels are detectable by the receiver. Bit rate is often related to BW. This relation depends on the digital code or format used.

6–2–3 Digital Formats (Coding)

Previously, you assumed that a "0" is represented by a low signal, and a "1" by a high signal. This is not the only **format** (coding) possible. When transmitting (or recording) digital data, the 1 and 0 can also be represented by transitions from level to level rather than by the level itself. A positive transition (0 to 1) may represent a 1 and a negative transition (1 to 0) a 0, or any transition might represent a 1 and no transition a 0. Figure 6–6 gives some of the formats in use. Note that for **RZ** (returns to zero), the number of transitions is two per bit. For **NRZ** (nonreturns to zero), it is bit pattern dependent; the largest number of transitions is produced by a 0101 . . . pattern and is one transition per bit. The other codes can be similarly classified. In the NRZ mark and **biphase (Manchester code),** a 1 or 0 is represented by a transition during **bit time,** rather than by a particular level.

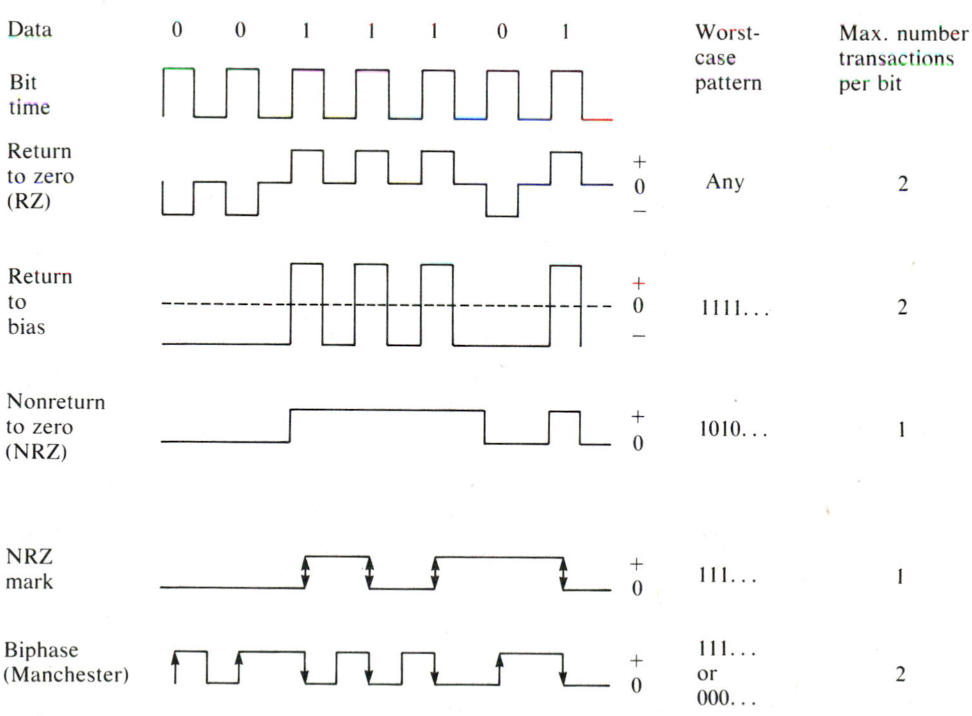

FIGURE 6–6 Digital data coding (formats).

The reason for considering the number of transitions involved is that the BW is related to the number of transitions per second. Clearly, to send an all-0000 message in NRZ requires only direct current; there are no transitions.

6–3 DATA RATE, BANDWIDTH, AND DISPERSION

Data rate, in bits per second, is the rate of information transfer and not necessarily the number of pulses or transitions per second. The number of signal transitions per second is given as the baud rate, which is a characteristic of the communication system. Telephone line systems operate at 9600 Bd (or lower), while very high speed systems may operate at rates in excess of 2 GBd. The baud rate specified for a particular system is related to the bit rate at which the system can transmit. For binary systems (systems recognizing only the levels 0 and 1), baud and bit rates are often interchanged.

Bandwidth is another term often used to describe the speed (rate of information transfer) of a system. BW refers to the highest sinusoidal frequency (not pulsed waveform) at which the system can operate with little amplitude loss in comparison with a much lower frequency. The BW of a system is directly related to the data rate (bits per second) that the system can handle. The dispersion in an optical fiber affects the BW and the maximum data rate that can be used in the system.

6–3–1 Bandwidth and Data Rate

The BW of a fiber or system is the range of frequencies that it can handle with minimal amplitude distortion. A more precise definition is that it is the range of frequencies between the two points, where the output optical power drops to 50% of its maximum (Figure 6–7). The drop to 50% means a loss of 3 dB.

$$10 \times \log(1/2) = -3 \text{ dB}$$

There are no low-frequency restrictions introduced by the fiber. The lowest frequency carried by the fiber is DC ($f = 0$). Consequently, BW is the upper 3-dB point, f_2 in Figure 6–7.

To understand the relation between data rate and BW, assume some upper frequency f_2, shown in Figure 6–8. As is demonstrated using an RZ code (11 . . . bit pattern), you can fit one bit of information into one cycle of the sine wave. This yields

$$B_r(\text{max}) \approx \text{BW} \tag{6–1}$$

where $B_r(\text{max})$ is the maximum bit rate. Because BW is the maximum frequency, the bit rate involved is the maximum bit rate. Equation 6–1 is only an approximation because the sine wave is a distortion of the original bit stream. In fact, most BWs calculated are only approximations. The rise and

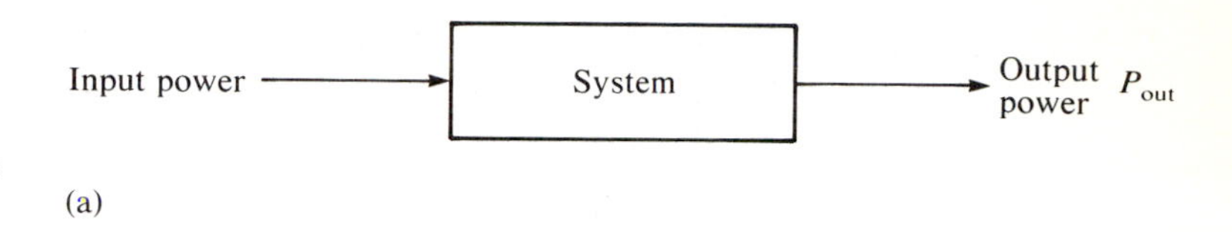

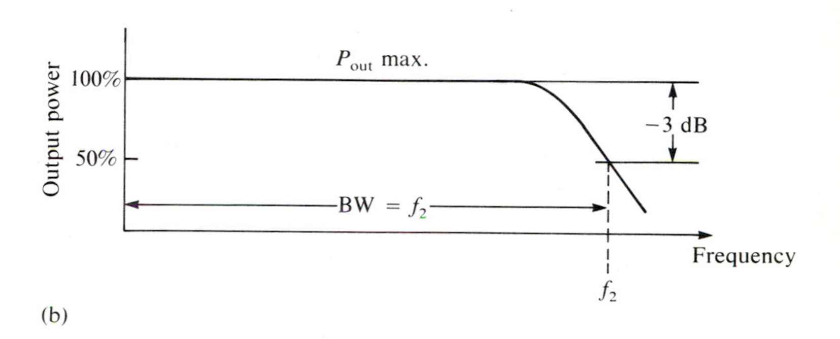

FIGURE 6–7 System bandwidths (BWs). (a) Block diagram. (b) Output power as a function of frequency (input power constant).

fall times are substantially larger for the sine wave. (To more faithfully reproduce the data bit stream of Figure 6–8, the BW must be much higher. In digital applications, however, it is necessary only to recognize levels or level changes, and hence, the BW given by Equation 6–1 is sufficient.) In Figure 6–9, the RZ code in Figure 6–8 is replaced by an NRZ code, with a 1010-bit pattern. Here,

$$B_r(\text{max}) = 2 \times \text{BW} \tag{6–2}$$

Each sine wave cycle can be used to represent two bits of information. The maximum bit rate is twice the BW.

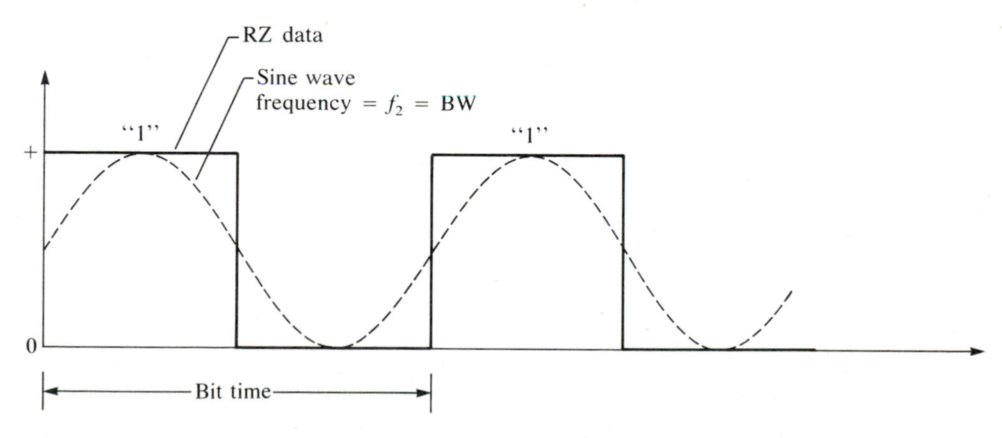

FIGURE 6–8 Bandwidth (BW) and data rate: return to zero (RZ) format. (Sine wave frequency shown is BW required for 1,1 data pattern.)

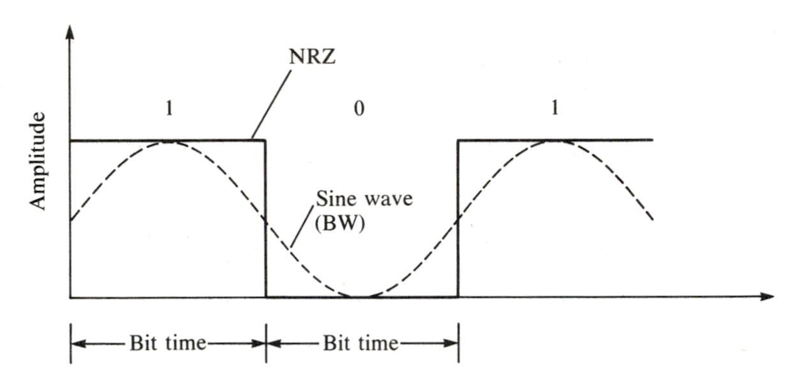

FIGURE 6–9 Bandwidth (BW) and data rate: nonreturn to zero (NRZ) format. (BW here is half that for return to zero code; data pattern is 1,0.)

6–3–2 Effects of Dispersion

Dispersion causes a broadening of the transmitted pulse. Figure 6–10 illustrates pulse broadening. The output amplitude is shown to be attenuated due to losses in the fiber. Note that the input pulse is idealized ($t_r = t_f = 0$.) As shown, the output pulse width is larger than the input pulse width, $t_{wo} > t_{wi}$.

The effects of pulse broadening are demonstrated in Figure 6–11. A 1010 pattern has been selected, and the output is shown under three dispersion

FIGURE 6–10 Pulse broadening.

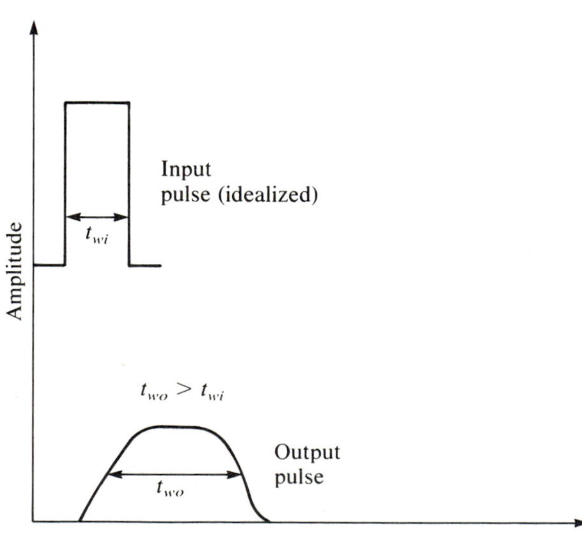

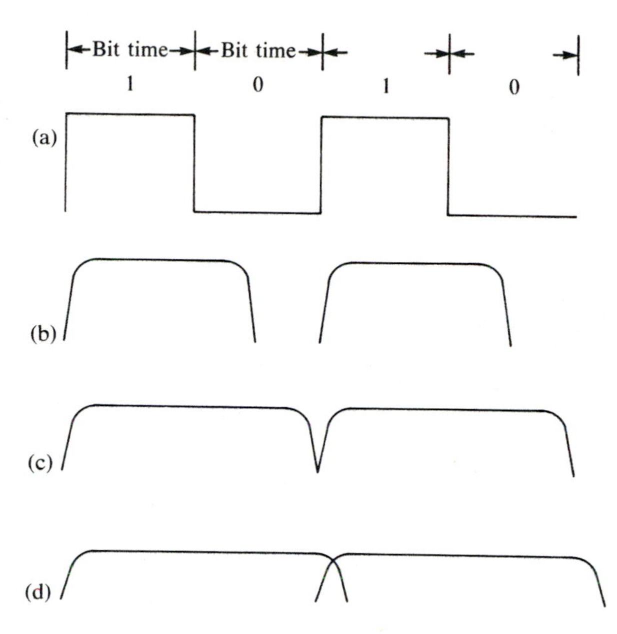

FIGURE 6–11 Effects of dispersion (pulse broadening). (a) Input data. (b) Output with pulse broadening and no intersymbol interference [ISI]). (c) Some ISI. (d) High degree of ISI—0 lost.

conditions. The term "bit time," used in Figure 6–11, refers to the time period during which the receiver examines the signal to determine whether a 1 or a 0 was received. For clarity, the propagation delay in the fiber was neglected. As you can see in Figures 6–11(b)–(d), the increase in dispersion eventually leads to errors in reception. The overlap between the symbols is called **intersymbol interference** (ISI). In Figure 6–11(d), the ISI is so substantial that the 0 is completely lost. A conservative estimate of the BW for a given dispersion Δt is given by

$$\text{BW} = 1/(2 \times \Delta t) \qquad \textbf{(6–3)}$$

The relationship between maximum bit rate of the fiber, its **capacity,** and BW is given in Equations 6–1 and 6–2 for different formats. All of the preceding discussion assumes that there is no ISI. In practice, the maximum rate can be larger, depending on the particular communication technique used and on the acceptable error rate. (In digital systems, error rate (the number of errors per some number of data bits), which is determined statistically, increases with increases in ISI. Allowing ISI to occur permits a higher data rate.)

Remember that the dispersion Δt is directly related to the type of fiber used. For step-index fibers, where intermodal dispersion dominates, BW is related to Δ and to numerical aperture (N.A.). BW is proportional to $1/(\Delta \times \Delta^{1/2})$ and $1/(\text{N.A.})^3$. Larger Δ or N.A. result in lower BW. Figure 6–12 shows the relation between the number of modes, Δ, N.A., and BW.

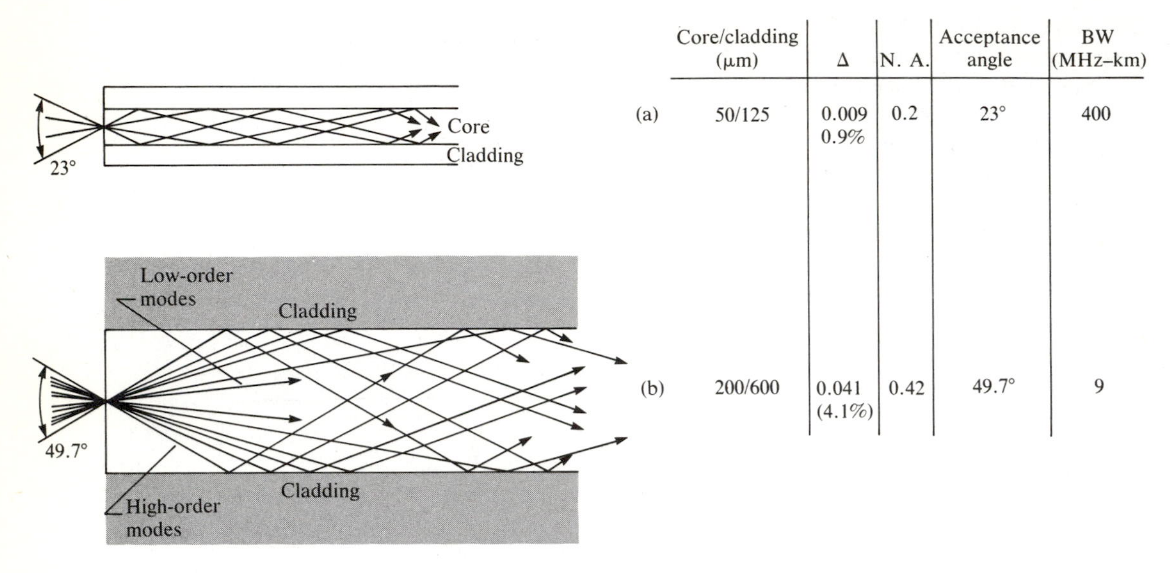

	Core/cladding (μm)	Δ	N. A.	Acceptance angle	BW (MHz–km)
(a)	50/125	0.009 0.9%	0.2	23°	400
(b)	200/600	0.041 (4.1%)	0.42	49.7°	9

FIGURE 6–12 Propagation modes. (a) Low Δ. (b) High Δ.

EXAMPLE 6–1

A graded-index fiber has a dispersion of 1.5 ns/km. An 8-km stretch of fiber was installed. Estimate the system BW due to the fiber alone.

Solution

$$\text{Total} = \Delta t_{tot} = 1.5 \times 10^{-9} \times 8 = 12 \text{ ns}$$
$$\text{BW} \simeq 1/(2 \times 12 \times 10^{-9}) = 42 \text{ MHz}$$

EXAMPLE 6–2

A communication system used a 1-km fiber with N.A. = 0.2 and BW = 50 MHz. The fiber was replaced by a fiber with N.A. = 0.25. Estimate the BW. (Neglect the effects of all components except the fiber.)

Solution

Because BW is related to $1/(\text{N.A.})^3$, the ratio

$$\text{BW}_{new}/\text{BW}_{old} = (\text{N.A.}_{old}/\text{N.A.}_{new})^3$$

The new BW becomes

$$\text{BW} = 50 \times (0.2/0.25)^3 = 25.6 \text{ MHz}$$

6–3–3 Bandwidth-Length Product

BW is length dependent. Longer fiber results in more pulse spreading and leads to lower BW. As a result, the fiber BW is often given in terms of the BW times kilometer product. A 1000 MHz × km fiber can usually operate with a 100-MHz BW if a 1-km fiber is used or with a 1000-MHz BW if a 1-km fiber is used.

EXAMPLE 6–3

---

1. Find the BW-length product in MHz × km for the problem in Example 6–1.
2. Find the BW for a 3-km length of the same fiber.

Solution

1. The BW for the 8-km stretch of fiber is 42 MHz.

$$\text{BW} \times L = 42 \times 8 = 336 \text{ MHz-km}$$

2. The expected BW for a 3-km fiber is

$$\text{BW} = (\text{BW length})/L = 336/3 = 112 \text{ MHz}$$

L denotes length.

6–3–4 Electrical and Optical Bandwidth

A distinction must be made between electrical and optical BW. **Electrical bandwidth** (BW_{el}) is defined as the frequency at which the ratio I_{out}/I_{in} (Figure 6–13) drops to 0.707. The **optical bandwidth** (BW_{opt}) is defined as the frequency at which the ratio, P_{Lo}/P_{Li} dropped to 1/2. (The ratios I_{out}/I_{in} and P_{Lo}/P_{Li} have maximum values of 1.) Because P_{Li} and P_{Lo} are directly proportional to I_{in} and I_{out}, respectively (and not to I_{in}^2 and I_{out}^2 as in an all-electrical system), the half-power point is equivalent to the half-current point. That is the point where I_{out}/I_{in} drops to 0.50, not to 0.707. This results in a BW_{opt} that is larger than the BW_{el}.

$$\begin{aligned} BW_{el} &= BW_{opt} \times (1/2)^{1/2} \\ BW_{el} &= 0.707 \times BW_{opt} \end{aligned} \qquad \textbf{(6–4)}$$

It is important to realize that these two parameters represent two ways of describing the same system. For example, a system can be said to have an optical BW of 10 MHz, which implies that its electrical BW is 7.07 MHz.

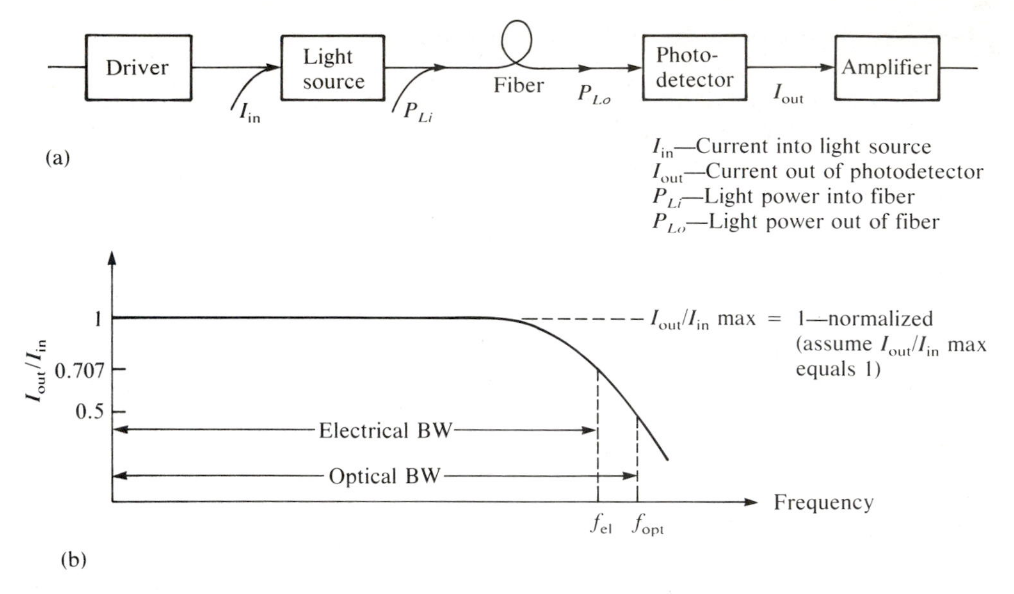

FIGURE 6–13 Electrical and optical bandwidth (BW). (a) Block diagram of system. (b) Frequency response.

SUMMARY AND GLOSSARY

The terms listed below are used for the first time in this chapter. Many other terms used in this chapter have been introduced in previous chapters and should be familiar to you. You should familiarize yourself with these terms and think of their significance and how they are connected to other terms and to the material covered in this chapter. Use these definitions to review the material covered.

ADC. Analog-to-digital converter.

Analog. A quantity that can vary continuously. It can attain any value (within minimum and maximum constraints.)

ASCII code. American Standards Committee for Information Interchange code. A 7-bit binary code, used to represent a variety of symbols, including the digits 0 through 9, the letters A through Z and a through z.

Baud. A line-signaling rate. Signal transitions per second.

Binary code. General meaning: a sequence of "1"s and "0"s used to represent symbols or numbers. Here, it refers to coding format, a method for representing data bits. For example, a 1 might be a high and a 0 a low, or a 1 might be a positive transition and a 0 a negative transition.

Biphase. A coding format for digital transmission (also called the Manchester code.) See Figure 6–6.

Bit time. The time slot in a transmission during which the data bit is present.

Capacity. When used in connection with a digital system or line, the highest data rate with which the system can operate. Maximum bit rate.

DAC. Digital-to-analog converter.

Digital. Discrete data quantities that vary in discrete steps.

Electrical bandwidth. Standard definition of bandwidth. For systems with direct current ($f_1 = 0$), it is the frequency for which the output amplitude drops to 0.707 of its maximum value, or by 3 dB.

External noise. Electrical noise generated outside the device or system (line interference, transients, etc.).

Fall time t_f. The time it takes for a waveform to change from 90 to 10% of its maximum amplitude.

Format. See definition of coding format under "binary code."

Intensity modulation. In connection with an optical system, changes in the intensity of the light used to represent data carried by the light.

Internal noise. Electrical noise generated by the device itself (thermal noise, resistor noise, etc.)

Intersymbol interference. The overlapping of binary symbols. For example, a 1 level might overlap a 0 that follows and thus increase the probability of error.

Light detector. Photodetector. A device that converts light intensity to current or voltage.

Linearity. A measure describing how close to a straight line a relation between two parameters is.

Manchester code. See "biphase."

NRZ. Nonreturn to zero format. See Figure 6–6.

Optical Bandwidth. The frequency for which the optical output power drops to half its maximum value.

Period T. The time of one cycle of a repetitive waveform.

Pulses per second. Pulse rate.

Response time. A system or device characteristic that describes how fast the system output can change from 10 to 90% of its maximum (like rise time for a waveform).

Rise time t_r. The time it takes for the voltage in a waveform to change from 10 to 90% of its maximum amplitude.

RZ. Return to zero format. See Figure 6–6.

Sample rate. The number of samples taken per second.

Sampling. Examining (measuring) a continuous analog signal at specific intervals of time.

Sampling theorem. At least two samples per cycle are required for the case of a noiseless channel.

Signal-to-noise ratio. The ratio of signal power to noise power in an electronic system.

FORMULAS

$$B_r(\max) = \text{BW} \tag{6–1}$$

For a RZ format, BW, equals the maximum data rate for the system $B_r(max)$.

$$B_r(\max) = 2 \times \text{BW} \tag{6–2}$$

Same as Equation 6–1, but for NRZ.

$$\text{BW} = 1/(2 \times \Delta t) \tag{6–3}$$

The relation between BW and total dispersion Δt.

$$\text{BW}_{\text{el}} = 0.707 \times \text{BW}_{\text{opt}} \tag{6–4}$$

The relation between electrical and optical BWs.

QUESTIONS

1. What is meant by intensity modulation?
2. Which of the following data, signals, or systems are digital and which are analog?
   a. Your home alarm system
   b. The water level in your pool
   c. The number of students in your class
   d. The intensity of light in your classroom
   e. Your high-fidelity system
   f. Your calculator
3. Typically, what is the relation between the frequency of an analog signal and the digital data rate used to transmit it?
4. What are the important characteristics of an analog transmission system?
5. Is the LED or the laser more suitable for analog transmission? Why?
6. What is meant by signal-to-noise ratio?
7. Is linearity important in digital transmission?

8. What is the difference among the RZ, NRZ, and NRZ mark formats?
9. What is intersymbol interference?
10. How is the system BW related to the data rate it can carry (its capacity)?
11. For a given data rate, how does dispersion affect intersymbol interference?
12. How do the following affect the fiber data rate (fiber capacity)?
    a. N.A.
    b. Fiber lengths
    c. Number of modes propagating in the fiber
    d. Relative refractive index difference Δ
13. Why is fiber BW specified in terms of Hz $\times$ km, and not simply in Hz?
14. What is the difference between electrical and optical BWs?

PROBLEMS

1. An analog signal with a frequency of 10 kHz is converted into a digital data stream. Each digital word is 8 bits, and 3 samples are taken for each cycle.
    a. What is the sampling rate?
    b. What is the data rate (bits per second)?
2. A digital system is said to have a BW of 100 MHz.
    a. Estimate the maximum data rate (capacity) of the system if RZ is used.
    b. Estimate the data rate if NRZ is used.
3. A fiber has a dispersion of 5 ns/km.
    a. What is the maximum data rate for a 10-km length using NRZ format? Consider only the effects of the fiber itself.
    b. Repeat step a for a 100-m length.
4. A fiber with a N.A. = 0.3 is replacing the original fiber used, which had N.A. = 0.2. If the original BW was 100 MHz, calculate the new BW. Consider only the effect of the fiber.
5. What is the BW of a 10-km length of the fiber in Figure 6–12(a)?
6. What is the BW of a 100-m length of the fiber in Figure 6—12(b)?
7. A fiber is specified to have a dispersion of 5 ns/km. Find the BW $\times$ km of the fiber.
8. In a system, a 10-km fiber with dispersion of 2 ns/km is replaced by a 15-km fiber with dispersion of 1.5 ns/km. Find the reduction or increase in BW (in terms of a ratio). Ignore all effects except those of the fibers.
9. A digital communication system is designed to operate at 100 Mb/s, and it requires a 10-km length of fiber. Which of the following fibers are suitable? (Assume the best choice of format and that only the fiber limits the data rate.)
    a. A fiber with a dispersion of 1.2 ns/km
    b. A 800-MHz $\times$ km fiber

10. A fiber optic communication system is said to have an optical BW of 100 MHz. What is its electrical BW?
11. The frequency response of a fiber optic system was taken by measuring the signal current of the photodetector I_{PD}, as a function of frequency. (The input amplitude remained constant.) At low frequencies, $I_{PD} = 120\ \mu A$ rms. At $f = 150$ MHz, $I_{PD} = 84.84\ \mu A$ rms. Find the optical BW.

7 Modulation and Multiplexing

CHAPTER OBJECTIVES

The subjects covered in this chapter relate to methods of data transmission, including various modulation schemes and multiplexing techniques. The objective is to give you working definitions of these methods in terms of how they are accomplished. There is no attempt to present detailed circuitry. The focus is on functional descriptions.

You will familiarize yourself with the standard terminology and be able to distinguish among the different techniques and when they can be applied. You will learn about amplitude, frequency, and pulse modulation schemes and three multiplexing systems: time division multiplexing (TDM), frequency division multiplexing (FDM), and wavelength division multiplexing (WDM).

7–1 INTRODUCTION

Modulation and multiplexing are techniques used in electronic communication and have been applied in varying degrees to fiber optic communication. The discussion of **modulation** concerns methods of transmitting information by a carrier. In this case, the carrier is light. **Multiplexing** involves transmitting data from many independent sources through a single channel. The channel used here is the optical fiber.

There are many schemes of modulation. In **frequency modulation** (FM), for example, the data to be transmitted modify the carrier frequency. The modulated carrier then carries the data, analog or digital, as frequency variations, which represent the input signal amplitude. For example, binary input signals, high and low, will be represented by two carrier frequencies. In other words, a shift in carrier frequency represents the change from "high" to "low" (or from "low" to "high"). In another scheme, **amplitude modulation** (AM), the changes in the carrier amplitude represent input signal amplitude changes. A high-input amplitude yields an increased carrier amplitude and vice versa. (This is a somewhat oversimplified definition of AM. Although the carrier amplitude appears to vary with input signal amplitude, the input information is actually carried by side bands.) In the discussion of the modulation of light, you are restricted to intensity modulation, where the intensity of the light represents the input signal amplitude (see Section 6–1). This simply means that the larger input signal amplitude, the higher the intensity of the light transmitted.

7–2 MODULATION

The discussion here focuses on another level of modulation. It is concerned with the relation between an analog input quantity and a pulse train that is used to represent it. The pulse train is then used to intensity modulate the light carrier. (Direct digital modulation, where binary numbers or codes are involved was discussed in Section 6–2–2.) Here, you have two levels of modulation. First, the input signal modulates a pulse train. The modulated pulse train then modulates the light.

The modulation of the pulse train may be pulse amplitude modulation (PAM), pulse code modulation (PCM), or other techniques such as pulse width modulation (PWM) and pulse position modulation (PPM). Remember that the input is an analog quantity, such as a voltage, which may vary with time (e.g., a sinusoidal voltage), which does not directly modulate the light carrier.

7–2–1 Pulse Amplitude Modulation

In **pulse amplitude modulation,** the amplitude of the modulating signal, the input, is made to control the amplitude of a pulse carrier. (Carrier here refers not to the light that ultimately carries the information, but to the pulse train that is modulated to represent the input signal.) Figure 7–1 gives the waveforms of such a system. You can view this as a sampling process. At pulse time (whenever the carrier pulse is high), the signal amplitude is observed (sampled) and the amplitude of the sample, with appropriate amplification, is forwarded to the drive circuitry of the light source. Appropriate filtering reconstructs the signal at the receiving end.

The pulse repetition rate of the carrier must be at least twice the signal frequency. To improve the immunity to noise, the sampling rate is often three

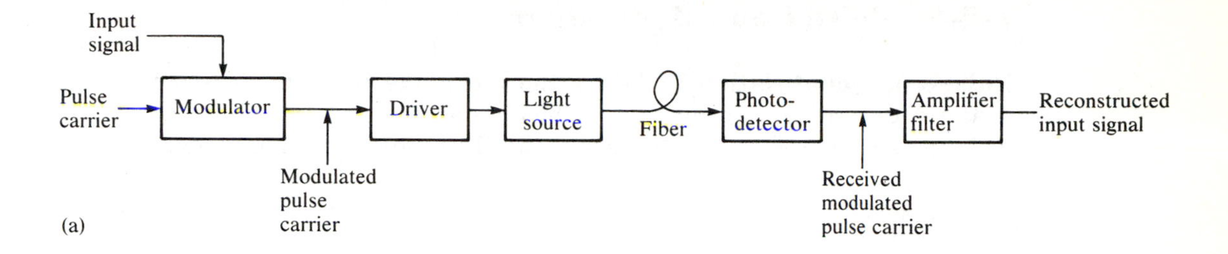

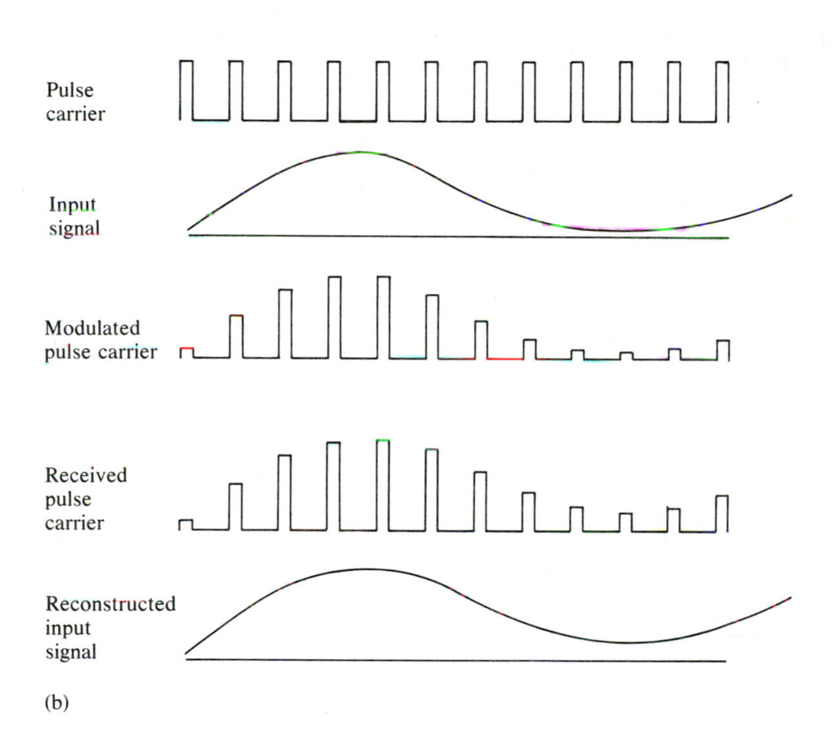

FIGURE 7–1 Pulse amplitude modulation. (a) Block diagram of system. (b) Waveforms.

or four times higher than the signal frequency. For a speaking voice (telephone voice), which has a bandwidth (BW) of about 4000 Hz, you can use a pulse carrier of about 16 kHz, four times higher than the highest frequency of the input signal. For a perfect (noiseless) input signal, you need only use a pulse train with a frequency double that of the signal frequency. This means that you need only two samples for each input signal cycle. PAM is sensitive to noise interference and consequently not used very often.

7–2–2 Pulse Code Modulation

Pulse code modulation (PCM) is the technique frequently used in long-distance telephone transmission. A block diagram and signal waveforms of PCM are shown in Figure 7–2. As in PAM, the input is sampled. This time, the

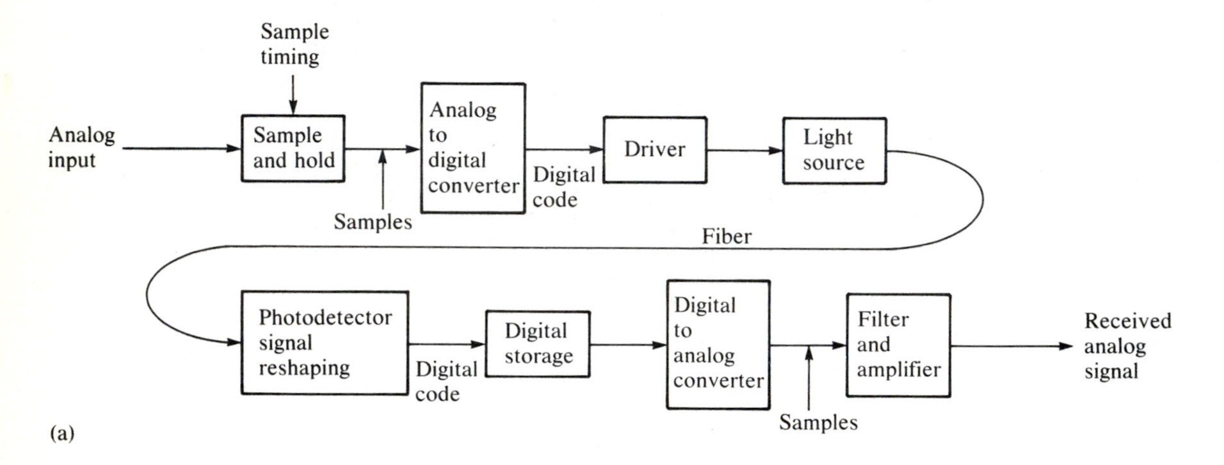

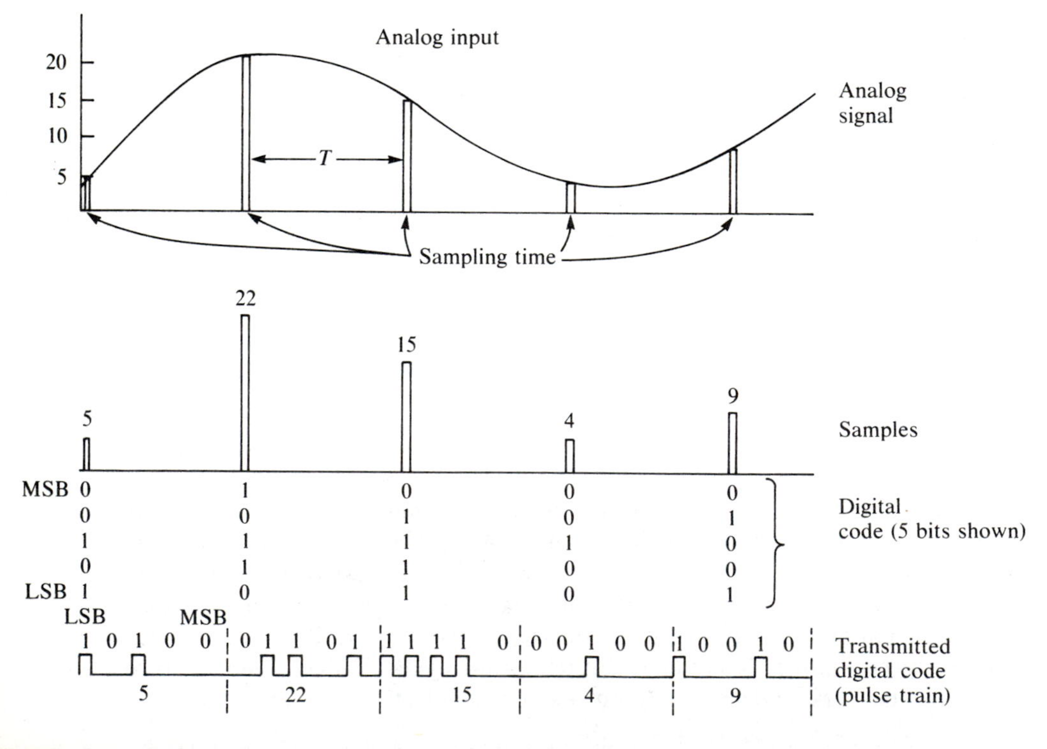

FIGURE 7–2 Pulse code modulation. (a) Block diagram of system. (b) Waveforms.

magnitude of each sample is converted to a binary code. In Figure 7–2, the binary code is simply the numerical binary value of the sample magnitude. This code is transmitted via the fiber. At the receiving end, the detected light is converted back to the binary code and then into the analog samples by a digital-to-analog converter. The filter shown is necessary to reproduce more closely the original input signal.

In this method, only "on-off" light signals are sent. It is easy to discriminate between a 0 and a 1, so noise is less of a problem. Note that in the PAM technique, the amplitude of each sample must be properly maintained, while here, you need only to discriminate between high and low levels.

Figure 7–2, shows five samples, with each sample converted to a 5-bit code. If, for example, you use four samples for each input cycle, with each sample converted into an 8-bit code, you end up with $4 \times 8 = 32$ bits for each input cycle. For a 4-kHz analog signal, you need $4000 \times 32 = 128$ kb/s. The minimum data rate, based on two samples per input cycle, for the 4000-Hz signal is then 64 kb/s. This rate is the standard used in the United States for PCM voice transmission. The general expression of the bit rate for PCM is

$$\text{Bit rate} = \text{BW} \times (\text{samples/cycle}) \times (\text{bits/sample}) \qquad \textbf{(7–1)}$$

EXAMPLE 7–1

The European standard for telephone-voice transmission via PCM is (1) voice BW: 3.4 kHz; (2) sampling rate: 8 kHz (8000 samples per second); and (3) code: 8 bits. Also, 16 timing bits are added every 30 bytes (240 bits).

Find the total bit rate.

Solution

Note that the sampling rate is slightly more than two samples/cycle (at the 3.4-kHz upper frequency).

$$(8000 \text{ samples/s})/(3400 \text{ cycles/s}) = 2.35 \text{ samples/cycle}$$

$$\text{Data rate} = 8000 \text{ samples/s} \times 8 \text{ bits/sample} = 64 \text{ kb/s}$$

$$\text{Timing rate (timing bits per second)} = (64{,}000/240) \times 16$$

64,000/240 is the number of times per second that 16 bits are added.

$$\text{Bit rate} = 64{,}000 + [16 \times (64{,}000/240)] = 68.267 \text{ kb/s}$$

7–2–3 Other Modulation Schemes

The following modulation schemes are not commonly used and are discussed here only briefly.

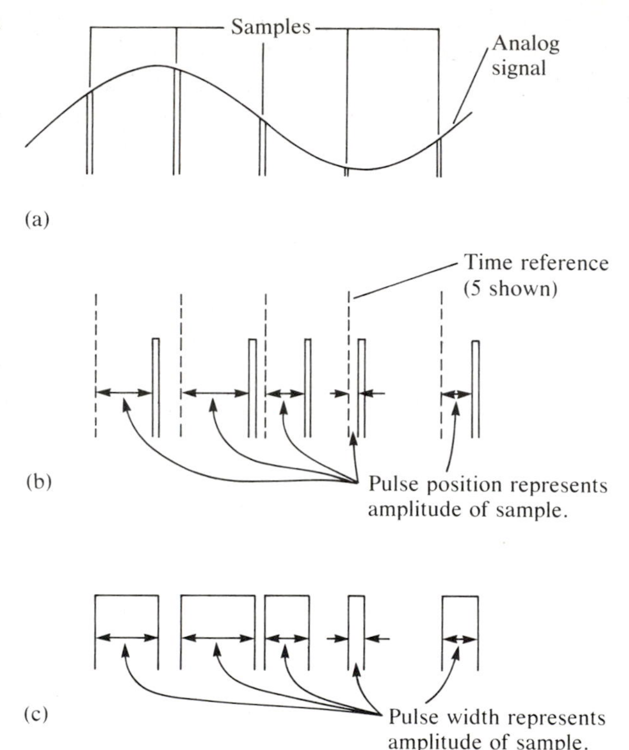

FIGURE 7–3 Pulse position modulation (PPM) and pulse width modulation (PWM). (a) Analog signal and samples. (b) PPM waveforms. (c) PWM waveforms.

- ***Pulse position modulation* (PPM).** The position of the pulse with respect to a reference time pulse (Figure 7–3(b)) represents the magnitude of the analog sample (Figure 7–3(a)).
- ***Pulse width modulation* (PWM).** The pulse width represents the amplitude of the sample. A sample with larger amplitude yields a wider pulse (Figure 7–3(c)).

For the PPM and PWM, the amplitude of the pulse is immaterial. It is an on-off pulse. Its position or width is the information. In both of these schemes, variations in time **(time jitter)** may cause errors. These are somewhat more susceptible to noise interference than is PCM.

7–3 MULTIPLEXING

The fundamental purpose of multiplexing is to share the BW of an information channel among many users. Here, fiber is the information channel; light is the carrier. We will assume that all independent data sources conform to some overall requirements, such as amplitude, rate, and BW.

7–3–1 Time Division Multiplexing

In **time division multiplexing** (TDM) transmission, time on the information channel is shared among many data sources. The source, in this context, may be data in any of the modulation formats discussed in the previous section. A source, for example, may be a PCM or a PAM pulse train that is used to represent the input analog signals. (Although there are multiplexing systems that multiplex analog signals directly, these will not be discussed here.) The data sources, one at a time, use the fiber to transmit their data. For clarity, assume that the data each source sends is a string of 1's and 0's, maybe PCM data.

Figure 7–4 gives a more detailed description of TDM when applied to digital data. Each source is allocated a time during which it transmits its data.

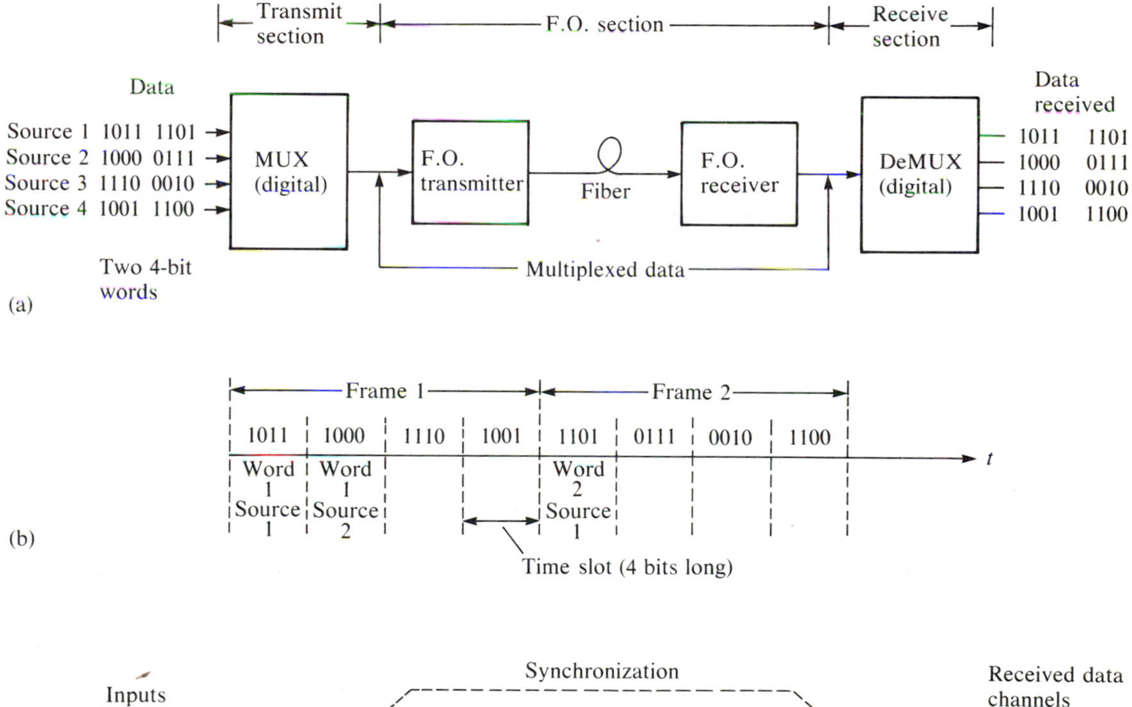

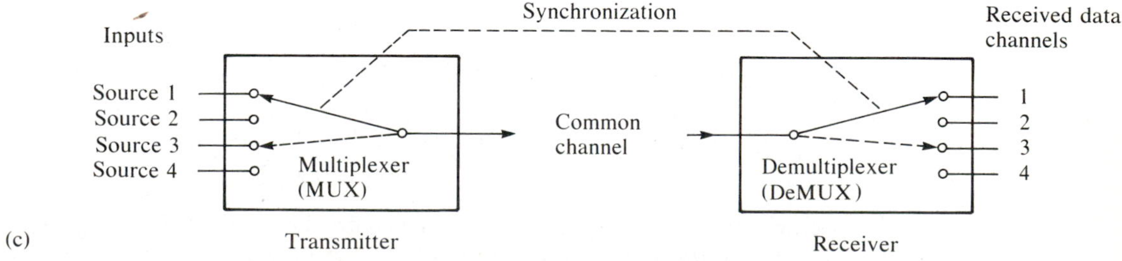

FIGURE 7–4 Time division multiplexing. (a) Block diagram of system. (b) Multiplexed data in time slots. (c) Simplified model of multiplexer (MUX) and demultiplexer (DeMUX).

One at a time, the four sources in Figure 7–4 transmit their data. When source 4 completes its transmission, the next time slot is assigned to source 1. (This scheme of multiplexing is called fixed TDM because the order of transmission is fixed. Other techniques allow sources with more data to transmit more often than do the others.) The sequence, source 1 through source 4 (or source 1 through the last source involved), constitutes a **frame.** Note that, in Figure 7–4, the data consists of 4-bit groupings.

Figure 7–4 is an oversimplification of the multiplexer (MUX) operation. In practice, each input channel feeds an input buffer used to temporarily store the input data. The data are then multiplexed onto the common channel, the optical fiber. In this way, a high-speed common channel can be used to transmit a large number of lower data rate inputs.

The relation among the number of multiplexed input channels N, the input channel data rate, and the output (or multiplexed) data rate is given by

$$\text{MUX output rate} = N \times \text{Input rate} \tag{7–2}$$

(Some timing and control signals that are usually required are ignored here.)

The capacity of the common channel connected to the MUX must be equal to or greater than the output rate in Equation 7–2. In other words, the common channel must be capable of operating at this output rate. In Equation 7–2, the data rate of all input channels is assumed to be the same. For varying input rates, the highest rate must be used to determine the common channel capacity (true for fixed TDM only).

Another parameter often mentioned in connection with TDM is the channel switching rate. This refers to the number of inputs visited (accessed) per second. (Each input is considered an input data channel.)

EXAMPLE 7–2

A digital MUX operates with 16 sources. The rate of data in each source is 1000 bytes/s. (Note that 1 byte is 8 bits.) Data are transmitted byte by byte.

1. What is the data rate of the MUX output?
2. What is the channel switching rate?

Solution

1. The data rate of each input channel is 8×1000 b/s. The output data rate becomes

$$\begin{aligned}\text{Output rate} &= N \times \text{Input rate}\\ &= 16 \times 8 \times 1000 = 128 \text{ kb/s}\end{aligned}$$

2. Each channel (source) must have access to the MUX 1000 times each second, transmitting 1 byte at a time. The channel switching rate is

$$16 \times 1000 = 16{,}000 \text{ channels per second}$$

The timing diagram shown in Figure 7–4(b) is incomplete. Timing markers (**synchronization** signals, which lock the signals to a time frame) are usually added at the end (or at the beginning) of each time slot and frame to identify clearly the time slots and frames. The receiving end uses these time markers to reconstruct the original data.

Figure 7–4(c) gives a simplified model for a MUX transmission system. Simple multiposition switches are used to simulate the multiplexer and **demultiplexer** (DeMUX). When the MUX is in position 1, so is the DeMUX switch. Consequently, for the duration that the MUX and DeMUX are in position 1, data from source 1 use the common channel for transmission. The next time slot is allocated to source 2. That means that both MUX and DeMUX are in positions 2, and so on. The MUX and DeMUX switches move in synchronism. Source 1 is reconnected to the common transmission channel following the transmission of data from the last multiplexed source, source 4 in Figure 7–4.

7–3–2 Frequency Division Multiplexing

In **frequency division multiplexing** (FDM), the information channel BW is divided into frequency slots (subcarriers) and each source is assigned a subcarrier. A block diagram of a typical FDM system is shown in Figure 7–5(a). Assume that a particular fiber system has a BW of 100 MHz. (That means that the light transmitted can be modulated by a carrier, or carriers, with frequencies of up to 100 MHz.) Then proceed to assign different carriers (in this context, referred to as **subcarriers**) to the different data sources. As shown in Figure 7–5(b) the allocations to the subcarriers are separated by **guard bands.** This makes it easier at the receiver end to separate the subcarriers.

As an example, in Figure 7–5(b), all subcarriers have a 5-MHz BW. In reality, the BW allocation would vary with the application. You do to the frequency scale in FDM, what you did to the time scale in TDM. Each subcarrier is modulated by its data source, which is PAM or PCM or any other data. You may have three levels of modulation. The analog data modulates a pulse train (PAM) that is then used to modulate a subcarrier (AM or FM). The combined modulated subcarriers modulate the light source.

A bank of **band pass filters** recover the original subcarriers at the receiver. These filters allow a limited range of frequencies to pass unattenuated. The filter bank, here, may consist of filters with BW of, for example, 5–10 MHz or 15–20 MHz.

In Figure 7–5(a), the mixing of the subcarriers is performed at the electronic signal levels. This is conducive to **crosstalk,** the interference by signals from one subcarrier with another subcarrier. The crosstalk can be eliminated by doing the mixing at the optical level. Figure 7–6 gives the block diagram of such a system. Note that the receiving end has not been changed at all. When multiplexing at the optical level, each data source modulates a

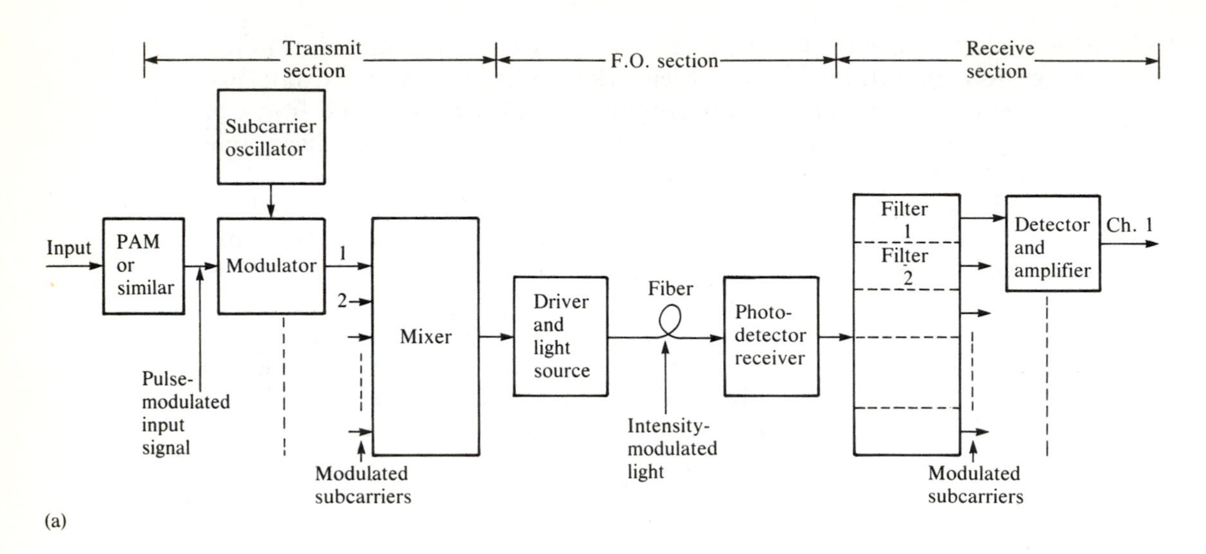

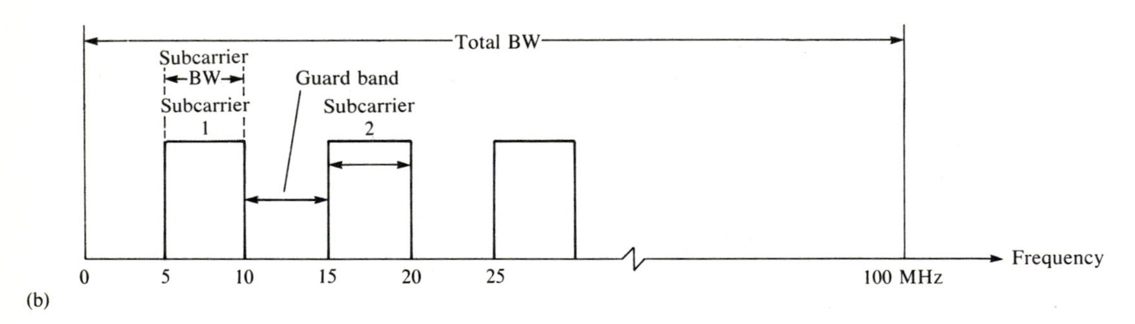

FIGURE 7–5 Frequency division multiplexing. (a) Block diagram of system. (b) Frequency bands of transmitted signal.

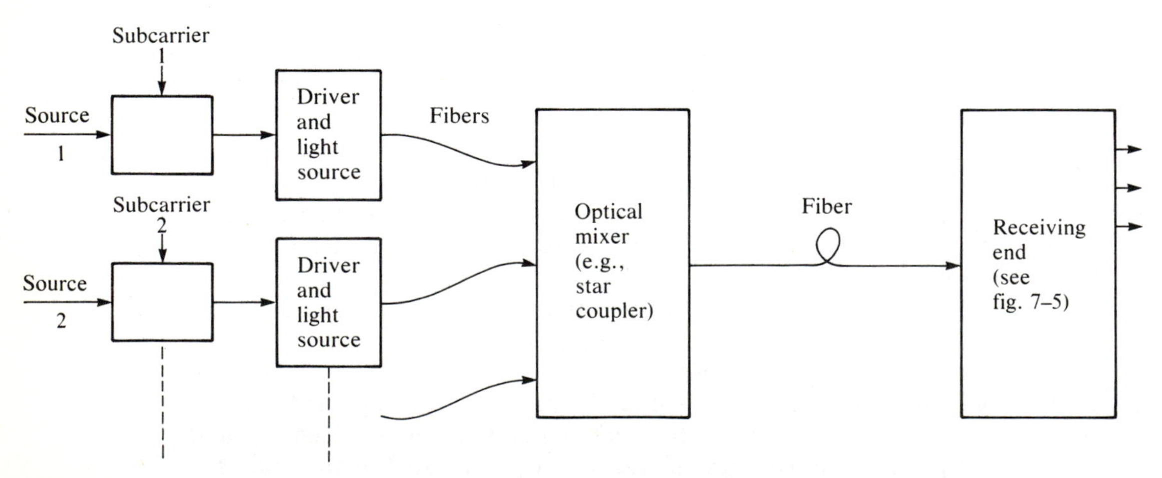

FIGURE 7–6 Frequency division multiplexing—mixing at the optical signal level.

subcarrier and drives its own light source. There will be as many light sources as multiplexed data sources. In the multiplexing scheme of Figure 7–5, there is only one light source connected to the fiber. The relatively large number of light sources introduce saturation problems at the photodetector. In other words, if a large number of inputs are involved, each with its own light source, and if you assume they may transmit simultaneously, the receiver photodetector (there is only one) is likely to saturate because of the cumulative light signal intensity. This will cause signal distortion and render the system inoperative. In this method, this means that there must be restrictions on the number of data sources involved and their maximum intensity.

7–3–3 Wavelength Division Multiplexing

In the **wavelength division multiplexing** (WDM) approach, each data source (or group of sources) is assigned a wavelength of transmission. (The different wavelengths are sometimes referred to as colors, hence, the term "color multiplexing".) A block diagram and wavelength plot are shown in Figure 7–7.

Typically, because of optical mixer complexities, only two or three wavelengths can be multiplexed presently. However, each wavelength carrier can be used with any of the previous multiplexing schemes. In this way, the data-carrying capacity (the overall data rate) can be increased many fold.

There is little interference between the multiplexed input signals when wavelength multiplexing is used. As is typical for all multiplexing at the optical level, the total optical power transmitted is increased as the number of data sources increases. This presents problems of saturation at the receiving optical detector. In WDM, however, the different light wavelengths (the different colors) are separated before the optical detector. Each wavelength is fed into a different detector, and saturation is unlikely. Both FDM and WDM allow simultaneous transmission by any or all data sources, while TDM requires that only one source transmits at a time.

7–3–4 Simplex and Duplex Systems

A communication channel connecting two nodes (node represents any device or system that receives and/or transmits data) may be one of two types. It may allow communications only in one direction (unidirectional channel), or it may provide for data flow in both directions (bidirectional channel). The unidirectional connection is called a **simplex channel;** the bidirectional is called a **duplex channel.**

The simplex system is used in data distribution where one node is a receiving node only. For example, the printer in a computer system only receives data; it does not transmit data (excluding some control signals) Figure 7–8 presents a block diagram of a typical fiber optic simplex system.

The two-way connection can be either **half duplex** or full duplex. The half-duplex connection allows two-way traffic, but not simultaneously. The

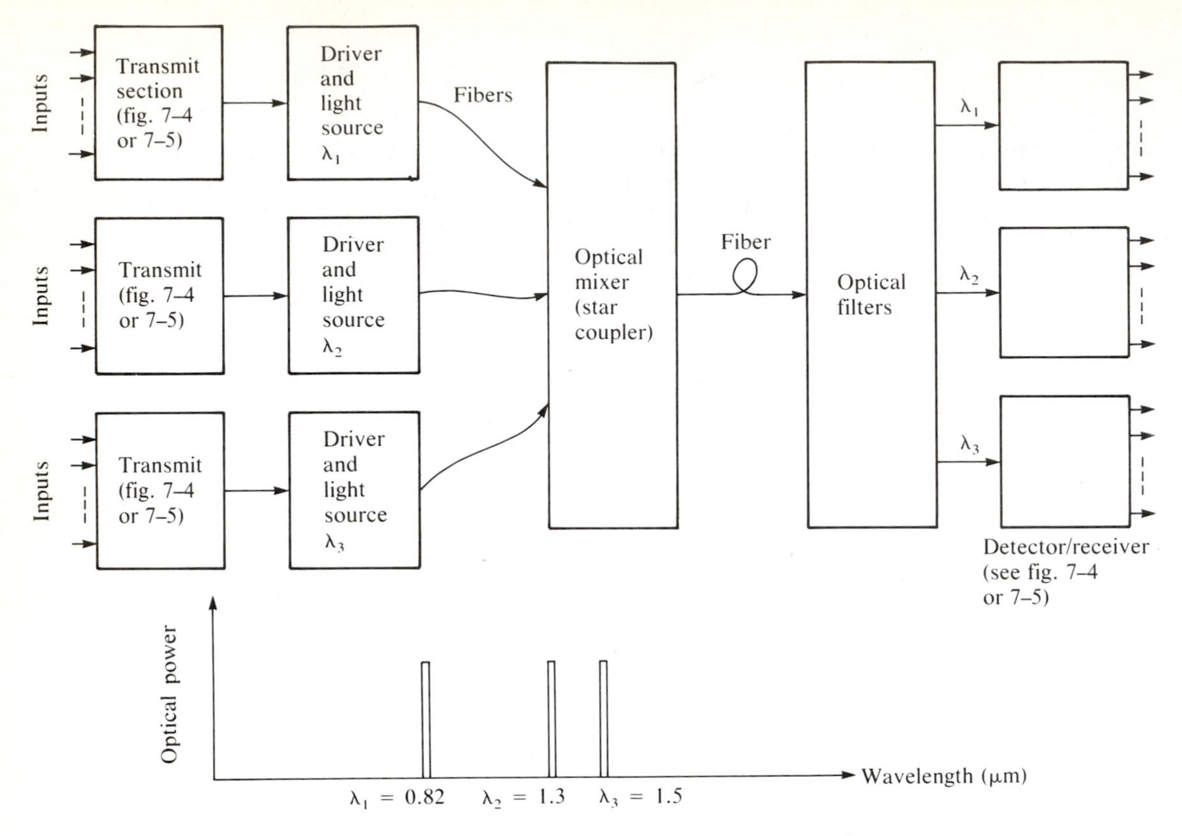

FIGURE 7–7 Wavelength division multiplexing. (a) Block diagram of system. (b) Typical wavelengths transmitted in main fiber.

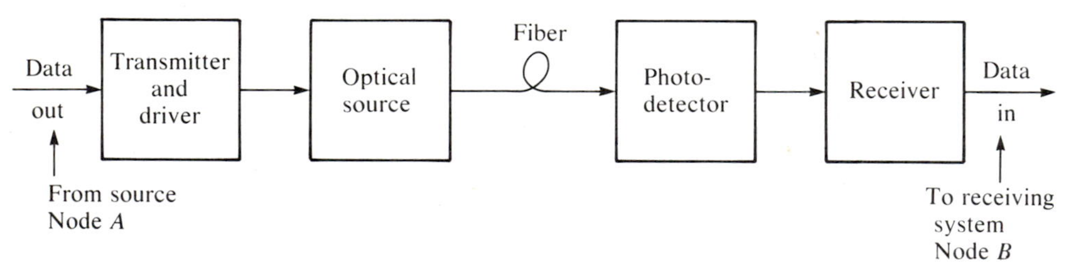

FIGURE 7–8 Simplex communication system.

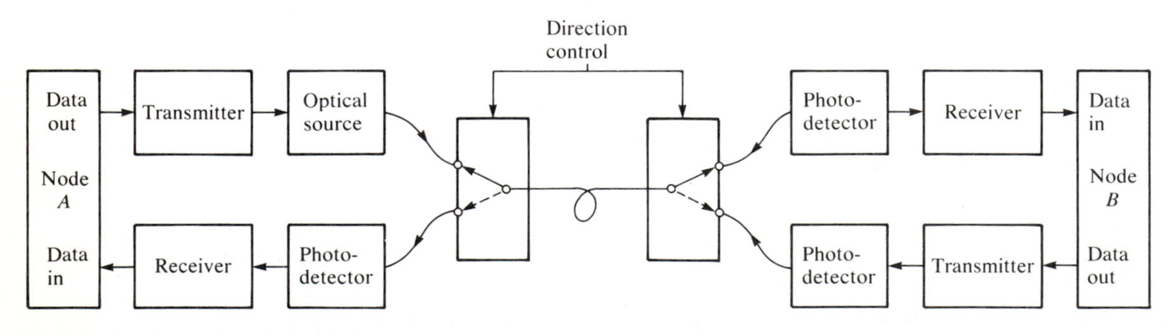

FIGURE 7–9 Half-duplex system.

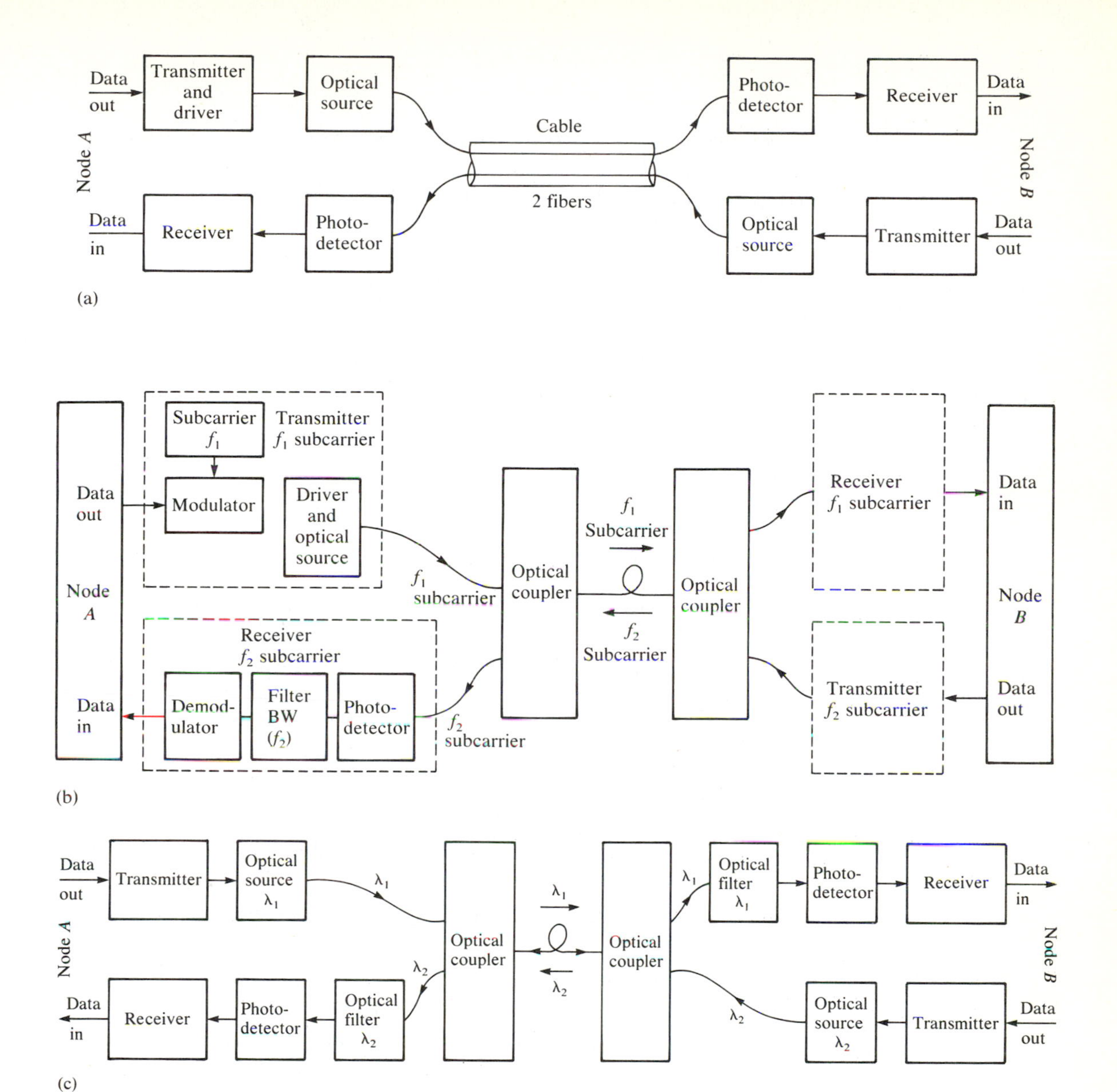

FIGURE 7–10 Full-duplex system. (a) Space multiplexing. (b) Frequency division multiplexing. (c) Wavelength division multiplexing.

block diagram in Figure 7–9 is typical of the half-duplex system. Note that data can move in either direction; however, it can move in only one direction at a time. When the control signal sets the switches up as shown by the solid lines, the flow is from node *A* to *B*. When the switches are set down as shown by the dashed lines, data move from *B* to *A*.

To provide full duplex, you must allow data transfer in both directions at the same time. This can be done a number of ways. Figure 7–10 shows three full duplex connections. In Figure 7–10(a), there are two fibers in the cable, each allowing traffic in one direction. This system is often called **space multiplexing.** Figure 7–10(b) shows how FDM can be used to provide full-duplex communications. Here a single fiber is used with two subcarriers. Each subcarrier is providing data flow in one direction. Subcarrier f_1 carries data from A to B, while subcarrier f_2 carries data from B to A. Figure 7–10(c) shows how WDM can be used to provide a full-duplex connection using a single fiber. Data flow from A to B is provided by the λ_1 wavelength light source, while λ_2 provides the reverse path. In both FDM and WDM, light is propagating in the fiber in opposite directions, simultaneously, without interference.

SUMMARY AND GLOSSARY

To be useful as a review tool, the terms and keywords in the glossary should be related to their respective contexts. It is not enough to memorize the definition, for example, of sample rate; you must understand the term in the context in which it appears in this text. To assist in this task, the definitions are sometimes deliberately narrow in scope, to be more closely related to the material covered in this chapter. Remember, only terms introduced in this chapter are covered here.

Amplitude modulation. A scheme involving changing the amplitude of a carrier in accordance with the amplitude of the input signal. Input information is ultimately carried by side bands.

Band pass filter. A circuit that attenuates all signal frequencies except a specific band.

Crosstalk. The interference of one multiplexer channel with another. The interference is equivalent to increased noise in the channel.

Demultiplexing. The retrieval of the original data from a multiplexed channel.

Duplex (full duplex). A communication that allows simultaneous transmission in both directions.

Error detection. Digital techniques used to detect the presence of digital errors (bit errors).

Error detection and correction. Digital techniques that identify the bit error and hence allow error correction.

Frame. One data set of multiplexed channels.

Frequency division multiplexing. The bandwidth of the common channel shared among a number of transmitting sources.

Frequency modulation. A scheme in which the carrier frequency is changed in accordance with the amplitude of the input signal.

Guard band. The range of frequencies separating the different subcarriers in an FDM system.

Half duplex. A communication channel that allows transmission in both directions, one direction at a time (not simultaneously).

Modulation. Modifying a carrier in some way to represent the input signal. (See "amplitude modulation" and "frequency modulation.")

Multiplexing. The use of one common channel by many transmitting sources. (See "frequency division multiplexing," "space multiplexing," "time division multiplexing," and "wavelength division multiplexing.")

Pulse amplitude modulation. The amplitude of a pulse train modified in accordance with an input signal amplitude.

Pulse code modulation. The input signal is sampled. Each sample is converted to a digital code used in the transmission.

Pulse position modulation. The position of the pulse in a pulse train, with respect to a fixed position, that changes in accordance with the input signal amplitude.

Pulse width modulation. Pulse width in the pulse train modified in accordance with the input signal amplitude.

Simplex channel. A communication channel that allows transmission in one direction only.

Space multiplexing. The use of different physical media (different optical fibers) for the different transmitting channels.

Subcarrier. When using frequency division multiplexing, each portion of the shared common channel bandwidth is a subcarrier.

Synchronization. Keeping a stream of data locked to a specific rate.

Time division multiplexing. Time on the common channels shared among a number of transmitting sources.

Time jitter. Rapid time variations in the signal, causing instantaneous loss of synchronization and increased probability of error.

Wavelength division multiplexing. Different wavelengths (colors) used for the different input transmitting sources.

FORMULAS

$$\text{Bit rate} = \text{BW} \times (\text{samples/cycle}) \times (\text{bits/sample}) \qquad \textbf{(7–1)}$$

Computation of PCM data rate.

$$\text{MUX output rate} = N \times \text{Input rate} \tag{7–2}$$

The output rate of a TDM MUX with equal input channel rates. (N is the number of input channels.)

QUESTIONS

1. What is a carrier?
2. What is amplitude modulation? Describe.
3. What is frequency modulation? Describe.
4. What is the carrier in PAM? How does PAM work?
5. What is one of the fundamental differences between PAM and PCM?
6. It is said that PCM is "bandwidth wasteful." Why? Relate the analog input BW to the bit rate transmitted.
7. What is meant by "the minimum sampling rate must be twice the highest frequency of the analog signal"?
8. What is the basic definition of
   a. PWM?
   b. PPM?
9. What is the fundamental function of multiplexing?
10. What is TDM? Explain.
11. What is FDM? Explain.
12. The data channel at the output of a time division MUX must have a BW that is equal to or greater than the sum of the input bandwidths. What does this mean?
13. What is the function of guard bands in FDM?
14. All multiplexed inputs can transmit simultaneously in FDM and WDM, but not in TDM. Why?
15. What are the differences among simplex, half-duplex, and duplex systems?
16. Is a simple telephone line simplex, half duplex, or full duplex? Explain.
17. In principle, can TDM be used to provide a full duplex? In practice, is it possible to simulate a nearly full duplex channel using TDM? How?
18. Draw a block diagram of a 16-channel multiplexer that uses 2-channel WDM and 8-channel FDM for each wavelength.
19. Of the three multiplexing schemes (FDM, TDM, and WDM), which is the least susceptible to crosstalk?
20. Under what conditions does multiplexing at the optical level give rise to saturation problems at the photodetector? Does the problem exist in WDM? Why?

PROBLEMS

1. PAM is used with an analog signal with a 3000-Hz BW. What is the minimum pulse rate involved?

2. In PCM systems, each input cycle is sampled 3.5 times. The word size is 8 bits. If the input BW is 3000 Hz, what is the bit rate?
3. Double the sampling rate in Problem 2 but do not change the bit rate. Calculate the input BW that could be accommodated.
4. What is the channel sampling rate in a MUX with 16 input channels if each input is allocated 1 ms at at time?
5. A MUX has 32 input channels. Each input channel operates at 16 kb/s. What is the output bit rate?
6. Sixteen analog inputs, each with a 4-kHz BW, are multiplexed. The minimum sampling rate is used. What is the pulse rate at the output if
   a. PAM is used?
   b. PCM is used with 8 bits per word?

8
Fiber Optic Components

CHAPTER OBJECTIVES

In this chapter you will become familiar with various types of fiber optic cables. You will gain insight into the mechanical structure of single-fiber and multiple-fiber cables and their characteristics.

In addition, you will gain an understanding of the way various splices are constructed and assembled. You should be able to follow specific splicing instructions given by splice manufacturers and to assemble a quality splice.

You will be introduced to various types of connectors and their assemblage onto the fiber. You will also learn how to calculate losses caused by misalignments and mismatches in fiber splices and connectors.

Part of this chapter presents a discussion of ways to distribute optical energy via fiber optic couplers and of how switching takes place in fiber optic systems.

8–1 INTRODUCTION

The typical glass fiber is about 100–250 μm in diameter. It is brittle and highly susceptible to damage such as scratches. Because of the small size of the fiber, this damage can appreciably increase radiation to the outside through the cladding and hence, increase losses. It is obvious that you cannot use the bare fiber. It must be protected.

The outside protection has the following features:

1. It gives the structure mechanical strength.

2. It protects the fiber against breakage and damage.
3. It does not introduce excessive pressure on the fiber. Such pressure could cause microbending and increase losses.
4. It allows easy field installation and maintenance.
5. It protects the fiber from environmental conditions, which is especially important in underwater installations.

8–2 SINGLE-FIBER CABLES

As part of the manufacturing process, the fiber gets a thin plastic coating, in addition to the fiber cladding, to protect it against damage during manufacture and assembly. Mechanical strength and additional protection are usually provided by heavy plastic jackets. It is important to realize that the mechanical structure varies with the application. For example, underwater cables have additional seals to prevent corrosion and special structures to withstand the high pressure. Also, the particular fiber encased in the mechanical structure may be step-index, graded-index, single-mode, or any other suitable fiber.

The single-fiber cables are constructed in two basic ways: **loose buffer (loose tube)** and **tight buffer.** Loose buffer means that the fiber is free of any stress. It is placed in an empty housing. Figures 8–1 and 8–2 show two loose buffer cables. (For reference, each figure includes the full data sheet, with all relevant specifications.) As you can see in the schematic cross-section, the fiber (the black dot in Figure 8–2) is free to move inside the buffer jacket (or primary tube). This pressure-free construction reduces the microbending effects.

Figure 8–3 presents a typical tight buffer cable. Again, the figure includes the full data sheet. Here, the protective jacket is tightly mounted around the buffer and fiber. Typically, a protective buffer coating covers the fiber, core, and cladding.

There are many different cable structures. Those included here are a small sample of what is available.

8–3 MULTIFIBER CABLES

In large installations, it is common to use many fibers contained in a reinforced cable, much like the multiwire copper cable in conventional use. The structure of these fiber optic cables is based on the use of the elementary single-fiber cable. Figure 8–4 shows a number of loose buffer (sometimes called loose tube) cables combined into a multifiber cable. A photograph of a cable, similar to the ones shown in Figure 8–4, is shown in Figure 8–5. The outside layers are necessary to protect the fibers. While Figures 8–4 and 8–5 show a loose buffer structure, the same basic configuration is used in a tight buffer cable structure. If the individual fiber tubes are replaced with tight buffer cables, you get a multifiber cable that uses a tight buffer structure.

1-Fiber PCS Single Fiber Cable

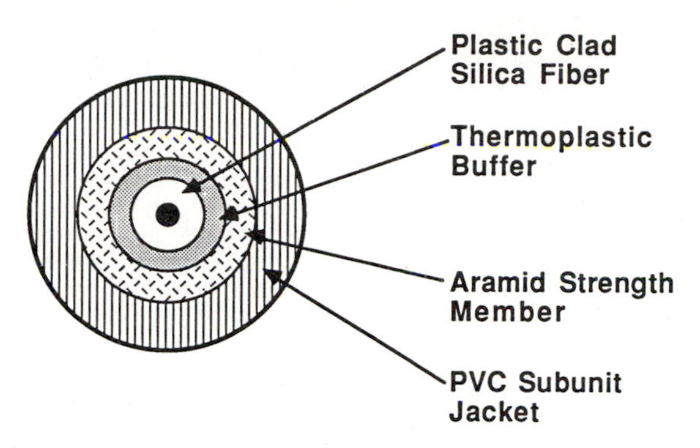

FIBER INFORMATION	Plastic Clad Silica Fiber
Core Diameter (μm)	200
Cladding Diameter (μm)	300
Buffering Diameter (μm)	900
Cladding Material	Silicone

CABLE INFORMATION	
Cable Diameter (mm)	2.4
(in.)	(0.09)
Cable Weight (kg/km)	5.5
(lbs/1000')	(3.7)

TRANSMISSION PERFORMANCE	
Cable Part Number	001Q31-A3027-00
Attenuation (dB/km)	
Maximum (@ 800 nm)	10.0
Minimum Bandwidth (MHz-km @ 800 nm)	10

MECHANICAL SPECIFICATIONS	
Maximum Tensile Load	
During Installation (short term)	300 N (67.5 lbs.)
After Installed (long term)	150 N (33.7 lbs.)
Minimum Bend Radius	
During Installation (loaded)	3.5 cm (1.4 in.)
After Installed (unloaded)	2.5 cm (1.0 in.)
Crush Resistance	2000 N/cm (1143 lbs./in.)
Impact Resistance @ 1.6 N-m	200 cycles
Cyclic Flex Resistance	2000 cycles
Maximum Vertical Rise	1000 m (3280 ft.)

ENVIRONMENTAL SPECIFICATIONS	
Storage Temperature	-40 to +70° C
Operating Temperature	-20 to +70° C
NEC Listing	OFNR
Flame Resistance	UL-1666 (for riser and general building applications)

FIGURE 8–1 Diagram and data sheet for single-fiber cable.
Courtesy of Siecor Corporation, Hickory, N.C.

20-01 Fiber Optic Cable

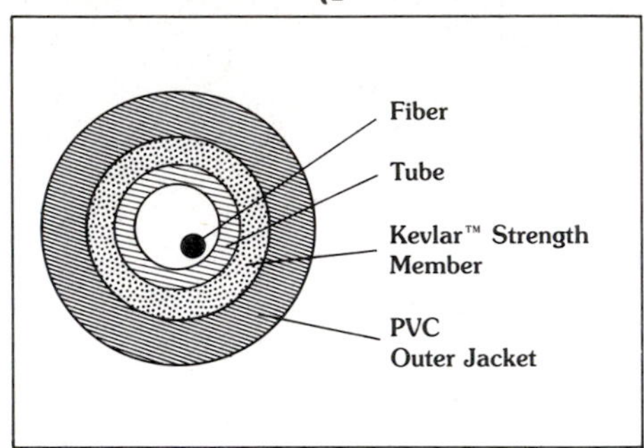

- Single Fiber Cable
- Loose Tube Structure
- Low Attenuation Over Wide Temperature Range
- High Tensile Strength
- Dielectric
- Jacket Pass UL VW-1 Flame Test

REMFO P/N	Fiber Diameter[1] Core/Clad	N.A.	Optical Performance[4]			
			@850 nm		@1300 nm	
	microns		dB/km[2]	MHz-km[3]	dB/km[2]	MHz-km[3]
20-01-1-354-154	50/125	0.20	3.5	400	1.5	400
20-01-1-000-106	50/125	0.20	—	—	1.0	600
20-01-3-402-202	85/125	0.26	4.0	200	2.0	200
20-01-4-501-301	100/140	0.29	5.0	100	3.0	100
20-01-4-503-303	100/140	0.29	5.0	300	3.0	300
20-01-5-800-000	200/380[5]	0.30	8.0	20	—	—

MECHANICAL SPECIFICATIONS[4]

No. of Fibers		1
Buffer diameter	microns	250/500
Tube diameter	mm	2.0
Storage temperature range, on original shipping reel	C	.40° to +70°
Operating temperature range, installed	C	.20° to +70°
Max. tensile load for installation	N (lbf)	500 (110)
Max. tensile load, long term installation	N (lbf)	100 (22)
Min. bend radius for installation	cm (in.)	10.0 (4.0)
Min. bend radius, unloaded (free) installation	cm (in.)	5.0 (2.0)
Crush resistance	N/cm	600
Impact resistance	times × Nm	50 × 1.0
Flexing ± 90°	times	10,000
Cable outside diameter, nominal	mm (in.)	4.0 (0.157)
Cable weight, nominal	kg/km (lbs/1000 ft)	13 (8.7)

Kevlar is a registered trademark of DuPont

1) Other fiber types and performance available.
2) Maximum attenuation.
3) Minimum bandwidth (−3 dB optical).
4) In accordance with industry testing standards.
5) Plastic-clad silica, typical 5.0 dB/km at 820 nm, 580 microns buffer diameter.

FIGURE 8–2 Data sheet for loose buffer step-index fiber cable.
Source: Remee Products Corporation, Florida, N.Y.

GenGuide-SX/Simplex Fiber Optic Cable

Cable Construction

- Optical Fiber–One multi-mode graded index glass fiber
- Core Diameter—50 μm (0.002″)
- Cladding Diameter–125 μm (0.005″)
- Protective Coating–Overall diameter of 500 μm (0.020″)
- Nylon Tight Buffer–diameter of 950 μm (0.037″)
- Strength Member–Kevlar®–a multi-end serving applied over the buffered fiber
- Outer Jacket–green flame retardent PVC jacket–diameter 2.8 mm (0.11″)

Optical Performance

- Numerical Aperture–NA = .20 ± .015
- Attenuation–4, 5, or 6 dB/km @ 850 nm wavelength performance
  2 dB/km @ 1300 nm wavelength
- Bandwidth–200, 400, 600, or 800 MHz/km

- Standard Cables

Standard Cables	Part No.
4dB–200MHz at 850 nm	1-801402 PC
2dB–400MHz at 1300 nm	3-801204 PC

Any combination of Attenuation and Bandwidth available upon request.

Mechanical Performance

- Maximum recommended tension 510N (100 lbs.)
- Minimum recommended bending radius is 25 mm (1 inch)
- GenGuide-SX Cable meets the standard acceptable tests for temperature, impact, flexibility, and crushing, as per General cable Specification 8006-NM latest issue.

Packaging

- Standard length–1 km
- Reel size–overall flange O.D. 16″ Traverse 6″–Drum 6″
- Shipping cartons–17″ x 17″ x 8″
- Shipping weight–approximately 18 lbs.

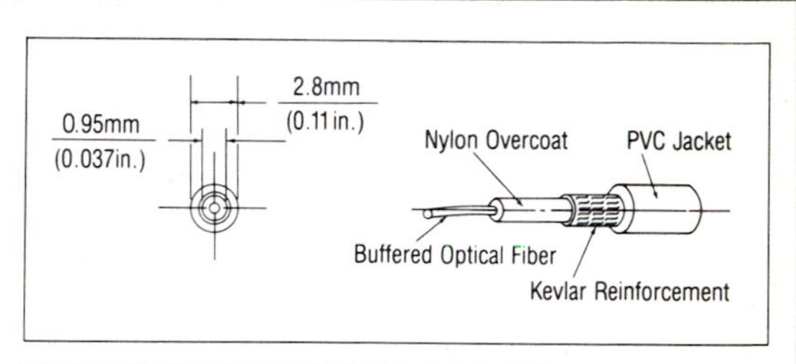

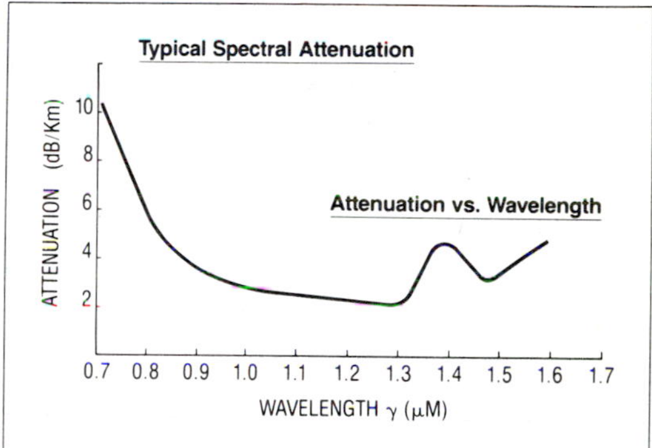

GenGuide Ordering Information

- Quantity–Attenuation–Bandwidth–Part Number
- Complete product data is available by requesting specifications–GCC 8004 and GCC 8006 NM
- For additional information, contact your local General Cable representative or the Fiber Optic Division at (201) 225-4780

Kevlar® is a registered trademark of Dupont Corporation.

FIGURE 8–3 Diagram and data sheet for the GenGuide-SX cable.
Source: General Cable Company, Edison, N.J.

SIECOR® 612/622 and 012/022 are general purpose fiber optic cables of proven design and large capacity for use in data transmission and telecommunications. These high-performance cables have improved specifications, simplified construction, and lower cost.

Low attenuation, high data rate

- Extended repeater spacing
- Low cost per channel
- Upgradable capacity

Size, weight, flexibility

- Low installation costs
- Minimal duct requirements

High-silica conductor

- Immunity to interference
- Isolation/Non-shorting
- Unaffected by water
- Chemically durable

SIECOR® 622 and 022 cables are rated at 6 dB/km and 400 MHz•km, or better, in the operating environment. Lower-cost 612 and 012 cables have attenuation less than 10 dB per kilometer and bandwidth greater than 200 MHz in one kilometer. SIECOR cables can handle high data rates including FT3 telephony (672 voice channels) over multi-kilometer unrepeatered distances.

These cables can be terminated with SIECOR and other commercially-available connectors. Splicing can be done in the field and takes about as long as splicing coaxial cables of similar size; however, fewer splices are required with SIECOR cables.

SIECOR cables are suitable for use indoors and outdoors, in conventional ducts, conduits, and trays. Installation is similar to copper cables, and is usually easier because optical cables are smaller and lighter. One kilometer or longer installation lengths are typically realized because of SIECOR cables' superior strength-to-weight ratio.
See SIECOR Technical Bulletin #9, "Installation Practices for SIECOR® General Purpose Optical Cables," for full details.

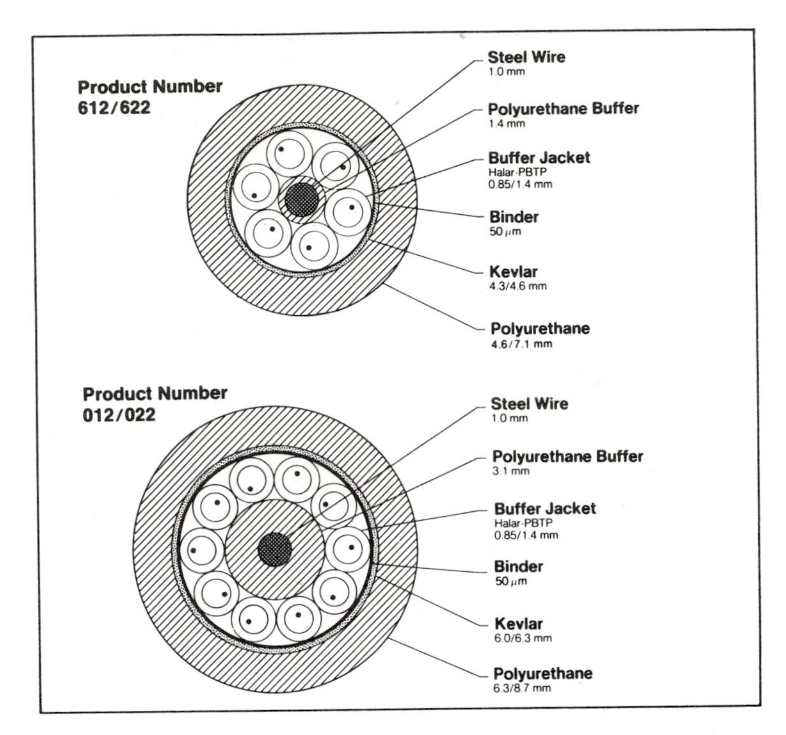

Cable design

Reliability of fiber optic cables is best achieved by maintaining the fibers in a stress-free condition. To achieve high reliability, the cables use a "loose buffer" design pioneered by SIECOR. Each optical fiber is mechanically isolated by spiraling it loosely inside an oversized buffer jacket. This permits considerable expansion and contraction of the cable from temperature change or mechanical loads before *any* stress is applied to the fibers, minimizing chances of increased attenuation or failure from stress fatigue. Also, when local stress is applied, the fiber accommodates it by sliding within the buffer jacket to reach the lowest possible stress level. Because of their design, SIECOR cables are rugged, stable in respect to tensile loads, and perform well over a wide temperature range.

Cable construction and materials

Each fiber is surrounded by a tough, double-layer buffer jacket that protects the fiber against impact and crushing forces. The buffer jacket is flexible, resists buckling, and can easily be removed mechanically for splicing and terminating.

The buffer jackets are stranded around a central member, a high-modulus steel wire coated with polyurethane, forming the cable's core.

A layer of aramid yarns surrounds the core for additional protection and cushioning, and provides the cable's tensile strength. An outer sheath of polyurethane completes the cable construction. Polyurethane is pliable but withstands abrasion and scuffing; it also resists chemical attack and cracking.

FIGURE 8–4 Diagram and data sheet for multifiber cable.
Courtesy of Siecor Corporation, Hickory, N.C.

Single Mode Fiber Optic Cable

Siecor single mode fiber optic cables offer high information carrying capacity over long distances between repeaters.

Transmission Benefits

- Current operation at 1300nm. Upgradability to 1550nm with no loss in performance.
- Capability for operation above 1 Gbit/s as electronics become available.
- Potential for greater information carrying capacity with wave division multiplexing (WDM) or use of narrow spectral width lasers.
- Typical cable attenuation from 0.4-0.7 dB/km at 1300nm
- Fiber dispersion less than 3.5 ps/nm-km (1285-1350nm).

Single mode cables have all the construction and installation advantages of the Siecor loose tube cable design. Single mode fiber with a mode field diameter of 10 μm, 125 μm cladding, and 250 μm CPC buffer composes the base from which the Micro-Bundle™, Mini-Bundle®, and Feeder-Bundle™ cable designs have evolved.

Siecor single mode cable carries several advantages for the user:

- Simple installation procedures for duct, aerial, or buried appli-
- Wide range of fiber counts available (2-144).
- Up to 6 kilometer continuous cable lengths. Longer reel lengths available soon upon special request.
- Custom reel lengths and cable designs available to meet specific applications.

Siecor optical cables are shipped currently in nominal 2, 3, or 6 kilometer lengths. Siecor offers a selection of splicing equipment, closures, and hardware as well as training site supervision and splicing services to ensure optimal system installation and operation.

FIGURE 8–5 Photograph of multifiber cabel and data sheet.
Courtesy of Siecor Corporation, Hickory, N.C.

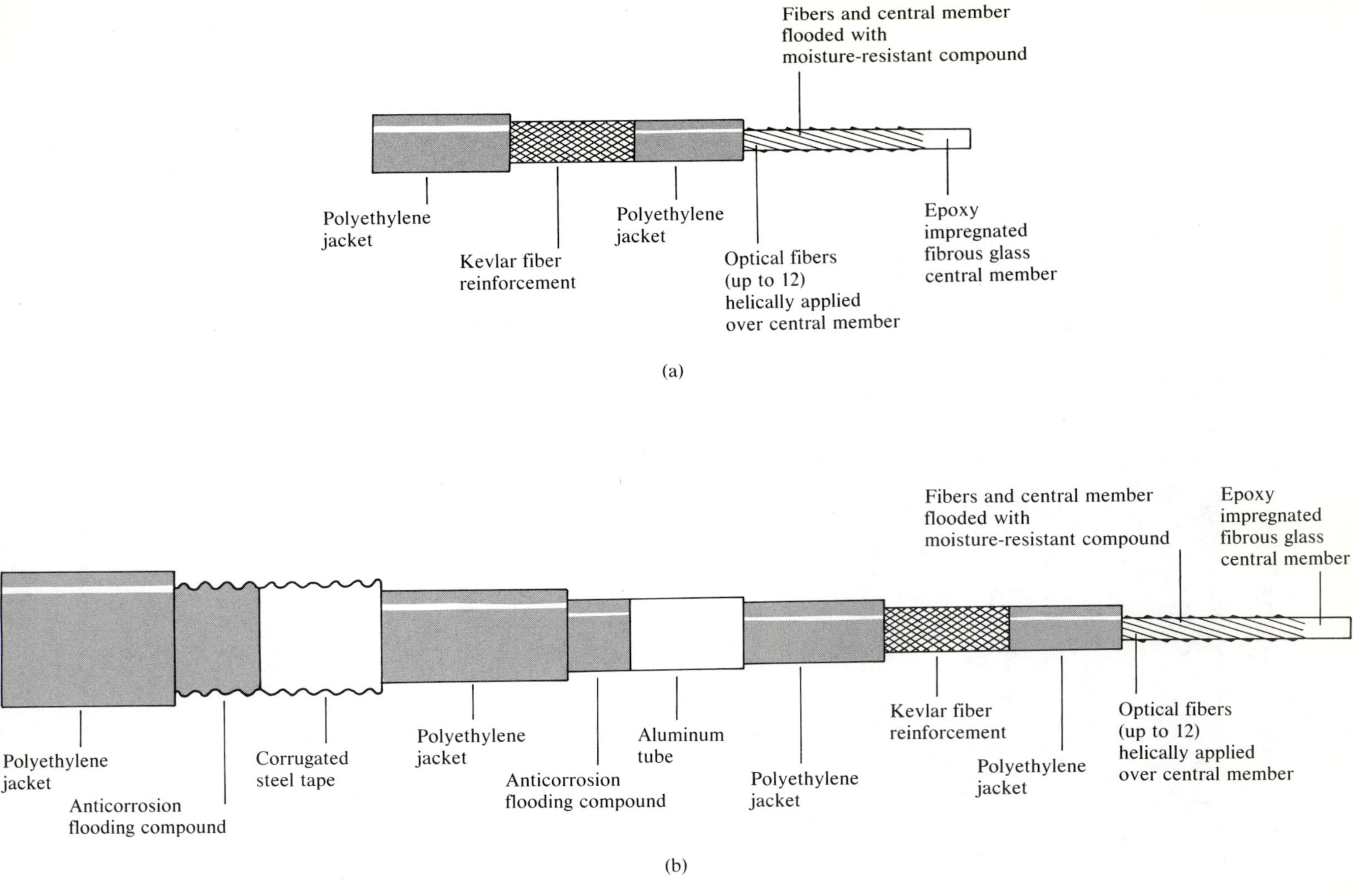

FIGURE 8–6 Cable structure. (a) Simple cable type NM. (b) Reinforced type GP.
Source: General Cable Corporation, Edison, N.J.

A somewhat different cable structure is shown in Figure 8–6. Here, the fibers are helically wound around a center support, with additional protective layers covering the fiber and the central member. The cable shown in Figure 8–6(b) is heavily reinforced and protected against corrosion. This extra protection is often necessary as a precaution against rodent attack. Rodents, such as gophers and squirrels, tend to chew on these cables and break through some of the plastic layers. The extra protection is also needed to protect against corrosive environments such as the seashore or underground ducts.

Typically, fiber optic systems require copper wires as well as fibers. These are sometimes incorporated into the fiber cable. Figure 8–7 presents some of the combinations of fiber-copper wires that are presently available in cable form.

The electrical and optical characteristics of a cable depend on the particular fibers used. Other characteristics such as mechanical strength, **minimum bending radius** (the radius of the smallest circular bend before the fiber breaks), and other mechanical properties are also important. (See the mechanical specifications listed in Figure 8–1.)

FIGURE 8–7 Photograph of a variety of fiber optic cable structures, some containing copper wires.
Source: The Rochester Corporation, Culpeper, Va.

8–4 SPLICING

When your copper intercom wire breaks, you either twist the broken edges or solder them. For optical fibers, you must **splice** the two fiber edges to repair a broken connection or to extend an optical link, thus permanently attaching the two fiber ends. Splicing, however, must be done in a way that does not introduce excessive losses. If the two edges are not properly aligned, light will escape at the junction and result in power loss. (The details of the alignment problems are covered in Section 8–6.) In addition, the connection must have long-term reliability. You should not have to redo a splice every 6 months.

To produce a quality splice, the fiber ends must be carefully cleaved. That is, they must be cut perpendicular to the fiber length. Some cleaving methods produce a clean and ready-to-use fiber end. Others require polishing of the fiber ends. All fiber preparation and joining of the fibers must be done as quickly as possible. Delays allow dust to accumulate and could cause additional light loss at the splice.

To reach the bare fiber, the outside jackets of the fiber must be carefully cut lengthwise without scoring the fiber. The longitudinal cut will allow the jacket to be pulled back to expose the buffered fiber. The buffer can then be removed chemically or with a stripping tool. (Chemicals, if not removed promptly, may damage the fiber.) The fiber is now ready for splicing.

Remember, when working with fiber, you must follow the specific splicing instructions available from the manufacturer of the splicing equipment and the cable manufacturer. *Be meticulous. Be careful.*

The two fiber edges can be connected by **fusion splice,** fusing the two

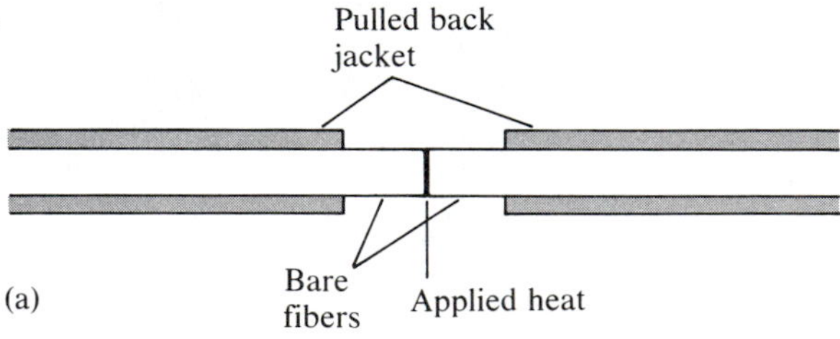

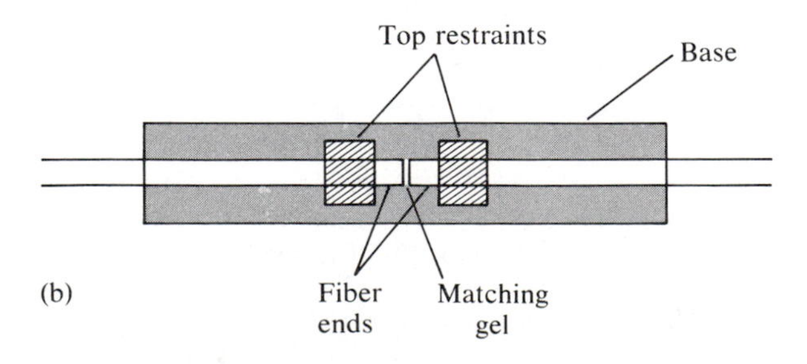

FIGURE 8–8 Splicing. (a) Fusion splicing. (b) Mechanical splicing (top view).

fibers (melting the glass), or by **mechanical splice,** mechanically lining up the two edges and permanently positioning them. The two methods are schematically shown in Figure 8–8. Figure 8–8(a) shows a fusion splice. The two edges are butted together and heated to the melting point. (You cannot use a fusion splice on an all-plastic fiber.) Figure 8–8(b) shows some mechanical means of locking the fibers in place. The **index-matching gel** added at the junction helps to eliminate any variation in refractive index from fiber to air to fiber. The gel is transparent and has the refractive index of the fiber core. Both methods shown require some way to align the two fibers.

8–4–1 Fusion Splices

Fusion splicing is performed by a sophisticated instrument that allows the alignment of fibers in all directions. First, the edges are aligned and appropriate butting pressure is applied. Then, heat is applied, usually by an electric arc. An electrical fusion splicer is shown in Figure 8–9. Once the fibers are fused, they become "one." You need only to cover the splice area to protect it from damage in handling. A plastic jacket is often used to protect the splice. The losses introduced by the fusion splice are about 0.1 dB.

Typically, fusion splicing is performed in the laboratory rather than in the field. It is often used in conjunction with a local injection and detection system. The purpose of this system is to allow better alignment of the fiber edges. An optical transmitter and a receiver are connected to the fiber sections to be fused. Alignment is accomplished by adjusting the two edges until maximum power is read by the receiver. The fibers are then locked in place, and fusion heat is applied.

8–4–2 Mechanical Splices

Mechanical splicing consists of aligning the fiber edges and locking them in place with the aid of various positioning devices and optical cement. A number of mechanical splice configurations are shown in Figure 8–10. The purpose of the various structures (the Vee groove, the three rods, and the ferrule) is to align the two edges. In addition to the structures shown, the splice is held in place by optical cement, tubing, or tape.

A detailed and complete splice structure is shown in Figure 8–11. A steel tubing, called a ferrule, with precise dimensions to fit the fiber and its jacket, is used to connect the two fibers. The funnels at the two ends are filled with optical adhesive to hold the fibers in place while the shrink tape gives mechanical strength to the overall splice.

Another mechanical splice structure is the **elastomeric splice,** a splice patented by General Telephone and Electronics, Inc. (GTE). Its details are

Mechanical Unit

Fiber adjustment method:	Orthogonal block/vacuum suction method Precision of axis discrepancy: less than ±1 μm
Clearance of suction slit:	50 μm ±5 μm
Jacket holder:	1.0 mmφ V-Groove/Magnet Clamp
Fiber moving method:	Manual stage (left side): Manually operated by micrometer (movable range: 0 ~ 3 mm) Automatic stage (right side): driven by DC motor (movable range: 0 ~ 4 mm)
Electrode holder:	Fixed V-Groove (1.0 mmφ)
Microscope:	100 power; 20 power (eyepiece) x 5 power (objective) Eyepiece microscale (0.1 mm pitch) included
Lighting lamp for microscope:	10V lamp fixed under the suction unit
Dimensions and Weight:	230 mm x 240 mm x 265 mm, 4.5 kg

Electrical Unit

• **Discharge Section**	
Power out-put method:	Constant current (10 step control)
Discharge current:	
High Current (silica fibers):	14 ~ 24 mA (10 step control)
Low Current (glass fibers):	1 ~ 5 mA (10 step control)
Out-put voltage:	AC 3500V (no load)
Discharge time:	0.1 ~ 9.9 sec. (0.1 sec. step control)
Discharge wave form:	50/60 Hz
• **Motor Control Section**	
Control function:	Forward, Reverse, Speed Control, Speed Shift, Time Control
Out-put voltage:	DC 2 ~ 8V (Helical Potentiometer Continual Control)
Rotation velocity:	0 ~ 6 r.p.m.
Feeding velocity:	0 ~ 45 μm/sec.
Time preset:	Forward, 0.1 ~ 9.9 sec. (0.1 sec. step control) Reverse, 0.1 ~ 9.9 sec. (0.1 sec. step control)
Dimensions and Weight:	230 mm x 245 mm x 150 mm, 7.0 kg

FIGURE 8–9 Electrical fusion splicer.
Source: Minitool Fiber Optic Inc., Campbell, Calif.

shown in Figures 8–12 and 8–13. The primary function of the splice structure is to align the fiber edges. Alignment is accomplished by placing the two fibers into a triangular tubing until they butt. To reduce reflection losses, matching gel is preinserted into the area where the fibers come together. The hexagonal outer surface of the triangular tubing is set into an outer sleeve to provide mechanical strength. The elastomeric splice is then assembled into top and bottom housings, as shown in Figure 8–13.

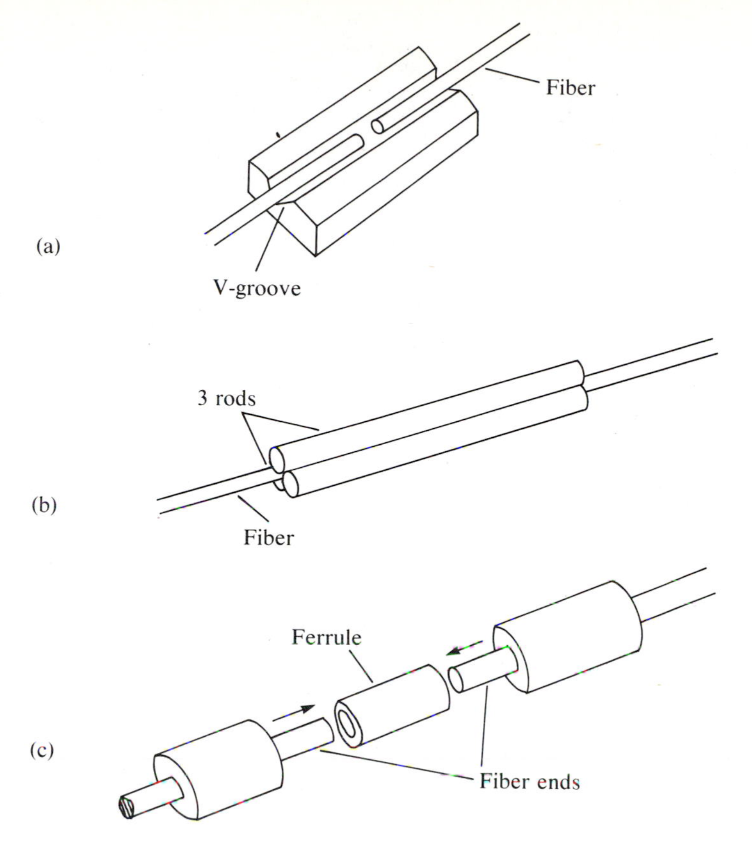

FIGURE 8–10 Mechanical splicing alignment. (a) Vee groove. (b) Three-rod alignment. (c) Ferrule.

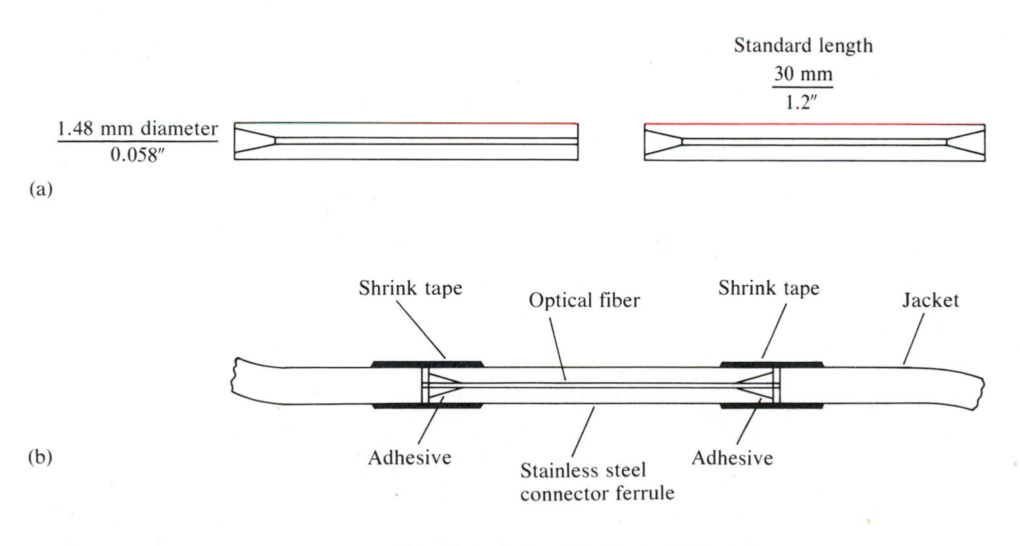

FIGURE 8–11 Mechanical splice. (a) Stainless steel ferrule. (b) Complete splice.
Source: Minitool Fiber Optic, Inc., Campbell, Calif.

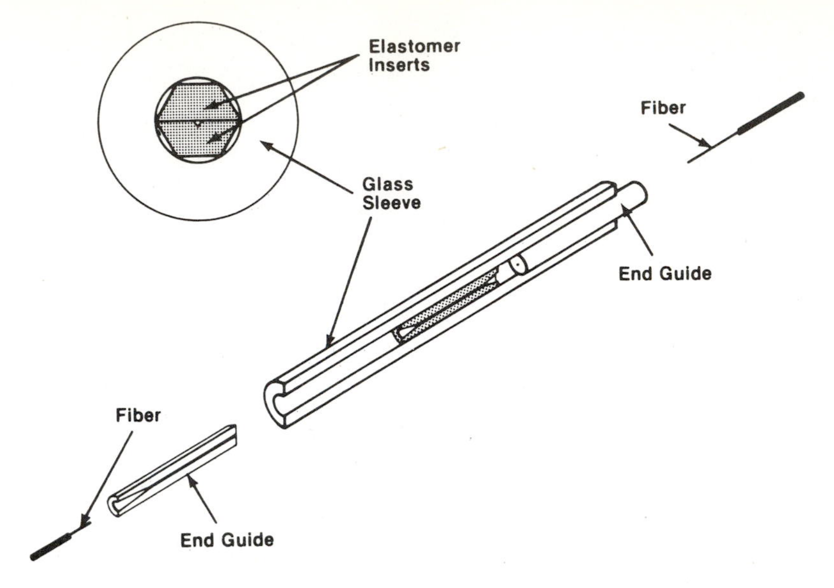

FIGURE 8–12 Optical fiber elastomeric splice.
Source: General Telephone and Electronics Products Corporation, Williamsport, Pa.

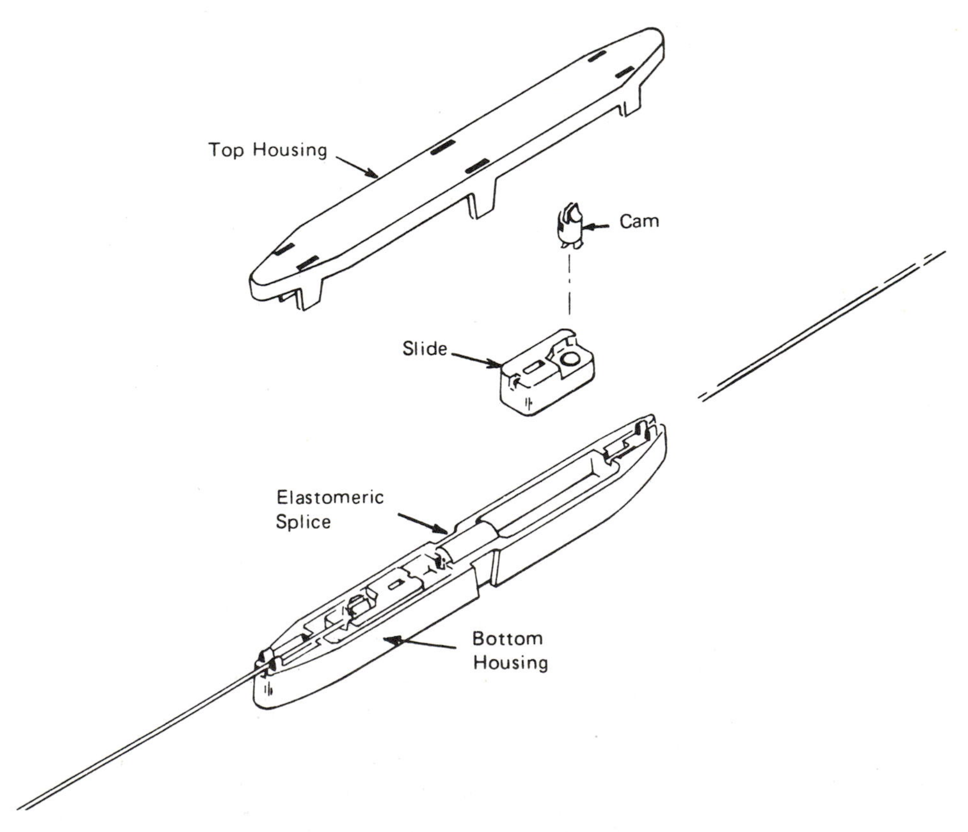

FIGURE 8–13 Splice assembly for elastomeric splice.
Source: General Telephone and Electronics Products Corporation, Williamsport, Pa.

8–4–3 Splicing Procedure

There are a number of steps involved in splicing.

1. Remove the outside fiber jacket without damaging the fiber using stripping tools, similar to the tools used for stripping copper cable. This can also be done simply by using a sharp blade to make a longitudinal cut in the jacket and then peeling the jacket off the buffered fiber. Remove only the length needed for the splice. Be sure to read and follow the specific instructions of the splice manufacturer.
2. Remove the buffer from the fiber. This plastic fiber covering can be removed by scoring the buffer carefully and peeling it off.
3. Cleave fiber edges to the desired length. (The term "cleave" is used instead of cut.) Well-cleaved edges should look as though they were polished. You must use a microscope to see if the edges are rough. If so, you should either cleave again or polish the fiber edges. It is important that the fiber edge be smooth and perpendicular to the fiber length. Various cleaving tools are available. The simplest tool is a carbide steel blade (Figure 8–14) or a diamond-coated one. The fiber is carefully scored with the cleaving tool and then bent until is breaks where scored. To obtain a clean cut with this method, you need practice.
4. When the fiber is ready, it can be inserted into the splice structure, clamped, or aligned in the fusion splice equipment.
5. For fusion splices, heat must now be applied to melt the edges of the fibers while they are butted.
6. The splice is completed by enclosing it in appropriate structural housing. Various cements are used in the final splicing stages. Some adhesives can be **cured** (allowed to harden) at room temperatures over a relatively long period of time; others must be cured at higher temperatures for a shorter duration. Still other adhesives can be cured by applying **ultraviolet** (UV) (invisible light below about 0.4×10^{15}

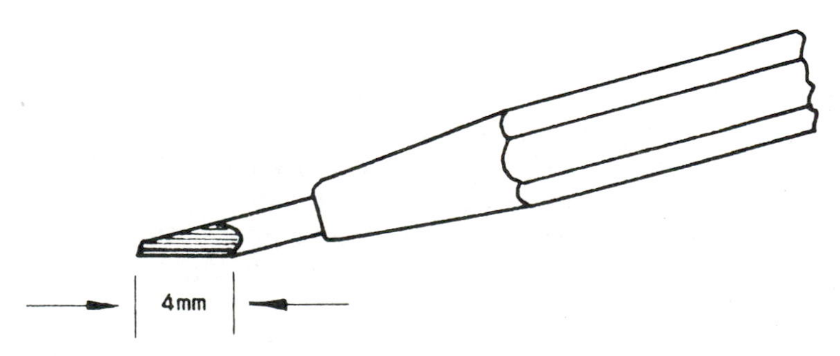

FIGURE 8–14 Optical fiber carbide cleaving tool, model MT–77.
Source: Minitool Fiber Optic, Inc., Campbell, Calif.

Hz) light. UV-cured adhesives are preferred because heat may damage the fiber jacket.

You know you have a good splice when the loss introduced by the splice is small. Fusion splices can be made with the losses ranging from 0.01 to 0.2 dB. Mechanical splices usually have losses of 0.15–0.5 dB.

The splicing of a multifiber cable involves, in addition to the splicing of each fiber, placing the spliced fibers in a **splice organizer,** which is designed to hold all splices and to give mechanical strength to the complete structure.

It is convenient to have all splicing tools organized and handy. The GTE Fiber Optic Splicing Kit is a briefcase-like housing containing all splicing tools. The kit and a list of tools included are shown in Figure 8–15.

8–5 CONNECTORS

The problems encountered with splicing are also applicable to fiber connectors. An additional and important difficulty is to provide for easy connect-disconnect. Fiber ends must come together, as close as possible, without damaging the fibers. The mounting of the connectors onto the fiber requires special care and precision. All fiber lengths and fiber jacketed lengths must be carefully measured according to the instructions of the connector manufacturers. Fiber ends must be polished. Often the polishing is done after the fiber is mounted in the connector.

The single-mode fiber connectors require a much higher degree of precision than do multimode connectors because even a very small **misalignment** is significant relative to the fiber diameter and will cause substantial loss.

A large variety of single-fiber connectors are available, ranging from plastic snap-on to precision steel screw-on connectors. There is a large selection of multifiber connectors as well. A low-cost, plastic single-fiber connector is shown in Figure 8–16. The connector shown is used mostly for larger diameter (core above 100 μm) fibers. Assembly instructions are also shown in Figure 8–16 and represent the typical assembly process. Note that the same structure is used for a variety of fiber diameters. You must pick the correct connector to fit the fiber dimensions.

A higher-precision, all-metal connector is shown is Figure 8–17. All relevant specifications, as well as some assembly instructions, are included. Note that this connector can be used for the small-diameter communication fibers as those shown in the table on ordering information. A collection of single-fiber and multifiber connectors is shown in Figure 8–18. The large linear connector (top) combines a printed circuit connector with eight fiber connectors.

Connecting the single-mode fibers requires a somewhat more sophisticated approach. The positioning precision required to align two cores of, say,

Complete P/N W0 1438
Basic P/N W0 1437
(without GTE Precision Cleaver)

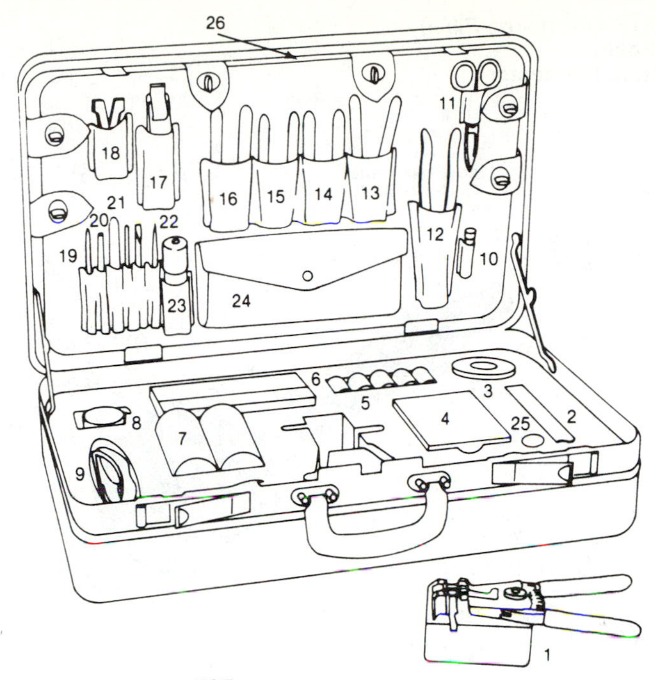

ITEM:	**USE:**
1. GTE Precision Cleaving Tool (included in complete kit only)	Produces consistent, high quality cleaves. Eliminates grinding & polishing of fiber ends.
2. Extra UV Lamp Bulb	Replacement use
3. Masking Tape	General use
4. Cardboard Box	Stores splices
5. "C" Batteries	For UV Lamp
6. UV Battery Operated Lamp	Cures UV adhesive
7. UV Adhesive	Bonds fibers permanently in splice
8. 10′ Retractable Tape Measure	General use
9. UV Safety Goggles	Eye protection
10. X-Acto™ Knife Blades	Replacement blades
11. Electrician's Scissors	Cut Kevlar® ; General use
12. Electrician's Pliers	General use
13. T-type Strippers (16 through 26 gauge)	Strips coating from central strength member; General use
14. Buffer Tube Strippers (red)	For up to 250μm buffer tube (.008)
15. Buffer Tube Strippers (white)	For up to 500μm buffer tube (.012)
16. Buffer Tube Strippers (green)	For up to 1mm buffer tube (.014)
17. Cable Jacket Stripper	Removes outer cable sheath 3/16″ to 3/4″ O.D.
18. Buffer Tube Strippers	For up to 1/8″ O.D. buffer tube
19. Tweezers	General use
20. Diamond Pen Scribe	To scribe-and-break fiber
21. X-Acto™ Knife	Removes polyurethane outer jacket; General use
22. Allen Wrench, 2 Screwdriver Heads	General use
23. Universal Screwdriver Head	For use with Allen wrench and two screwdriver heads
24. Storage Pouch	General storage
25. TB Grooved Bottom Clamp (included in complete kit only)	Adapts cleaver for tight buffer fiber
26. Instructions	

FIGURE 8–15 Splicing kit. (a) Photograph of kit. (b) List of components.
Source: General Telephone and Electronics Products Corporation, Williamsport, Pa.

OPTIMATE LFR (Large Fiber)
Single Position
Fiber Optic Connector

Dimensioning:
All dimensions in millimetres and inches. Values in brackets are equivalent U.S. customary units.
Chart contains dimensions in millimetres over inches.

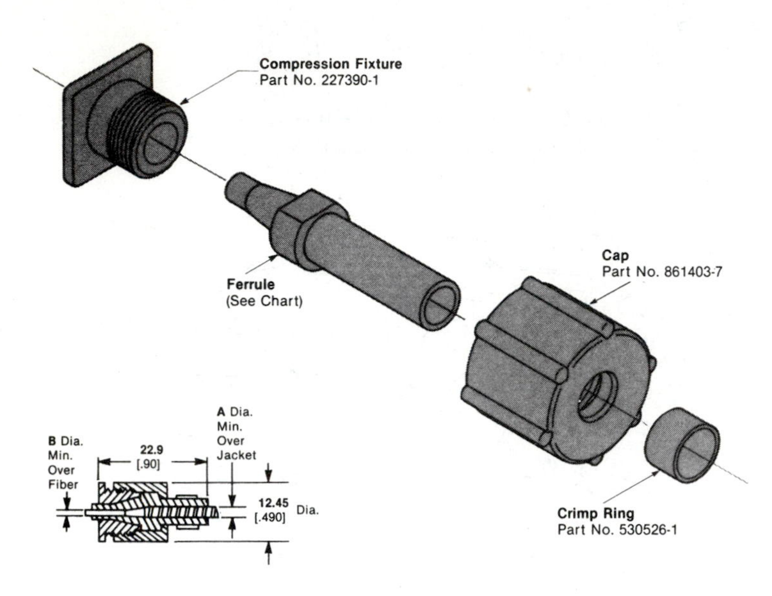

Note: A connector assembly includes all components illustrated above. Individual components can be supplied.

Dimensions		Part Numbers		Ferrule Color
A	B	Assembly	Ferrule	
3.15 .124	**2.16** .085	1-530530-7	1-530954-7	Violet
3.15 .124	**0.61** .024	1-530530-4	1-530954-4	Yellow
3.12 .123	**1.6** .063	530530-8	530954-8	Gray
3.12 .123	**1.17** .046	530530-1	530954-1	Brown
3.12 .123	**0.41** .016	1-530530-1	1-530954-1	Brown
2.79 .110	**0.64** .025	1-530530-2	1-530954-2	Red
2.49 .098	**0.64** .025	530530-5	530954-5	Green
2.62 .103	**0.58** .023	530530-7	530954-7	Violet
3.12 .123	**1.07** .042	228671-1**	227686-1*	Brown
2.29 .090	**1.17** .046	530530-2	530954-2	Red
2.29 .090	**0.81** .032	1-530530-5	1-530954-5	Green
2.29 .090	**0.53** .021	530530-4	530954-4	Yellow
2.29 .090	**0.43** .017	530530-9	530954-9	White
1.98 .078	**0.43** .017	1-530530-0	1-530954-0	Black
1.96 .077	**1.96** .077	530530-3	530954-3	Orange
1.57 .062	**0.64** .025	1-530530-3	1-530954-3	Orange
1.57 .062	**0.41** .016	530530-6	530954-6	Blue

*This ferrule is designed to be used with 1 mm plastic fiber, jacketed cable. Use of epoxy is not required. Termination is done by hot blade technique using AMP cut-off fixture No. 227386-2.

**Compression fixture not included.

Typical Assembly

Step 1
Strip cable.

Step 2
Epoxy fiber (where required).

Step 3
Slide on connector assembly.

Step 4
Crimp with hand tool.

Step 5
Mate with compression bushing before epoxy cures.

Step 6
Polish fibers.

Note: Use epoxy as recommended by fiber manufacturer, and polishing procedure described in AMP Instruction Sheet IS 2878-2.

Hand Tools

CERTI-CRIMP Hand Tool
Part No. 90364-2

CHAMP Hand Tool
Part No. 220193-1

FIGURE 8–16 Optical fiber connector (with assembly instructions).
Source: AMP, Inc., Harrisburg, Pa. Reprinted with permission.

FEATURES

- Lower cost yet high quality
- Guaranteed repeatability over 500 insertions
- Low loss (at least 1.0 dB, typically .6 dB*)
- All metal construction
- Precision tungsten carbide tipped ferrule
- Easy field assembly
- Operating temperature from -20° C to +80° C
- Accommodates 50/125, 100/140 or 200/230 μm fiber

*50/125 μm fiber

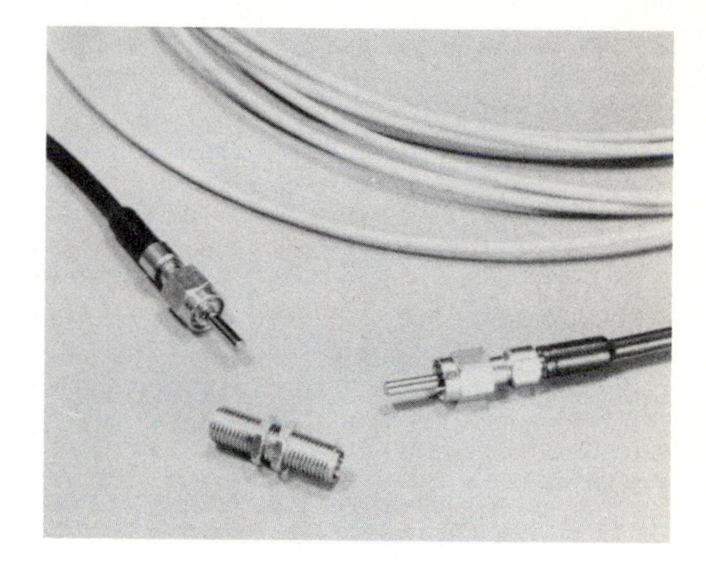

APPLICATIONS

- Telecommunications
- High rate data links
- Medical
- High precision instrumentation

DESCRIPTION

INTEROPTICS$_{TM}$ GFS-24/A fiber optic connectors are designed to be used to terminate 50/125, 100/140 or 200/230 μm diameter multimode fibers.

The ferrule tips are made out of tungsten carbide, an extremely hard metal with a very low coefficient of expansion.

The rugged and precise design allows a moderate number of repeated insertions with losses no worse than .8 dB within a temperature range of -20° C to +80° C.

The connectors can be assembled in a short period of time using INTEROPTICS$_{TM}$ Z213 Connector Assembly Tooling Kit and by following a very simple step-by-step procedure. No special skills are required.

MECHANICAL DETAILS

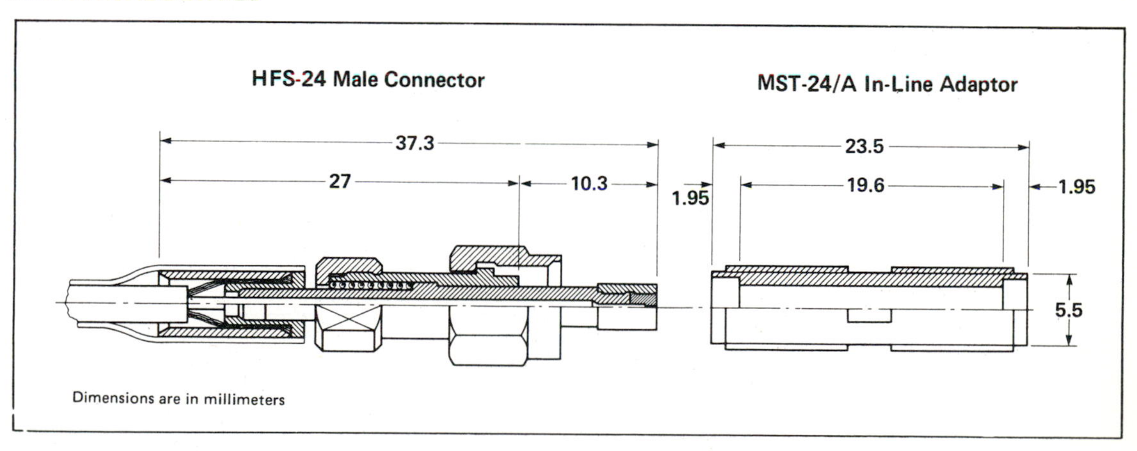

FIGURE 8–17 Data sheet for fiber optic in-line connector.
Source: Interoptics, Burlingame, Calif.

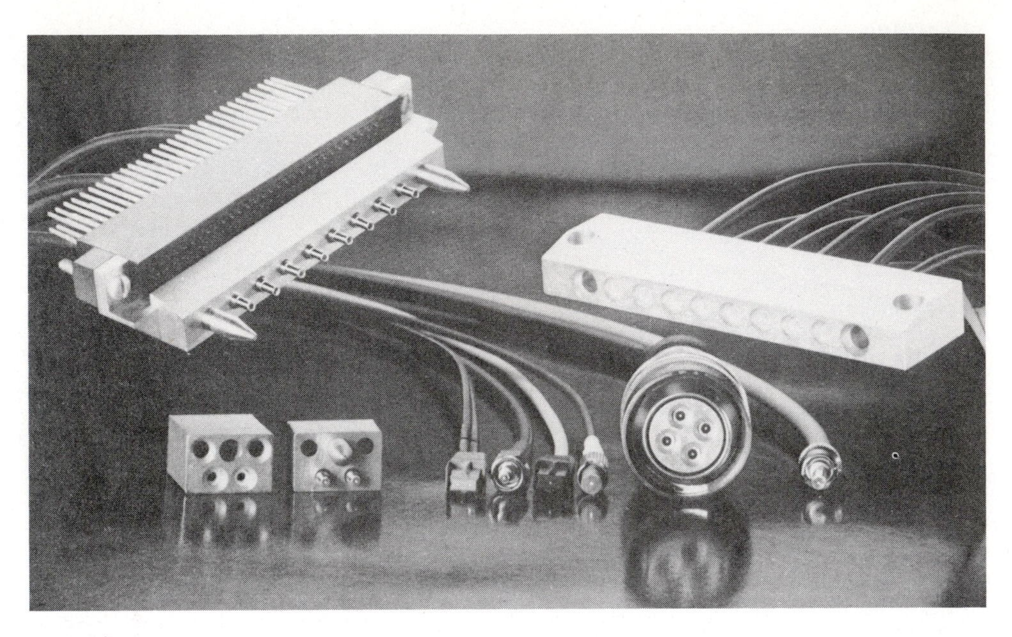

FIGURE 8–18 Single-fiber and multifiber connectors.
Source: Amphenol Fiber Optic Products, Lisle, Ill.

10-μm diameter becomes nearly impossible to attain. The solution lies in the use of lenses in the **expanded beam connector** (a connector that uses lenses to broaden the beam). The approach is shown schematically in Figure 8–19. The basic concept is to enlarge the effective size of the beam so that lateral misalignments (see Section 8–6) do not produce as much loss as a direct fiber-to-fiber connection. A detailed diagram of a connector using the expanded beam approach is shown in Figure 8–20.

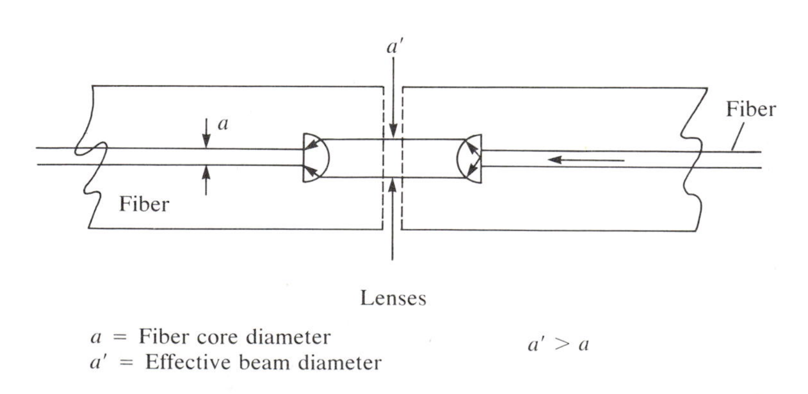

FIGURE 8–19 Expanded beam principle.

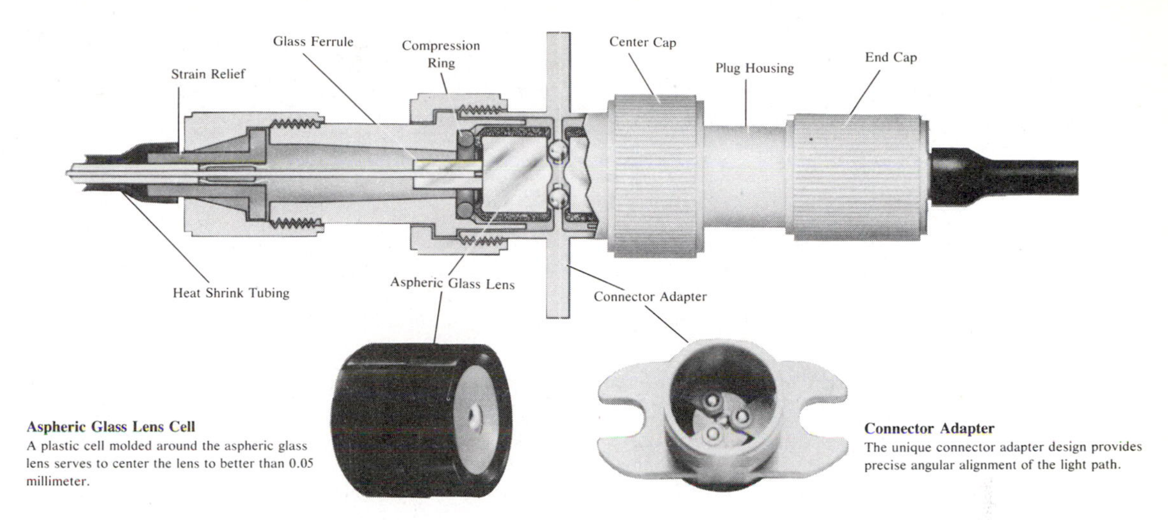

FIGURE 8–20 Photograph of single-mode expanded beam connector.
Source: Lamdek Fiber Optics, Division of Eastman Kodak, Rochester, N.Y.

8–6 CONNECTION LOSSES

The causes of optical loss in a fiber-to-fiber connection by splice or by connector fall into two classes: misalignment losses and losses resulting from mismatches in fiber characteristics, such as core area, numerical aperture (N.A.), and core profile. In addition, you may have **Fresnel reflection losses** when there is an air gap between the two fiber ends. Note that the discussion that follows largely pertains to multimode step-index fibers. Single-mode fibers are much more sensitive to both misalignment and fiber mismatches.

8–6–1 Misalignment Losses

This text covers three types of misalignment loss, all shown in Figure 8–21. Also shown are graphs relating the misalignment to the resulting loss for multimode and single mode fibers. You must look at each type of loss separately, as though the others did not exist.

Lateral Misalignment. **Lateral misalignment** loss (the loss caused by the misalignment of the center lines of the fiber), the most problematic misalignment, can be estimated mathematically (in decibels) by[1]

$$\text{Loss}_{\text{lat}} = -10 \times \log[1 - (1.28 \times l/d)] \quad \textbf{(8–1)}$$

1. The equation was derived from D. Glog, "Offset and Tilt in Optical Fiber Splices," *Bell Systems Technical Journal* 57 (7):905–16 (1976). It is a representation of the curve shown in Figure 8–12(a).

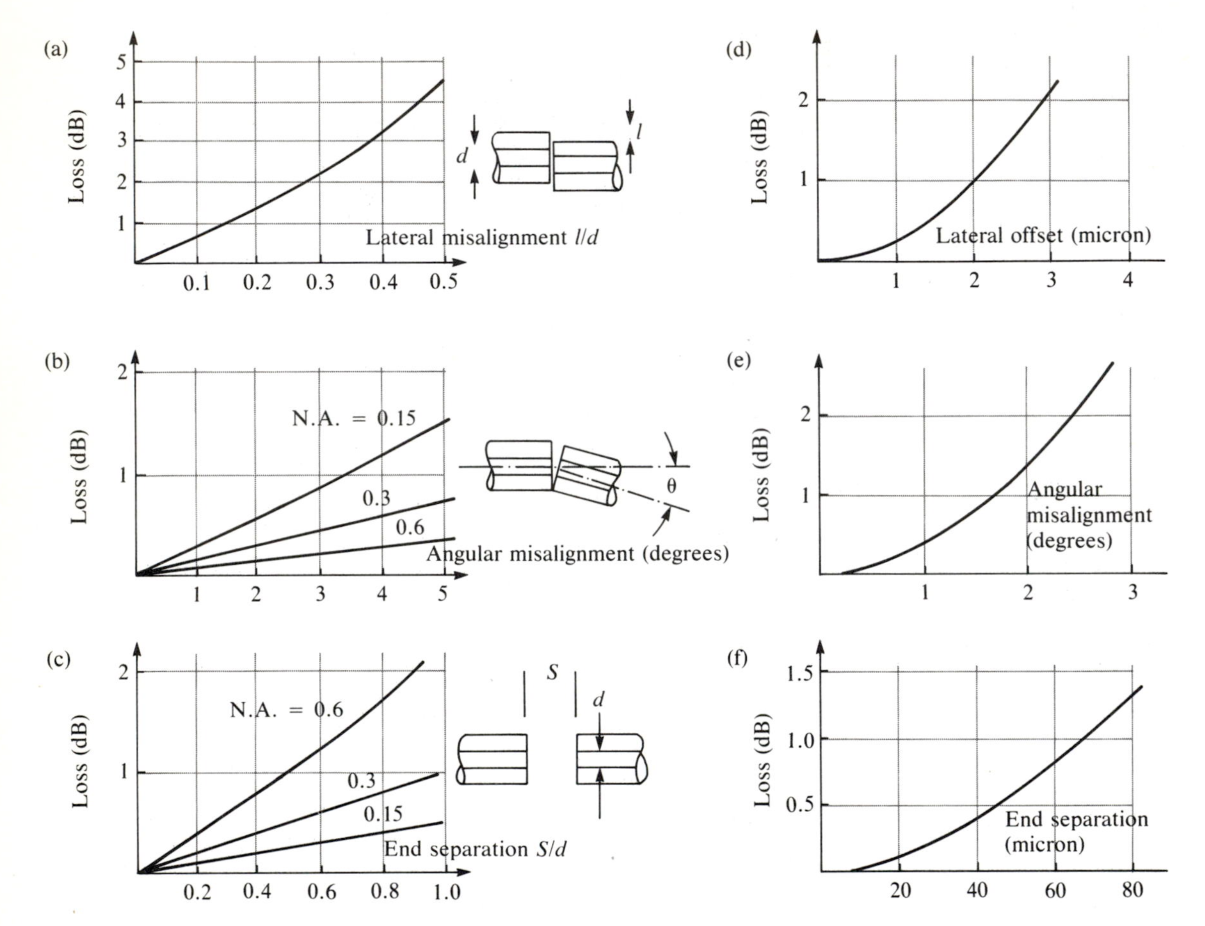

FIGURE 8–21 Mechanical misalignment losses. (a, b, and c) For multimode 100-μm core. (d, e, and f) For 8-μm single-mode fiber.
Source: Amphenol Fiber Optic Products, Lisle, Ill.

where l is the misalignment and d is the core diameter. The minus sign is introduced so that the resulting calculated loss is positive. Equation 8–1 is valid for l/d less than about 0.1. The term l/d represents the relative lateral displacement. It does not include other loss effects. In other words, assume no other misalignment and no Fresnel reflections.

EXAMPLE 8–1

Two 50/125-μm fibers are connected (no air gap between the fiber ends) with a 4-μm misalignment. Find the loss introduced.

Solution

50/125 μm means the core diameter is 50 μm, hence,

$$l/d \times 4/50 = 0.08$$

and

$$\begin{aligned} \text{Loss}_{\text{lat}} &= -10 \times \log[1 - (1.28 \times 0.08)] \\ &= 0.47 \text{ dB} \end{aligned}$$

End Separation Misalignment. Two factors cause optical loss in the case of **end separation misalignment.** The first is Fresnel reflections, as given by Equation 2–10. This is a result of the variation in refractive index from fiber to air in the gap and again to the fiber core. The total loss involved is double the Fresnel reflection from each fiber end. To conform to the definition of loss as $P_{\text{out}}/P_{\text{in}}$, the Fresnel loss becomes

$$\text{Loss}_{Fr} = -10 \times \log(1 - Fr) \tag{8–2}$$

where Fr, the Fresnel reflection, is given by Equation 2–10.

EXAMPLE 8–2

A fiber with core index of 1.45 is spliced (or connected) allowing an air gap between the fiber ends. Calculate the total loss because of Fresnel reflections.

Solution

From Equation 2–10, shown again here, you have

$$\begin{aligned} (n_1 - n_2)^2/(n_1 + n_2)^2 &= (1.45 - 1)^2/(1.45 + 1)^2 \\ &= 0.0337 \end{aligned}$$

$$\text{Loss}_{Fr} = -10 \times \log(1 - 0.0337) = 0.15 \text{ dB}$$

For the two fiber ends, the total Fresnel loss is 0.3 dB.

Example 8–2 yields a typical Fresnel loss of 0.3 dB. This loss can be eliminated by filling the gap between the fiber ends with an optical gel that has a matching refractive index. When using optical gel, beware of its tendency to attract dust and thus cause unexpected losses.

The second loss mechanism introduced by fiber end separation is demonstrated in Figure 8–22. Some of the rays from the transmitting fiber escape the fiber and are not incident on the receiving fiber end. For separations (S) that are less than 10% of the core diameter, the loss is less than 0.1 dB (for N.A. as large as 0.4) and not nearly as significant as other losses. Note that end separation loss increases with increases in N.A. (See Figure 8–21(c).)

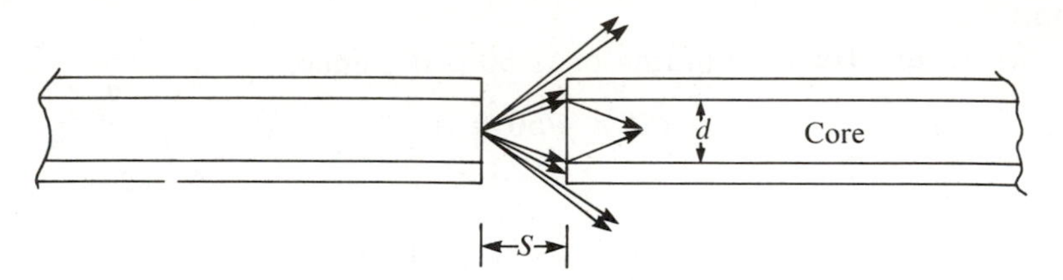

FIGURE 8–22 End separation losses.

Angular Misalignment. **Angular misalignment** is present when the two fiber ends are not lined up along the same axis. Even a small misalignment angle, θ (Figure 8–21(b)), can cause substantial losses, as shown in the graph of Figure 8–21(b). A mathematical expression that gives a good approximation for the angular misalignment loss is

$$\text{Loss}_{\text{ang}} = -10 \times \log[1 - (n \times \theta)/(180 \times \text{N.A.})] \tag{8–3}$$

where θ is in degrees and n is the refractive index in the gap;[2] n is usually either 1 for air or the same as the core index when index-matching gel is used. Again, this expression does not include any other losses.

EXAMPLE 8–3

A fiber with N.A. of 0.2 and n_{core} of 1.45 is connected with an angular misalignment of 5°. Find the loss. An index-matching gel is used.

Solution

$$\begin{aligned}\text{Loss}_{\text{ang}} &= -10 \times \log[1 - (1.45 \times 5)/(180 \times 0.2)] \\ &= 0.98 \text{ dB}\end{aligned}$$

EXAMPLE 8–4

For problem 8–3, no index-matching gel is used. Calculate the total Fresnel and angular misalignment losses.

Solution

The misalignment loss with n of 1 (air) is

$$\begin{aligned}\text{Loss}_{\text{ang}} &= -10 \times \log[1 - 5/(180 \times 0.2)] \\ &= 0.65 \text{ dB}\end{aligned}$$

2. The equation is from H. Tsuchia, H. Nagakone, N. Shimiza, and S. O'hare, "Double eccentric connectors for optical fibers," *Applied Optics* 16 (5):1323–31 (1977).

For each end,

$$\begin{aligned} \text{Loss}_{Fr} &= -10 \times \log[(1.45 - 1)^2/(1.45 + 1)^2] \\ &= 0.15 \text{ dB} \\ \text{Total Fresnel loss} &= 0.3 \text{ dB} \\ \text{Total loss} &= 0.95 \text{ dB} \end{aligned}$$

Examples 8–3 and 8–4 show that for angular misalignment it is not always useful to use index-matching material. Here, the losses are the same whether gel is used or not. However, gel tends to accumulate dust, which may cause additional losses. Note that for angular misalignments, an increase in N.A. decreases the loss; the opposite is true for end separation losses.

8–6–2 Mismatches of Fiber Characteristics

Here you will learn about three types of mismatches in fiber characteristics: cross-section mismatch, N.A. mismatch, and profile mismatch.

Cross-Section Mismatch. Energy loss can take place only if the cross-section of the transmitting fiber core is larger than that of the receiving fiber. The reverse situation introduces no losses. This **area mismatch** (two fiber ends with different areas) loss is directly proportional to the ratio of the core areas:

$$\begin{aligned} \text{Loss}_{\text{area}} &= -10 \times \log(a_2/a_1) \\ &= -10 \times \log(d_2/d_1)^2 \\ &= -20 \times \log(d_2/d_1) \end{aligned} \qquad \textbf{(8–4)}$$

where a_1 and a_2 are the areas of the transmitting and receiving fibers, respectively, and d_1 and d_2 are their respective diameters.[3]

EXAMPLE 8–5

Find the area mismatch losses when the transmitting fiber has a diameter, d_{core}, of 62 μm and the receiving fiber has a d_{core} of 50 μm.

Solution

$$\text{Loss}_{\text{area}} = -20 \times \log(50/62) = 1.87 \text{ dB}$$

The variation in core area for the same fiber is very small, making area mismatch loss very small when splicing two ends of the same type of fiber.

3. See footnote 2.

N.A. Mismatch. As is true for area mismatch, the loss caused by the variations in N.A. along a fiber is very small. This loss is neglible when connecting the ends of the same type of fiber. This type of loss exists only when the N.A. of the transmitting fiber is larger than that of the receiving fiber. Equation 8–5 gives an approximation of the loss:

$$\text{Loss}_{\text{N.A.}} \cong -20 \times \log(\text{N.A.}_2/\text{N.A.}_1) \qquad \textbf{(8–5)}$$

where N.A._1 and N.A._2 are, respectively, the N.A. of the transmitting fibers and that of the receiving fibers, and $\text{N.A.}_2 < \text{N.A.}_1$.

Profile Mismatch. **Profile mismatch** means that the shape of the two fiber ends is not the same. For example, one end might be elliptic and the other circular. The losses are basically area mismatch losses and are very small in most standard fibers.

Remember that when dealing with single-mode fibers, the problems introduced by misalignment and mismatch are much more severe than those with multimode fibers and the resulting losses are more substantial. The effect of the three types of misalignment losses on a single-mode fiber are shown in Figures 8–21(d) through 8–21(f).

The average losses introduced by a connector vary from 0.5 to 1.5 dB, and they vary with repeated use. When a connector is reconnected, its loss may not be what it was before reconnection. It may vary from connection to connection. Often rotating the connector can improve coupling and reduce loss or vice versa.

The average splice loss ranges from 0.1 to 0.5 dB. The same procedure using the same type of splicing method may not produce splices with the same loss. The loss will vary from splice to splice.

The large difference between splice and connector losses results from the nature of the connection. With a splice, which is a permanent, fixed connection, it is relatively easy to position the fibers accurately and to minimize all losses. Splice characteristics do not change with time, particularly, the fusion splice. The characteristics of the "connectorized" (lengths of fiber containing connectors) fibers deteriorate with repeated disconnect-connect operations.

8–7 FIBER OPTIC COUPLERS

It is relatively simple to distribute the voice in your intercom system to as many stations as you desire. Most of the time, it is necessary only to connect a few wires. The distribution of optical signals to many stations is somewhat more complex. *You cannot simply connect a few fibers.* The devices that are used in distribution (one-to-many or many-to-one) of optic signals are fiber optic **couplers** (Figure 8–23). These are special devices with one or many input fibers and one or many output fibers.

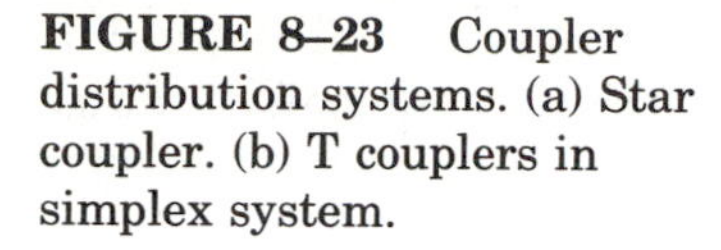

FIGURE 8–23 Coupler distribution systems. (a) Star coupler. (b) T couplers in simplex system.

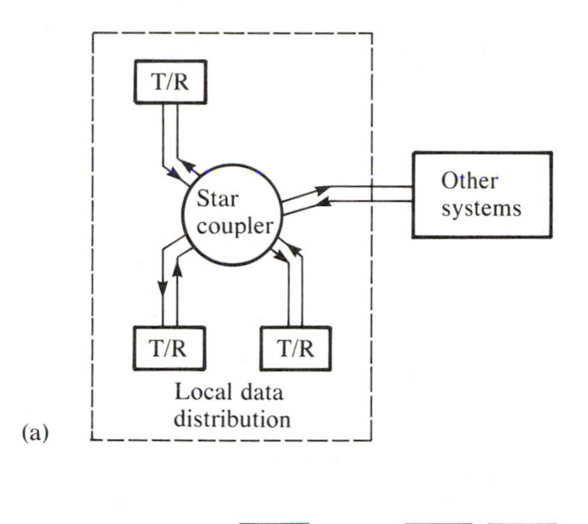

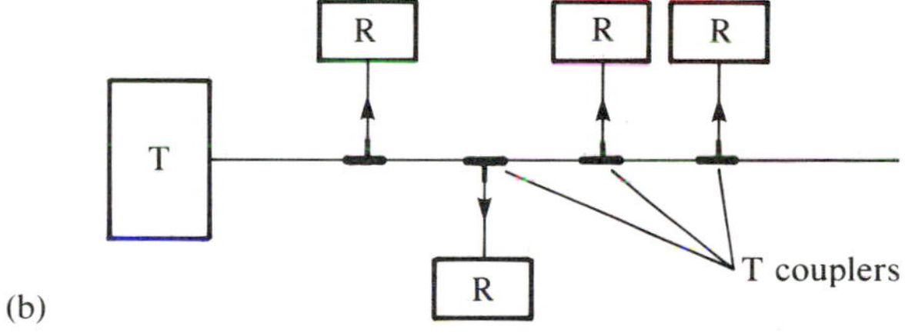

8–7–1 Star Couplers

Two basic structures of **star couplers** are shown in Figure 8–24. In the transmissive type (Figure 8–24(a)), optical signals that are sent into the mixing block from any single input fiber are available at all output fibers. Typically, the number of input fibers is the same as that of the output fibers.

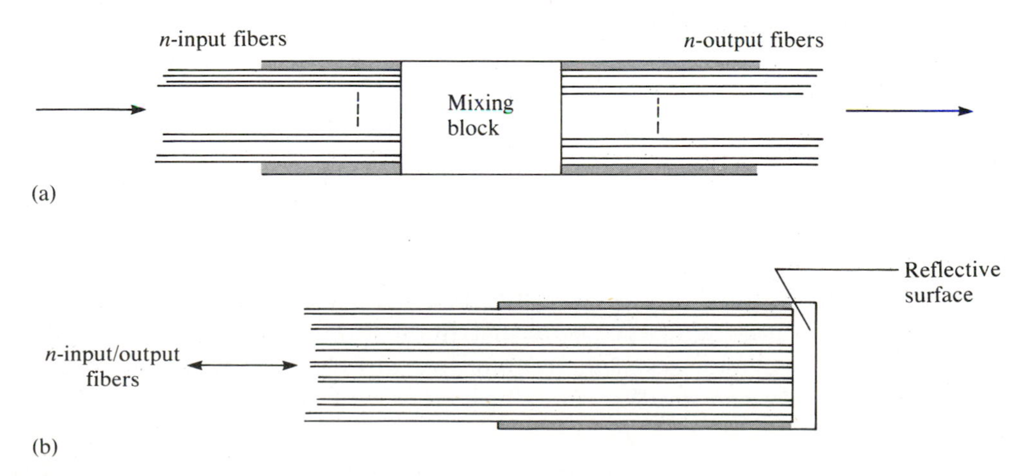

FIGURE 8–24 Star couplers. (a) Transmissive. (b) Reflective.

Power is evenly distributed. For an $n \times n$ star coupler (n input fibers and n output fibers), the power available at each output fiber is $1/n$ the power of any single input fiber:

$$P_o = P_{in}/n \tag{8–6}$$

P_o is the power at a single output fiber. The **power division (power splitting ratio)** in decibels associated with a star coupler is then given by

$$PD_{ST} = -10 \times \log(1/n) \tag{8–7}$$

(The minus sign is added so the result will be positive.)

EXAMPLE 8–6

An 8×8 star coupler is used in a fiber optic distribution system to connect the signal from one computer to eight terminals. If the power at an input fiber to the star coupler is 100 μW, find (1) the power at each output fiber and (2) the power division in decibels.

Solution

1. The 100-μW input is distributed to eight fibers. Each has 100/8 (or 12.5) μW.
2. The power division is

$$PD_{ST} = -10 \times \log(1/8) = 9.03 \text{ dB}$$

The power division in decibels gives the number of decibels apparently lost in the coupler from input fiber to single-fiber output. This is not a real loss because the total output power on all n output fibers may be equal to the input power. **Excess power loss** is defined as the power lost from input to total output, not related to the power division:

$$\text{Loss}_{ex} = P_{ot}/P_{in} \tag{8–8}$$

$$\text{Loss}_{ex}|_{dB} = -10 \times \log(P_{ot}/P_{in}) \tag{8–8a}$$

where P_{ot} is the total power on all output fibers. Equation 8–8a is Equation 8–8 expressed in decibels.

EXAMPLE 8–7

A 10×10 star coupler is used to distribute 3-dBm power of a laser diode to 10 fibers. The excess loss of the coupler is 2 dB. Find the power at each output fiber in dBm and in μW.

Solution

The power division in decibels is

$$PD_{ST} = -10 \times \log(1/10) = 10 \text{ dB}$$

To find the P_{out} for each fiber, subtract PD_{ST} and $Loss_{ex}$ from P_{in} in dBm:

$$3 \text{ dBm} - 10 \text{ dB} - 2 \text{ dB} = -9 \text{ dBm}$$

Note that, if you start with dBm, you get an answer in dBm.

$$P_o = -9 \text{ dBm}$$

To find P_{out} in watts (or μW), use

$$\begin{aligned} -9 &= 10 \times \log(P_{out}/1 \text{ mW}) \\ P_o &= \text{antilog}(-0.9) = 0.126 \text{ mW} \\ &= 126 \ \mu\text{W} \end{aligned}$$

The calculations of P_o, the power in each fiber, are important in power budget considerations (see Chapter 11).

An important characteristic of the transmissive star coupler is **cross coupling,** the extent to which fibers on the input side couple power to each other. In Figure 8–24(a), fibers from the left are expected to couple power to fibers on the right but not to other fibers on the left. Cross coupling (sometimes called **crosstalk**) is given in decibels. For example, the Model SFC, made by ADC Magnetic Controls has −40-dB crosstalk (see Figure 8–25). This means that the power coupled to any input fiber is 40 dB below the power entering any other input fiber.

The transmissive star coupler can be manufactured by fusing a number of fibers under tension. The process is shown in Figure 8–26.

The reflective star coupler (Figure 8–24(b)) has the same power division as the transmissive coupler. It is meaningless, however, to talk about crosstalk because, here, all fibers are on one side. That is, power from any fiber is distributed to all fibers.

8–7–2 T Couplers

A schematic representation of a **T coupler** is given in Figure 8–27. As shown, power launched into **port** 1 is split between ports 2 and 3. The power split does not have to be equal. Port 2 may receive more or less power than port 3, depending on the detailed coupler structure. The power division (the splitting ratio or power-splitting ratio) is given in decibels or in percent. For example, a 80/20 split, 80% to port 2 and 20% to port 3, translated into 0.97 dB for port 2 and 6.9 dB [about 7.0 dB] for port 3. (Both of these numbers are actually negative, but the negative meaning is self-evident, so the minus sign is often dropped.) In terms of the **power ratios;**

$$\begin{aligned} P_2/P_1 &= 0.8 = -0.97 \text{ dB} \\ P_3/P_1 &= 0.2 = -6.9 \text{ dB} \end{aligned}$$

Another important term relating to T couplers is **directivity.** Transmission between the ports in a T coupler usually has directivity. If P_3/P_1 is 0.2, you

The SFC series of passive transmissive star couplers make possible the efficient distribution of analog and digital signals for voice, video, and data communication systems. Because of its unique optical characteristics, the Star coupler has wide acceptance as the key component for applications such as Local Area Networks (LAN's) and process control systems. The important properties of Star couplers are: low excess loss, low optical crosstalk (high directivity) and good output uniformity.

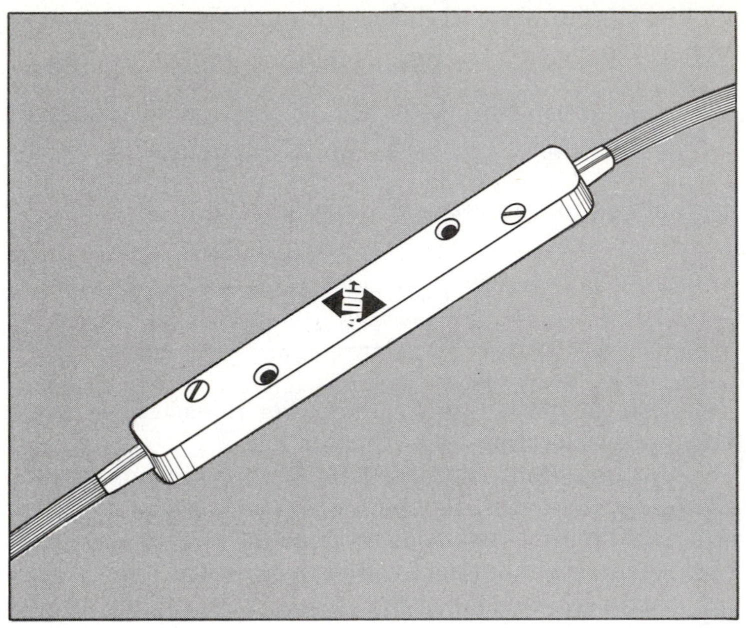

Standard Features

3x3 to 20x20 ports
Fiber types: 100/140 G.I. or S.I. and 200/250 S.I.
Low excess loss: less than 3dB
Low optical crosstalk (high directivity): −40dB typical
Output uniformity: ±4dB typical
High performance, reliable, rugged: totally passive component
Small package size:
76×10×8 mm (3×3 to 7×7 ports)
100×12×10 mm (8×8 to 20×20 ports)
1 meter long fiber at each port
Unidirectional matrix provided
Temperature and vibration tested

FIGURE 8–25 Description and specification of transmissive star coupler.
Source: ADC Magnetic Controls Company.

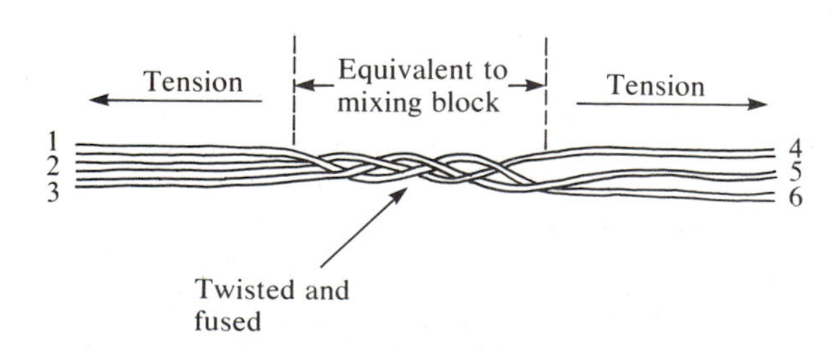

FIGURE 8–26 Fused transmissive star coupler.

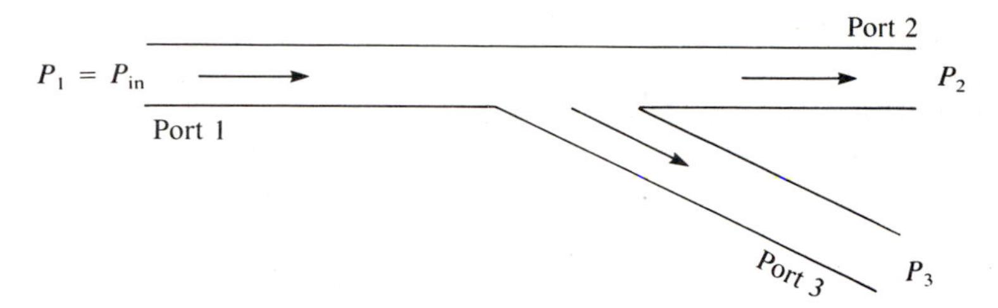

FIGURE 8–27 Schematic diagram of T coupler.

cannot assume that P_3/P_2 is also 0.2. In other words, port 3 may receive 20% of the power entering port 1 but not 20% entering at port 2. In highly directive T couplers, P_3/P_2 (Figure 8–27) is very small. Only a small part of power entering port 2 (if port 2 is the input port) reaches port 3. The directivity can be defined as the power ratio between one port and another, which is expected to be zero (ideal directivity). No power is expected to be transferred from port 2 to port 3 in Figure 8–27. Typically, no power is expected to be transferred between any two ports on the same side of the coupler.

Excess loss has the same meaning here as for the star coupler. It is the ratio of total output power (all output ports) to input power.

EXAMPLE 8–8

A T coupler with four ports, essentially two T couplers combined, is shown in Figure 8–28. The table in the figure gives the power division data. Use the worst case values for coupler TC3–TC4 model A (first column in the table).

1. If P_1 is power in at the input port and has −3 dBm, calculate P_2, P_3, and P_4.
2. Find the directivity in decibels. (Read directly from the table.)

Solution

1. To calculate P_2, P_3, and P_4, use the first portion of the table and excess loss of 1 dB (the last grouping in the table). This is the worst case (maximum) excess loss.

   - P_2 is 75% of P_1, or 1.2 dB lower. (See table under A, first two rows.)
   - P_2 = −3 dBm − 1.2 dB − 1 dB = −5.2 dBm
   - P_3 = −3 dBm − 13 dB − 1 dB = −17 dBm (13 dB found under A, fourth row.)
   - P_4 = −3 dBm − 40 dB − 1 dB = −44 dBm (40 dB found under A, sixth row.)

   The effects of excess loss, 1 dBm, are included in each case.
2. The coupling between port 1 and port 4 (both on the same side) is expected to be very small. This is the directivity. Reading from the

Bidirectional Couplers

Performance Characteristics TC3-TC4

input port	output port	Model A	Model B	Model C	NOMINAL COUPLING
①	②	75 1.2	60 2.2	40 4	% dB
	③	5 13	20 7	40 4	% dB
	④	.01 40	.01 40	.01 40	% MAX dB MIN
②	①	75 1.2	60 2.2	40 4	% dB
	③	.01 40	.01 40	.01 40	% MAX dB MIN
	④	5 13	20 7	40 4	% dB

Directivity performance is measured at the adjacent port with the output port terminated using index matching fluid.
Directivity of greater than 50 dB can be supplied on special order.
Excess loss is the difference between the input and the sum of the outputs.

input port	output port	Model A	Model B	Model C	NOMINAL COUPLING
③	①	5 13	20 7	40 4	% dB
	②	.01 40	.01 40	.01 40	% MAX dB MIN
	④	75 1.2	60 2.2	40 4	% dB
④	①	.01 40	.01 40	.01 40	% MAX dB MIN
	②	5 13	20 7	40 4	% dB
	③	75 1.2	60 2.2	40 4	% dB
Excess Loss		20 1 0.5	20 1 0.5	20 1 0.5	% MAX dB MAX dB TYP
Output port ratios		15:1	3:1	1:1	Main Port: Secondary Port

Applications

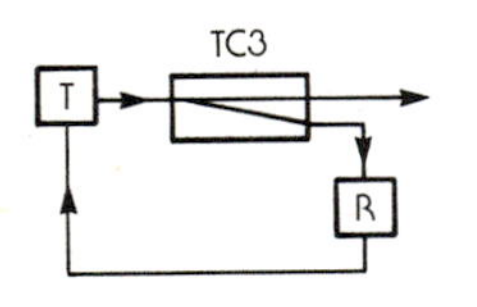

Feedback Loop

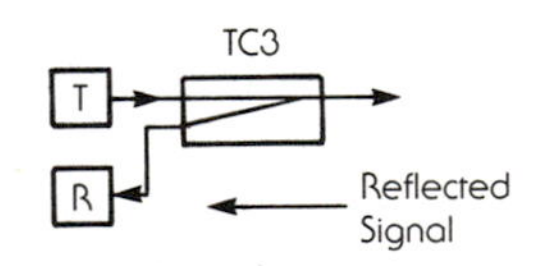

Reflectometer

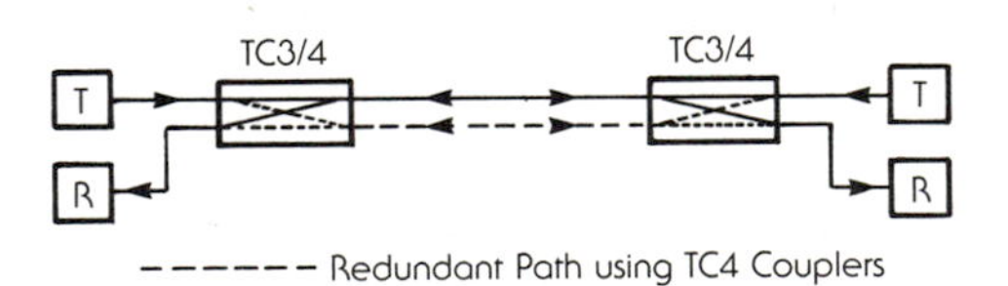

Full Duplex and Bidirectional Transmission on a Single Fiber or Separate Fibers for a Protected Route.

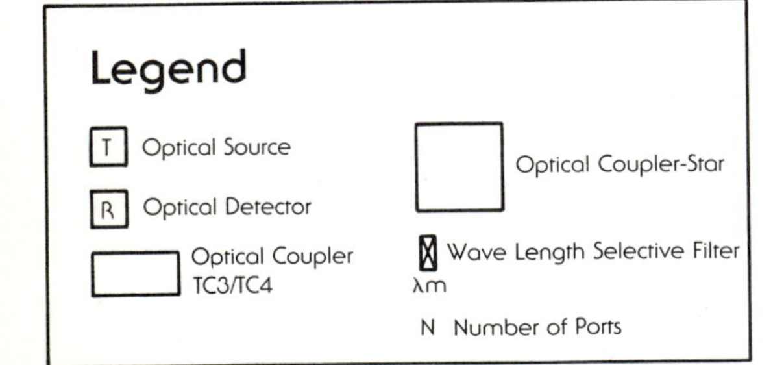

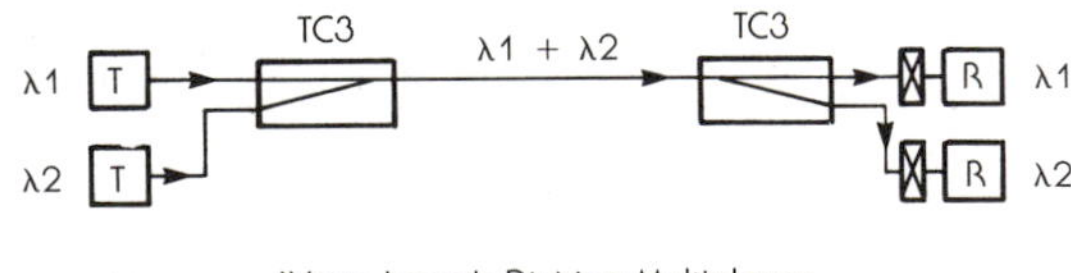

Wave Length Division Multiplexer

FIGURE 8–28 Bidirectional coupler specifications.
Source: Canstar Communications, Division of Canada Wire and Cable, Ltd.

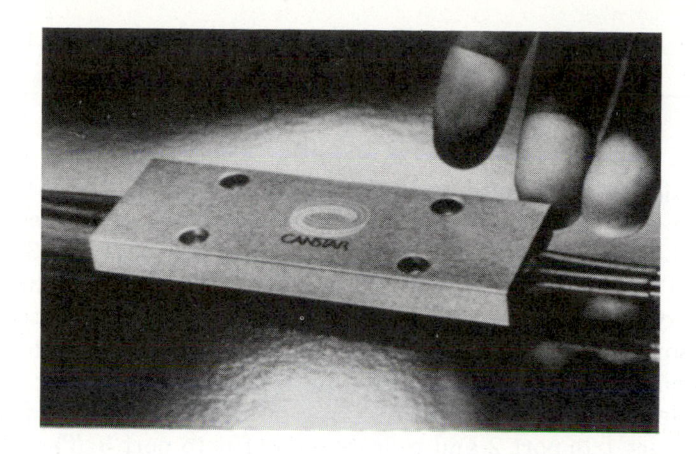

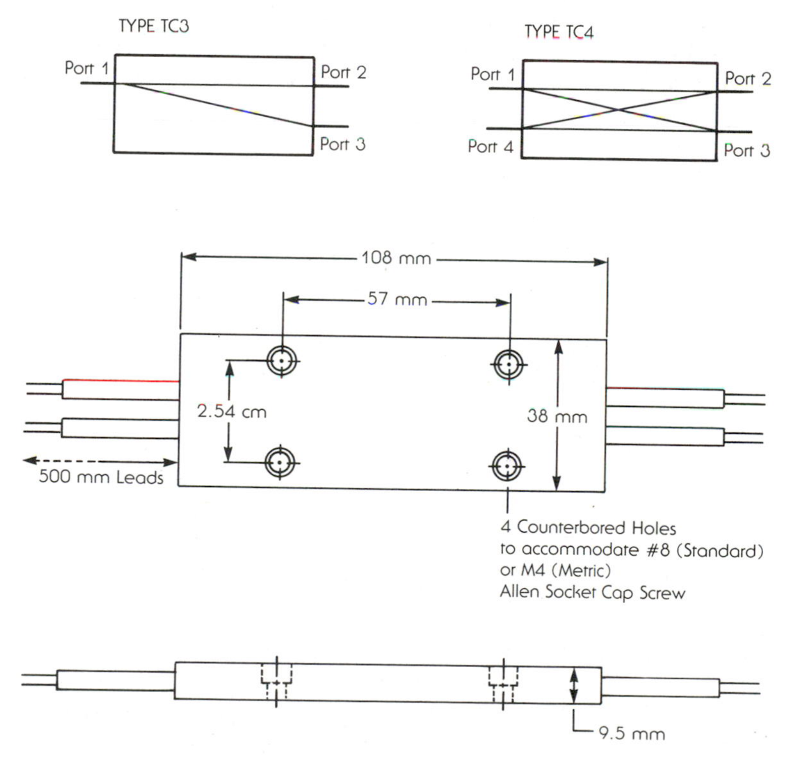

FIGURE 8–28 *(continued)*

table, you have 40 dB, or 0.01%. Note that 0.01% is a ratio (0.01 to 100 or 0.0001), which is −40 dB (10 × log 0.0001).

The table in Figure 8–28 gives the output port ratios (last row), which are the ratios of the two receiving ports, ports 2 and 3, when the input is port 1 or port 4. For example, the port 1 to port 2 ratio is 75% from model A. Port 1 to port 3 is 5%, so that the output port ratio of port 2 to port 3 is 75 to 5 (or 15:1).

Two types of T couplers are shown in Figure 8–29, the fused coupler and the expanded beam coupler. The fused type is constructed similarly to the transmissive star coupler. The expanded beam coupler is basically an expanded beam connector with a partial reflecting mirror placed between the fiber lenses. The partially reflecting mirror allows part of the light to travel from port 1 to port 2 and reflects part of it to port 3. The power division is a function of the reflectivity of the mirror.

Another type of T coupler uses a GRIN lens and a partially reflective surface to accomplish the coupling. A **GRIN lens** (graded-index lens) is constructed like a graded-index fiber. Its length is carefully selected to accomplish a particular focusing (aiming) function. Here, a quarter pitch (or quarter period) lens is used. (The GRIN lens is also known as a **SELFOC® lens,** SELFOC® being a registered trademark of NEC Corporation.) The

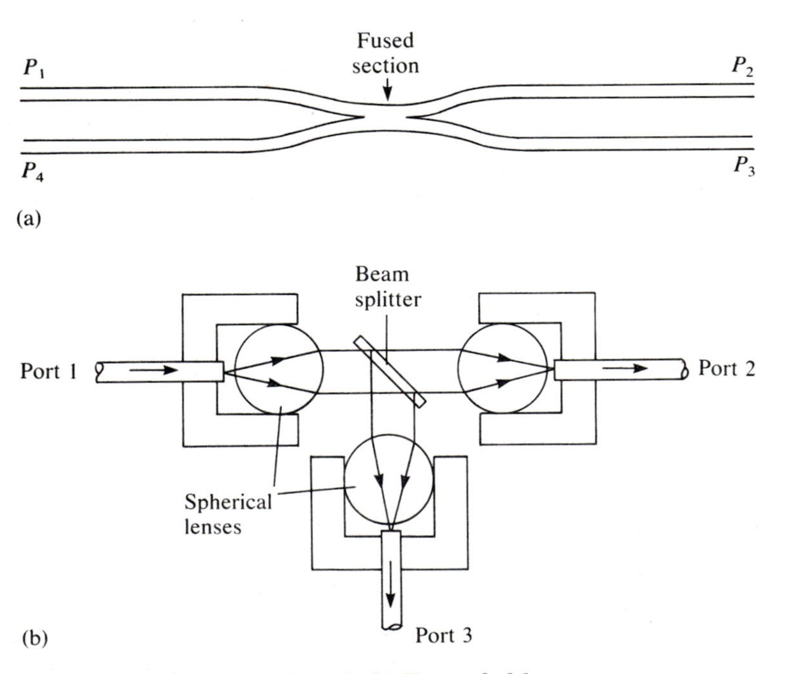

FIGURE 8–29 T couplers. (a) Fused. (b) Expanded beam.
Source for part b: I. C. Callahan, "Expanded Beam Connectors," *Proceedings of SPIE* 374: 64 (1983).

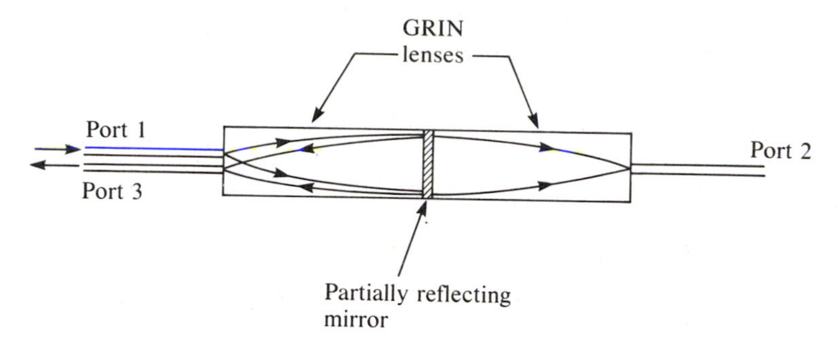

FIGURE 8–30 Expanded beam coupler of graded-index (GRIN) lens.
Source: G. A. Gasperian, "Expanded Beam Directional Couplers for LAN," *Proceedings of SPIE* 434: 18 (1983).

coupler arrangement is shown in Figure 8–30. The beam from port 1 is split as shown and is partially reflected to port 3, while most of the light is refracted onto port 2. The power division is a function of the reflecting mirror. This coupler is often used to monitor optical power in a fiber optic line. Port 3 is the monitoring port.

8–7–3 Other Couplers

The couplers discussed so far have been designed to be insensitive, as much as possible, to wavelength. The characteristics of the couplers were relatively independent of the wavelength within the operating region. The couplers used in wavelength division multiplexing (WDM) (discussed in Chapter 7) are designed specifically to make the coupling between ports a function of wavelength. The purpose of these couplers is to separate (or combine) signals transmitted in different wavelengths. A functional description of couplers used in WDM is given in Figure 8–31. The λ_1 and λ_2 signals (two separate data sources) are combined in the transmitting coupler and converted back into two separate lines λ_1 and λ_2, by the receiving coupler. The combined signals, $\lambda_1 + \lambda_2$, are carried by the main transmission fiber. Essentially, the transmitting coupler is a mixer, while the receiving coupler acts as a wavelength filter.

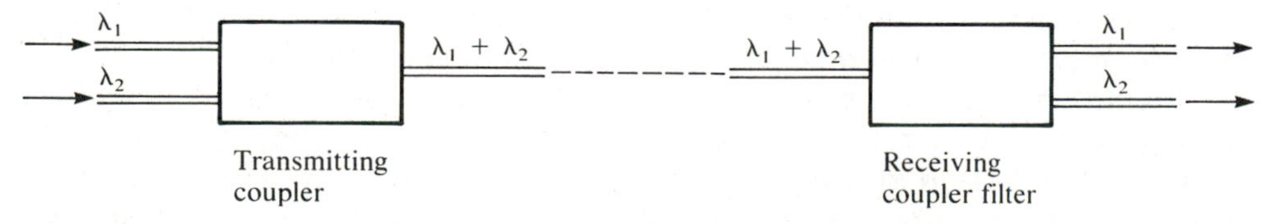

FIGURE 8–31 Diagram of wavelength division multiplexing system.

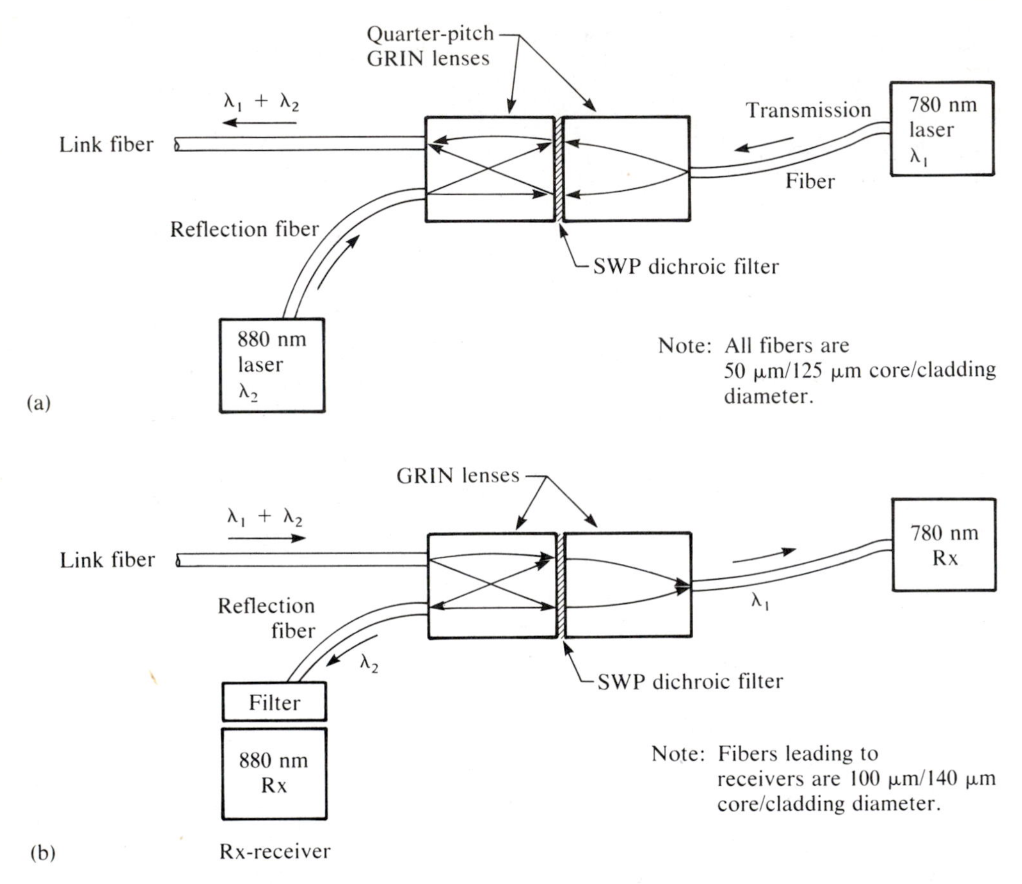

FIGURE 8–32 Wavelength division multiplexing using couplers (780–880 nm). (a) Multiplexed. (b) Demultiplexed.
Source: J. C. Williams and C. L. Kozibowski, "A Two-Wavelength Fiber Optic Multiplexing System for Analog Data Links," *Proceedings of SPIE* 417: 19 (1983).

Figure 8–32 shows how a special GRIN lens is used in the design of a WDM coupler. In Figure 8–32(a), a special wavelength filter, a dichrotic filter, is used to pass the 780-nm wavelength to the link fiber and to reflect the 880-nm signal into the same link fiber. This is a multiplexing scheme that combines λ_1 and λ_2 into the link fiber. The design in Figure 8–32(b) functions as a demultiplexer. The mixed signal enters from the link fiber. The 880-nm wavelength is reflected to one receiver, while the 780-nm signal is passed through to the other receiver. A WDM demultiplexing coupler, using a GRIN lens and a reflection grating (reflecting diffraction grating) is shown in Figure 8–33. The grating refracts the different wavelengths at different angles, and the GRIN lens focuses the wavelengths onto the three receiving fibers.

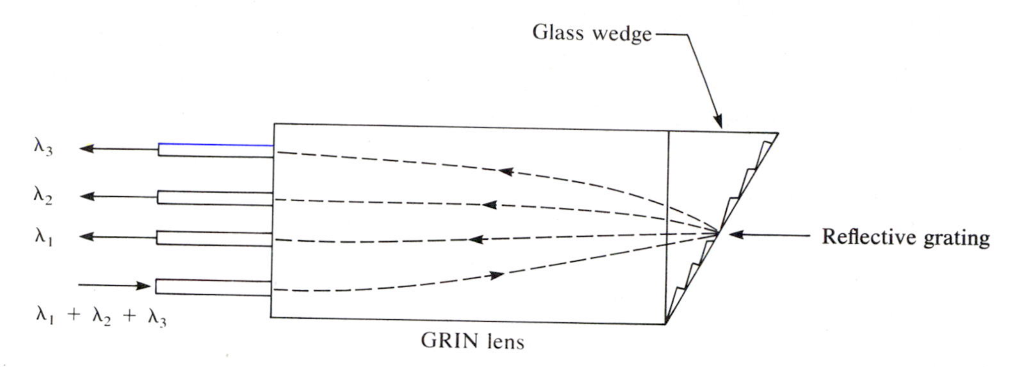

FIGURE 8–33 Wavelength division multiplexer, using reflective gratings.
Source: Liz Jou and Bruce Metcalf, "Wavelength Division Multiplexing," *Proceedings of SPIE* 340: 69 (1982).

8–8 FIBER OPTIC SWITCHES

Until the development of optical switches, the functions of switching and controlling data flow in a communication network were carried out at the electronic level. Electrical rather than optical signals were switched. Since optical switches became available commercially in the late 1970s, the control of data flow in fiber optic communication has been accomplished by switching optical signals between fibers.

There are numerous advantages to optical rather than electrical switching. The first advantage is the cost. It is simpler and less costly to switch optical signals in a fiber optic system than to convert the optical signals to electrical signals, execute the switching, and convert back to optical signals. Second, a much higher bandwidth and higher data rate can be accommodated. Third, it gives the fiber optic system the flexibility that only electronic systems had. Line testing can be easily implemented; nodes (transmitting and receiving stations) can be bypassed when they fail and thus can be prevented from disturbing the communication system connected to the node; and signal routing in the network can easily be accomplished.

8–8–1 Mechanical Switches

The fiber optic mechanical switching is accomplished by moving fibers, or prisms, to connect or to disconnect the optical circuit. A simplified version of a mechanical fiber optic switch is shown in Figure 8–34. By sliding the assembly that holds fibers 2 and 3, port 1 can be connected to port 2 or port 3. Remember that the fibers are very small, so it is not easy to line them up to obtain a useful switch.

A switching technique that relies on beam expansion (similar to the expanded beam connector) is shown in Figure 8–35. Moving the prisms as

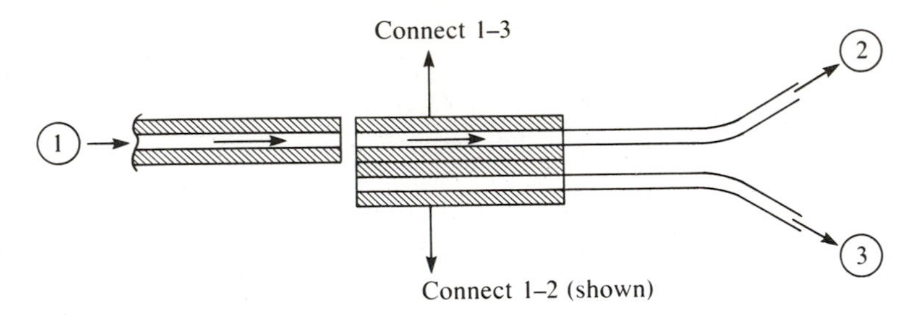

FIGURE 8–34 Simplified schematic of fiber optic switch.

shown by the arrows would connect port 2 to port 1 (prism up) or port 2 to port 3 (prism down). Here, because the beams are expanded, positional accuracy is somewhat less stringent.

The fibers, or the prisms, in Figures 8–34 and 8–35 can be moved magnetically. A data sheet describing some electromagnetic switches is shown in Figure 8–36. Note that switching time is measured in milliseconds. Here, it is 5–8 ms.

8–8–2 Direct Optical Switches

The electromagnetically activated switches are relatively slow. It takes many milliseconds to do the switching. In contrast, the direct optical switches can switch in a matter of nanoseconds. Their insertion loss (the power lost by inserting the switch into the optical circuit) is relatively high, about 3–10 dB, in comparison with about 1 dB for the mechanical switch. The cost of a direct optical switch is also substantially higher than that of the mechanical switch.

The underlying principle of direct optical switching is shown in Figure 8–37. The refractive index of the crystal is altered by the applied voltage. This results in a different angle of refraction in the crystal. In Figure 8–37(a), there is no applied voltage. The input is connected to output 1. With voltage

FIGURE 8–35 Prism switch.

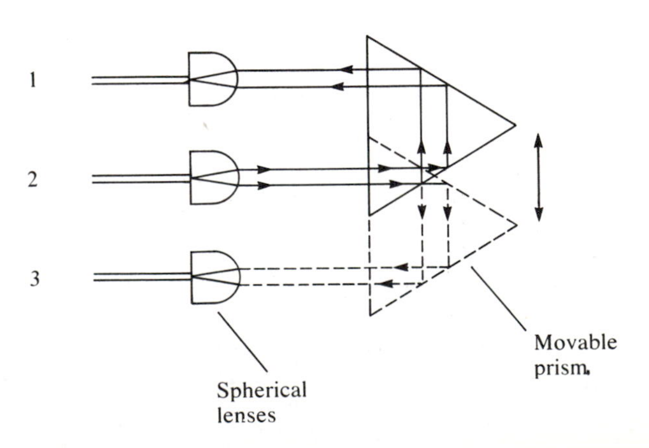

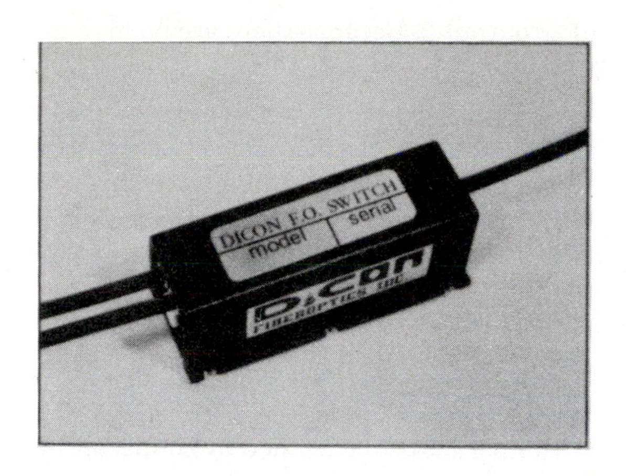

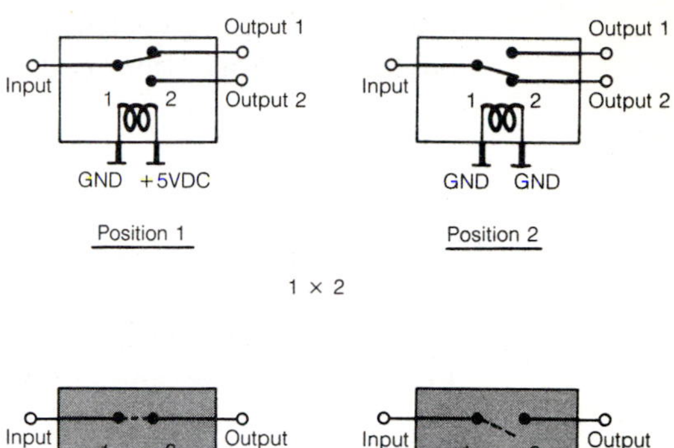

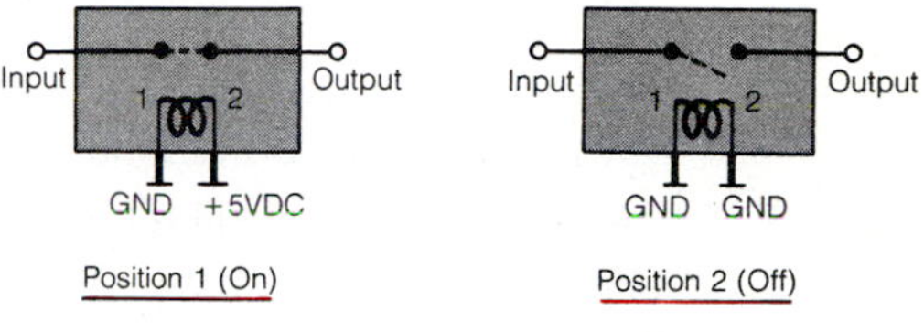

Item	Rating/Specifications
Insertion loss	0.6 dB typical, 1.2 dB max
Switching time	5-8 ms latching style; 8-14 ms non-latching style
Wavelength range	no restriction
Cross-talk	less than −60 dB
Repeatability	less than 0.05 dB
Durability	greater than 10^6 cycles
Operation temperature	−20°C to +80°C
Storage temperature	−40°C to +100°C
Actuation method	latching or non-latching style
Fiber types	50/125, 62.5/125, 85/125, 100/140

FIGURE 8–36 Data sheet for 1 × 2 and on-off fiber optic switches.
Source: Dicon Fiber Optics, Inc., Fremont, Calif.

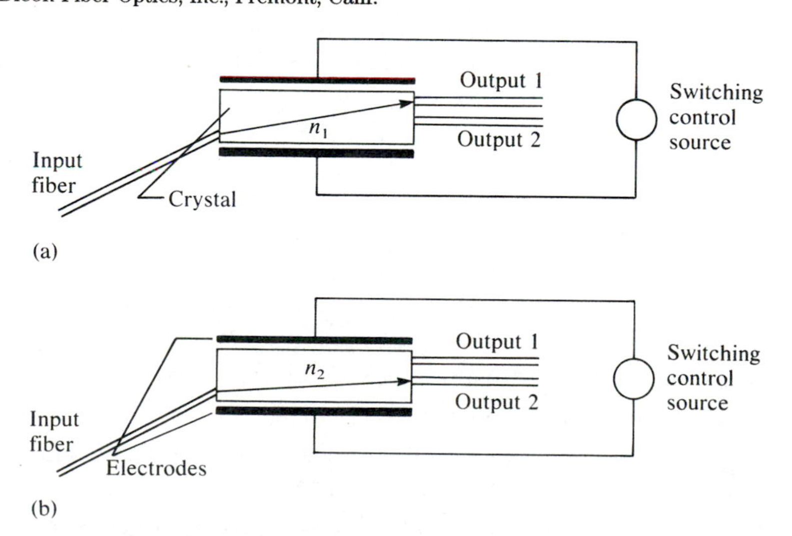

FIGURE 8–37 Direct optical switch. (a) Without applied voltage (refractive index n_1). (b) With applied voltage (refractive index n_2).

applied, the input becomes connected to output 2. Because there are no moving parts, switching times can be very short, a few nanoseconds.

SUMMARY AND GLOSSARY

The glossary terms and their definitions are specifically related to the material you have just covered in this chapter. They represent a good portion of this material. Understanding and memorizing this list will reinforce what you have just learned and will generally broaden your technical vocabulary.

Angular misalignment. The angle by which one of the fibers to be joined deviates from the axis of the other fiber.

Area mismatch. The fact that the two fiber ends to be joined by connector or by splice may not have the same area.

Coupler. A means of transmitting optical power among different fibers. (See "star coupler" and "T coupler.")

Cross coupling. The power coupled between fibers that are not expected to be coupled at all, usually given in decibels and usually caused by imperfections in construction.

Cross-section mismatch. See "area mismatch."

Crosstalk. See "cross coupling."

Cure. Allow or cause an adhesive to harden.

Directivity. The coupling in a direction opposite to the expected direction of optical power flow, usually given in decibels (reverse coupling in decibels).

Elastomeric splice. A type of mechanical splice developed by GTE.

End separation misalignment. The distance between fiber edges S (separation), usually given in relation to the fiber core diameter d, as S/d.

Excess power loss. Loss in a coupler in excess of the power division. The ratio of total output power at all output ports to the total input power at all input ports.

Expanded beam connector. A fiber connector that uses lenses so that the beam between two fiber ends which are to be joined is broader (has larger cross-section). This type of connector is usually used with single-mode fibers.

Fresnel reflection losses. Losses caused by the reflection at refractive index discontinuities, such as the interface between a fiber core and air. Important for connectors and splices not using index matching.

Fusion splice. A splice accomplished by melting the fiber ends, using heat.

GRIN lens. A lens made like a graded-index fiber (but with a substantially larger diameter) with carefully selected length.

Index-matching gel. Optical material placed in the gap between two fiber ends being connected. Its refractive index matches that of the fiber core.

Lateral misalignment. The distance between the center lines of the two fiber edges.

Loose buffer (loose tube). A fiber cable construction in which the fiber is free to move (loose) inside a buffer jacket.

Mechanical splice. A splice where the two fiber ends are held together mechanically without fusing the fiber edges.

Minimum bending radius. Represents the sharpest possible bend the cable or bare fiber can tolerate before it breaks. It is the radius of the smallest possible circle into which the fiber can be formed.

Misalignment. See "angular misalignment," "end separation misalignment," and "lateral misalignment."

N.A. mismatch. The fact that the numerical aperture (N.A.) of the two fiber edges to be joined may not be equal.

Port. A point of input or output to a system or a circuit. Fiber ends that allow access to a system.

Power division. The fraction of input power that is coupled to an output fiber.

Power ratio. The ratio of power between two output fibers in a T coupler, where the power for each output is expressed as a percentage of the input.

Power-splitting ratio. See "power division."

Profile mismatch. The cross-sections of two fibers may not have the same shape. For example, one may be elliptic and the other circular.

SELFOC® lens. See "GRIN lens."

Splice. A method of permanently connecting two fibers. (See "elastomeric splice," "fusion splice," "mechanical splice," and "splice organizer.")

Splice organizer. When dealing with splices in multifiber cables, the splice organizer is the housing for the many individual splices.

Star coupler. A star-like structure that allows coupling of optical power among a large number of fibers.

T coupler. Typically, a coupler with three ports (sometimes four for a dual T structure).

Tight buffer. A fiber cable where the outside jacket is tightly wrapped around the fiber.

Ultraviolet. Invisible light of short wavelength (below about 0.4×10^{15} Hz) used to cure certain optical adhesives.

FORMULAS

$$\text{Loss}_{\text{lat}} = -10 \times \log[1 - (1.28 \times l/d)] \quad \textbf{(8–1)}$$

Loss caused by lateral misalignment of fibers at splice or connector. (The relative lateral misalignment is l/d.)

$$\text{Loss}_{Fr} = -10 \times \log(1 - Fr) \tag{8–2}$$

Loss due to Fresnel reflection at butted fiber surfaces. (Fr is the Fresnel reflection.)

$$\text{Loss}_{\text{ang}} = -10 \times \log[1 - (n \times \theta)/(180 \times \text{N.A.})] \tag{8–3}$$

Loss due to angular misalignment of connected fibers. (N.A. is the numerical aperture of the fiber, n is the refractive index in the gap between fiber ends, and θ is the angular misalignment in degrees.)

$$\text{Loss}_{\text{area}} = -10 \times \log(a_2/a_1) = -20 \times \log(d_2/d_1) \tag{8–4}$$

Loss due to area mismatch between connected fibers. (a_1 and a_2 are the areas of the respective fibers and d_1 and d_2 are their diameters.)

$$\text{Loss}_{\text{N.A.}} \cong -20 \times \log(\text{N.A.}_2/\text{N.A.}_1) \tag{8–5}$$

Loss due to mismatch in numerical aperture N.A. ($N.A._1$ and $N.A._2$ are the numerical apertures of the two fibers.)

$$P_o = P_{\text{in}}/n \tag{8–6}$$

Power at each output fiber for a star coupler with n output fibers.

$$PD_{\text{ST}} = -10 \times \log(1/n) \tag{8–7}$$

Power division in a star coupler in decibels, assuming no loss in the coupler.

$$\text{Loss}_{\text{ex}} = P_{\text{ot}}/P_{\text{in}} \tag{8–8}$$

Excess loss in coupler, the power loss in terms of the ratio of the total output power P_{ot} to the input power P_{in}.

$$\text{Loss}_{\text{ex}}|_{\text{dB}} = -10 \times \log(P_{\text{ot}}/P_{\text{in}}) \tag{8–8a}$$

Excess loss expressed in decibels.

QUESTIONS

1. What are the basic design goals of fiber optic cables?
2. What is one of the main advantages of the loose buffer structure?
3. Does the cable structure contribute to the optical characteristics of the fiber?
4. What is meant by minimum bending radius? Why is this feature important?
5. What is the difference between a spliced fiber and a connected fiber? When are the two used?
6. Does fusion splice or mechanical splice have power loss? Explain.
7. What, in addition to an optical joint, must the splice provide?
8. For what is a splice organizer used?
9. Why must the fiber ends in a connector or a mechanical splice be highly polished?
10. What are the three misalignment loss mechanisms? Explain.
11. Describe the three mismatch types of losses. Which is least important? Which is most important?
12. Why are losses in single-mode fibers much more sensitive to connector or splice misalignments than are multimode fibers?
13. What is one of the most important disadvantages of the use of index-matching gels?
14. What is the purpose of index-matching gel used in splices or connectors?
15. How are end separation and angular misalignment losses affected by the N.A.?
16. How does the use of index matching affect Fresnel and angular misalignment losses?
17. Would you expect to encounter Fresnel losses in a fusion splice? Explain.
18. How is the power of a laser monitored using a coupler?
19. Why is it useful to have fiber optic couplers?
20. How does the star coupler connect one fiber to many and many fibers to one?
21. What is the basic functional difference between a reflective and a transmissive star coupler? Do both have separate input and output fibers?
22. Is a star coupler characterized by a constant power division (the same for all fibers)? Explain.
23. What is the basic function of a T coupler?
24. Can you think of a way to use T couplers to interconnect many nodes?
25. Describe, briefly, a fused T coupler and an expanded beam T coupler.
26. What is meant by a WDM coupler? Give an example.
27. What are the two basic methods of switching optical power? Give advantages and disadvantages of each.
28. How does the electromagnetically driven optical switch operate like a relay?
29. What is the principle of operation of a direct optical switch?

PROBLEMS

1. A fiber with the following characteristics is connectorized with a lateral misalignment of 15 μm. Find its loss due to lateral misalignment alone. The core diameter is 62 μm, core refractive index is 1.4, and N.A. is 0.25. Assume a small air gap ($n = 1$).
2. For Problem 1, calculate
   a. The Fresnel losses (no index matching)
   b. The total lateral misalignment and Fresnel losses. (Add the losses in decibels.)
3. If the connector in Problem 1 has *only* an angular misalignment of 4°, calculate
   a. The loss with index matching
   b. The loss without index matching
   c. The difference between the first and second parts of this problem and determine which is better
4. Find the maximum angular tolerance for the connector in Problem 1 if the angular misalignment loss must be kept below 0.5 dB.
5. Repeat Problem 1 for a fiber with core diameter of 200 μm, core refractive index of 1.45, and N.A. of 0.3.
6. The fiber in Problem 5 is connectorized with an angular misalignment of 6°. Find the angular misalignment loss
   a. With index matching
   b. Without index matching
7. The fiber in Problem 5 is spliced with a gap of 50 μm. Find
   a. The end separation loss
   b. The Fresnel loss (no index matching)
   Which is more significant? Should you use index matching?
8. The fiber in Problem 1 is connectorized with the following misalignments: lateral, 10 μm; angular, 5°; and end separation, 15 μm.
   a. Estimate the total connector loss, with no index matching. (Include Fresnel losses.)
   b. Find the total loss with index matching. (Remember that the angular loss is affected by index matching.)
9. A transmitting fiber with core diameter of 62 μm, core index of 1.4, and N.A. of 0.2 is connected to another fiber with core diameter of 52 μm, core index of 1.4, and N.A. of 0.25. Find the loss due to
   a. Area mismatch alone
   b. N.A. mismatch alone
   (*Hint:* Be careful to determine which types of losses actually exist.)
10. Repeat Problem 9 with the transmitting and receiving fibers interchanged. (*Hint:* Be careful to determine which types of losses actually exist.)
11. Assume, in the connection in Problem 9, that the refractive index of one of the fibers (only one) is 1.5 instead of 1.4. Calculate the Fresnel losses

(assume no gap) and add them to the losses in Problems 9 and 10. This is the total connector loss, assuming no misalignments. (*Hint:* Since you assume no end separation, there is only one Fresnel reflection between the $n = 1.4$ and $n = 1.5$ materials.)

12. From the graphs given in Figures 8–21(d) through (f), estimate the total connection misalignment loss for a single-mode fiber connection with end separation of 40 μm. lateral misalignment of 2 μm, and angular tilt of 2°.

13. A star coupler has 18 fibers used as inputs and 9 fibers used as outputs. Calculate the power division if
  a. A reflective star is used
  b. A transmissive star is used

14. The input into a fiber in a reflective star coupler is -10 dBm (0.1 mW). The excess loss of the coupler is 2 dB. Calculate the power in each fiber (in dBm) for a coupler with
  a. 10 fibers
  b. 20 fibers

15. A T coupler has a 80% to 20% output power ratio. Find the power division in decibels for each output fiber (three ports).

16. A T coupler has a power splitting ratio of -1 and -6.86 dB. Excess loss is negligible.
  a. If the input power is 100 μW, what is the power at each of the two output fibers?
  b. Calculate the output power ratio in percent.

17. Repeat Problem 16 with an excess loss of 1 dB.

9 Optical Sources for Communication

CHAPTER OBJECTIVES

This chapter will provide you with an understanding of the physics of light sources, light-emitting diodes (LEDs) and lasers, as well as some of their important operating characteristics. You will learn about various LED and laser structures and how they affect performance. After you have studied this chapter, you will be able to evaluate specifications of LEDs and lasers to determine their suitability for specific applications. In this chapter, you will also be introduced to various electronic drive circuits and some of their design parameters.

Many terms and concepts mentioned in this chapter have been discussed in detail in previous chapters.

9–1 GENERAL CHARACTERISTICS

An optical source for use in fiber optic applications must have the following general features:

1. For communication, the light intensity (optical power) must be large enough to make long-distance fiber communication feasible. This feature will also be affected by the next feature given.
2. The structure must permit effective coupling of light into a fiber.
3. The wavelength of the light must be compatible with the wavelength most suitable for fiber optic use. Today, the most suitable wavelength in fiber optic communication is about 1.55 μm. Other wavelengths used are 1.3 and 0.82 μm.

4. The line width must be narrow to allow a high data rate to be used.
5. The response time must be short (the bandwidth [BW] very high) to allow a high data rate. The circuitry to drive the optical source must be reasonably simple.
6. The device must produce stable power that does not vary with temperature and other environmental conditions, and it must be reliable.

9–1–1 Power and Efficiency

For the purpose of this discussion, assume that for a signal to be properly received, its power P_R must be about 100 nW. To make the calculations easier, convert this to -40 dBm. (Note that 0 dBm = 1 mW and $P|_{\text{dBm}} = 10 \log(P_R/10^{-3})$. Here, $P|_{\text{dBm}} = 10 \times \log[(100 \times 10^{-9})/10^{-3}] = -40$ dBm.) Further assume that the losses along the path of transmission are 30 dB (20 km of 1.2-dB/km fiber and various other losses of 6 dB). This means that the power at the source is -40 dBm + 30 dB = -10 dBm. The power sent out by the source must be 30 dB larger than the power received to account for the losses. A source that must couple -10 dBm (0.1 mW) into the fiber is required.

Typical lasers produce a few milliwatts of optical power, while LEDs usually produce no more than about 1 mW. As you will see soon, you incur substantial losses in the coupling of the power to the fiber. The need for -10 dBm at the fiber input may require a high-efficiency light source that can effectively be coupled to the fiber.

Even though the optical power is relatively small (only milliwatts) because the typical total efficiency of these devices is also small, it is not uncommon to expect the electrical power to be 100 times the optical power (of the order of 100–500 mW). This may not mean much compared to your 100-W high-fidelity system, but it complicates the circuitry and increases costs.

Efficiency can be defined in terms of photons generated per electron **(quantum efficiency):**

$$\eta_Q = \text{Photons generated/Electrons injected} \tag{9–1}$$

η_Q is the quantum efficiency, sometimes called the **internal efficiency.** The **external efficiency** (also called **power efficiency**) η_{PT} is defined by

$$\eta_{\text{PT}} = P_{\text{out}}/P_{\text{el}} \tag{9–2}$$

where P_{out} is the emitted optical power and P_{el} is the electrical power input to the light source. η_{PT} is distinguished from the total efficiency η_T. It expresses the system efficiency. In addition to η_{PT}, you must consider the **coupling efficiency** η_c, the ratio of power coupled to the fiber to the power output of the device:

$$\begin{aligned}\eta_c &= P_F/P_{\text{out}} \\ &= \text{Power into fiber/Power out of device}\end{aligned} \tag{9–3}$$

The total efficiency of the device relates the power into the fiber to the electrical power input.

$$\eta_T = \eta_{PT} \times \eta_c = P_F/P_{el} \tag{9–4}$$

The coupling efficiency is a function of the structure of the source and its light intensity distribution, as well as the size and the numerical aperture (N.A.) of the fiber.

9–1–2 Numerical Aperture, Active Area, and Intensity Distribution

Internal and external efficiencies are substantially different for LEDs and lasers. Coupling efficiency, however, has some general characteristics that apply to all types of light sources. Coupling efficiency is related to the active area and distribution of the source and to the cross-section and N.A. of the fiber. To simplify the discussion, you can make the following assumptions:

1. The active area of the source (that is, the surface which produces the light) is substantially smaller than the core cross-section.
2. The beam from the source covers the core.
3. The intensity distribution is Lambertian.

The second and third assumptions are shown in Figure 9–1. In Figure 9–1(b),

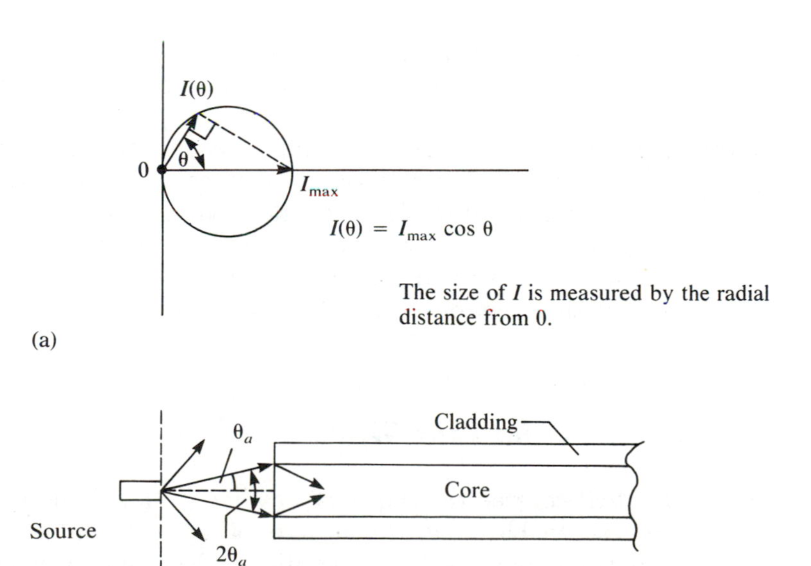

FIGURE 9–1 Light source to fiber coupling. (a) Lambertian source. (b) Coupling to fiber.

$2\theta_a$ is the acceptance angle of the fiber. The rays emanating from the source overflow this angle. Clearly, a portion of the light will radiate beyond the acceptance angle of the fiber. The distribution of the light power is shown in Figure 9–1(a). The light intensity at a direction θ is a function of cos θ times I_{max}, the intensity in the direction perpendicular to the radiating surface. At $\pm 60°$, the intensity is 0.50 I_{max}. The coupling efficiency for this case is given by

$$\eta_c = \sin^2 \theta_a = (\text{N.A.})^2 \tag{9–5}$$

EXAMPLE 9–1

A Lambertian light source, with a total power output of 1.2 mW is coupled to a fiber. Assume the active light source area is smaller than the fiber core. Find the power coupled to the fiber with (1) N.A. = 0.2 and (2) N.A. = 0.4.

Solution
First find the coupling efficiency η_c, and then find the power into the fiber P_F.

1. $\eta_c = 0.2^2 = 0.04$ (or 4%)
$\eta_c = P_F/P_{out}$ $0.04 = P_F/1.2$ mW
$P_F = 0.04 \times 1.2$ mW $= 48$ μW

2. $\eta_c = 0.4^2 = 0.16$ (or 16%)
$P_F = 0.16 \times 1.2$ mW $= 192$ μW

As Example 9–1 shows, larger N.A. means more power coupled into the fiber. (Remember, however, that larger N.A. means higher dispersion and lower data rate.)

With the introduction of single-mode fibers with core diameters of 3–10 μm, special high-intensity and low-area sources were developed. As you see later, the laser is more suitable for single-mode operation. The laser can reasonably simulate a point source with a very narrow beam, which can efficiently be coupled to the single-mode fiber. Its line width is also very narrow, which further improves its performance in single-mode transmission.

9–1–3 Wavelength and Line Width

In Section 4–4, this text covered the choice of wavelength for communication systems. It seems reasonable that the design of optical sources should concentrate on producing devices that operate at desirable wavelengths. Historically, devices were first available for 820- to 850-nm wavelengths. This range represents the first low-loss window. Subsequently, operation was shifted to the 1.3 μm wavelength area, where dispersion in single-mode fibers is minimal. With the introduction of the dispersion-shifted fibers, which operate at about 1.55 μm, suitable sources were (and still are) being developed. The 1.55-μm wavelength represents a low-loss window with losses near the theoretical limit set by Rayleigh scattering, about 0.2 dB/km.

To operate at different wavelengths, the material and the structure must both be carefully selected. Single-mode operation requires a narrow line width source. Lower $\Delta\lambda$ (line width) yields a wider BW and thus higher data rates. In this context, higher data rates are rates of the order of 100–1000 GHz (and even higher). Note that 1 GHz = 1000 MHz = 10^9 Hz.

9–1–4 Modulation and Response Time

In the **intensity modulation** scheme, which is the only method used commercially to modulate data, data are represented by a variation in the intensity (or power output) of the source. Digital data are transmitted by turning the source "on" and "off," which corresponds to 1 and 0, respectively. A useful source is one that can be turned on and off easily and rapidly.

Rise time t_r and fall time t_f are defined the same way they were defined in the section on electronic pulse theory. Figure 9–2 shows an idealized input pulse and the resulting "turn on" and "turn off" of the source. t_r is the time the source takes to rise from 0.1 (10%) to 0.9 (90%) of full power out. t_f is similarly defined for turn-off time.

For high data rates, the device must have low t_r and t_f. These are usually specified by the manufacturer of the source. The relation between the optical BW and the rise time (or the fall time) is given by

$$BW_{opt} = 0.35/t_r \qquad \textbf{(9–6)}$$

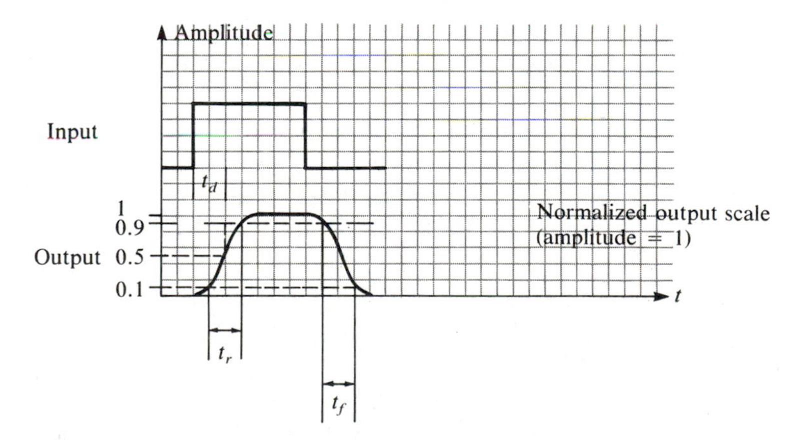

FIGURE 9–2 Rise, fall, and delay times.

EXAMPLE 9–2

The rise time of an LED is specified to be 25 ns. Find (1) its BW and (2) the corresponding electrical data rate when using a return to zero code.

Solution

1. From Equation 9–6, $BW_{opt} = 0.35/(25 \times 10^{-9}) = 14$ MHz
2. Because the electrical BW is $0.707 \times BW_{opt}$ (Section 6–3–4), $BW_{el} = 0.707 \times 14 = 9.9$ MHz. From Equation 6–1,

$$B_r(\text{max}) = \text{BW} = 9.9 \text{ Mb/s}$$

The circuitry to drive (modulate) the source should be as simple as possible. Generally, this is true for the LED but not for the laser. Note, however, that the laser has many other advantages.

For analog intensity modulation, the light intensity represents the amplitude of the transmitted signal. This means that to be able to faithfully reproduce the transmitted signal, the relation between light energy and input amplitude should be as linear as possible. This effect is often specified in terms of the "harmonic distortion" of the device. The nonlinearity of the device distorts the original input sinusoidal waveform. The extent of the distortion is measured by the amount of harmonic frequency content introduced. If the input is 1000 Hz, the output may contain 2000 Hz, 3000 Hz, etc., in varying amplitudes. These are the harmonics that represent the distortion introduced.

9–1–5 Stability and Reliability

Light energy produced by a source must depend only on the input signal. It should not vary with ambient temperature or other environmental conditions. Some of the original lasers were constructed in a temperature-controlled (cooled) chamber to minimize temperature effects. Lasers are still quite sensitive to temperature variations.

In some cases, the sources that are installed underwater should have a long life of uninterrupted stable performance. The typical life expectancy in communication systems is more than 10^5 hours of continuous operation. The incandescent lamp does not last that long. Note that 10^5 hours is about 11 years.

9–1–6 Safety

Light energy, particularly laser beams, is considered highly destructive and dangerous. However, this is true only when relatively high power is involved. Mostly, the light energy involved in communication will do no damage. Remember, however, that even a small amount of light energy can damage the human eye. The communication laser beam can do substantial damage to the eye. (See Appendix II.)

NEVER LOOK DIRECTLY INTO A SOURCE OF LIGHT—ESPECIALLY A LASER! Remember, you may see nothing, and yet the damage may be substantial. The wavelengths involved are not necessarily visible. To verify the existence of a beam, you must use photo-optical devices, **NOT YOUR EYES!**

9–2 PRINCIPLES OF LIGHT EMISSION

This text refers to light emission as the generation (radiation) of photons. A photon is a particle of light (as described by the quantum theory) or a packet of energy. The energy of a photon is given by

$$E_p = h \times f \qquad h = 6.626 \times 10^{-34} \text{ J-s} \quad \text{Planck's constant} \tag{9–7}$$

f is the frequency of the optical energy.

A photon may be generated when an electron moves from a higher to a lower energy level in the atomic structure. Note that not every such electron transition produces a photon. Electron energy levels exist in bands (energy bands) corresponding to the location of the electron in the shell structure of the atom. Electrons in outer shells are of higher energy. The energy bands associated with light generation are the conduction and the valence bands. The conduction energy band corresponds to the conduction electrons, the free electrons that form the electric current flow. The valence band is associated with the electrons in the outer shell of the atom. Note that current flow involves the motion of both electrons and holes (the term "hole" refers to the absence of an electron where an electron should have been).

Energy bands are separated by energy gaps, or forbidden energy gaps. Electrons or holes do not occupy these energy gaps. Figure 9–3 schematically shows electrons occupying the conduction band (E_2) and holes occupying the valence band(E_1). Think of holes in terms of electrons that have left their places in the valence band, become free electrons, and left behind an equal number of holes.

9–2–1 Spontaneous Emission

The **recombination** of an electron with a hole (that is, the transition of an electron from the conduction band back to the valence band) (Figure 9–3(b)) may be associated with the generation of a photon. The recombination involves an electron that has undergone an energy change of $E_2 - E_1$. If a photon is generated, it must have an energy equal to that change:

$$E_p = E_2 - E_1 = h \times f \tag{9–7a}$$

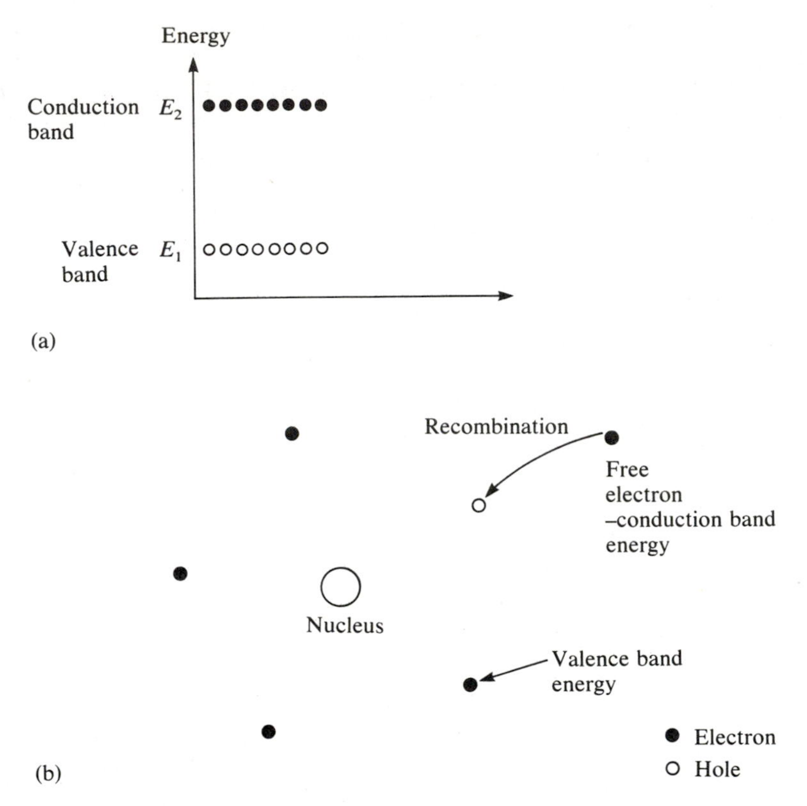

FIGURE 9–3 Electron-hole energy levels. (a) Energy band representation. (b) Atomic shell (only valence shell shown).

Because the light frequency f is given by $f = c\lambda$, you get, for this type of radiation,

$$\lambda = (h \times c)/(E_2 - E_1) = (h \times c)/E_g \quad \textbf{(9–8)}$$

where E_g is gap energy $E_2 - E_1$. The wavelength of the emitted radiation is related to the energy difference between the conduction band and the valence band. Because different materials have different energy gaps, they produce radiation of different wavelengths. (See Appendix III.) This type of photoemission is called **spontaneous emission** and is the mode of light generation in LEDs. In contrast, the photon emission in lasers is largely stimulated emission, which will be discussed in Section 9–2–2.

EXAMPLE 9–3

A material with an energy band gap of 1.2 eV is used in photon emission.[1] Find the frequency and wavelength of the emitted light.

[1] 1 eV = 1.602×10^{-19} J. 1 Coulomb (C) = 6.242×10^{18} electrons. 1 eV = $1/(6.242 \times 10^{18})$ J.

Solution

The 1.2-eV band gap means that $E_2 - E_1 = 1.2$ eV. From Equation 9–7,

$$1.2 \text{ eV} = h \times f$$

$$f = (1.2 \times 1.6021 \times 10^{-19})/(6.626 \times 10^{-34})$$
$$= 2.90 \times 10^{14} \text{ Hz}$$

Also,

$$\lambda = c/f = (300 \times 10)^6/(2.9 \times 10)^{14}$$
$$= 1.03 \times 10^{-6} = 1.03 \ \mu\text{m}$$

and

$$\lambda = 1.03 \ \mu\text{m}$$

In Equation 9–8, E_g is usually given in electronvolts. Combining all constants, Equation 9–8 can be rewritten as

$$\lambda = 1.24/E_g \qquad \textbf{(9–8a)}$$

λ is in μm and E_g is in eV. In Example 9–3, wavelength can be obtained from

$$\lambda = 1.24/1.2 = 1.03 \ \mu\text{m}$$

To obtain an efficient light source, a large number of recombinations must take place and a substantial number of these must be radiative. (Not all recombinations result in radiation.) This is accomplished by (1) selecting suitable materials where recombinations tend to be radiative, (2) producing a structure that has a large number of electrons in the conductive band, which may undergo a transition to the valence band, and (3) confining the generated light so that it is not radiated in all directions.

Figure 9–4 shows a P-N junction with and without applied voltage. In Figure 9–4(a), no voltage is applied, and there is a wide depletion region. When voltage is applied, a large number of electrons is pushed, or injected, into the N region (or a large number of holes into the P region). This situation, where there is a large concentration of high-energy electrons, is referred to as **population inversion.** (In normal materials, there are many more low-energy (valence) electrons than high-energy electrons. The reverse is thus termed population inversion.) The electrons and holes penetrate into the depletion region and recombine (Figure 9–4(b)). This recombination, represented in Figure 9–4(c) as an electron energy transition, produces the radiation. The injected electrons (or holes) constitute electric current. Light emission takes place at the recombination centers, that is, very close to the junction. Figure 9–4(b) shows the active area, as the area close to the junction.

It is worth noting that the energy gap for gallium arsenide (GaAs), for example, is about 1.43 eV, which will yield a light wavelength of $\lambda = 0.87$ μm. This is the infrared region, typical for communication applications. **Infrared-**

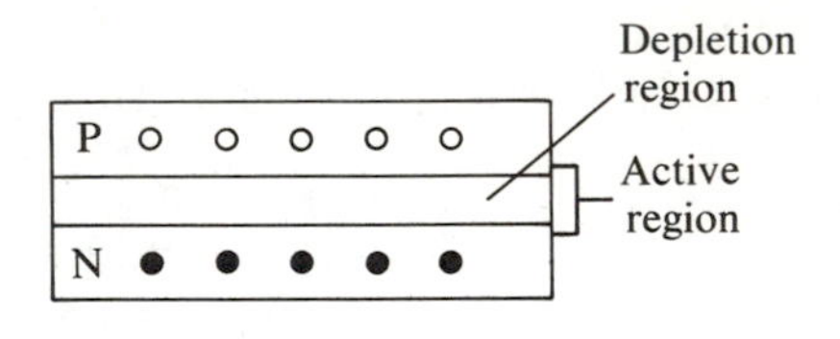

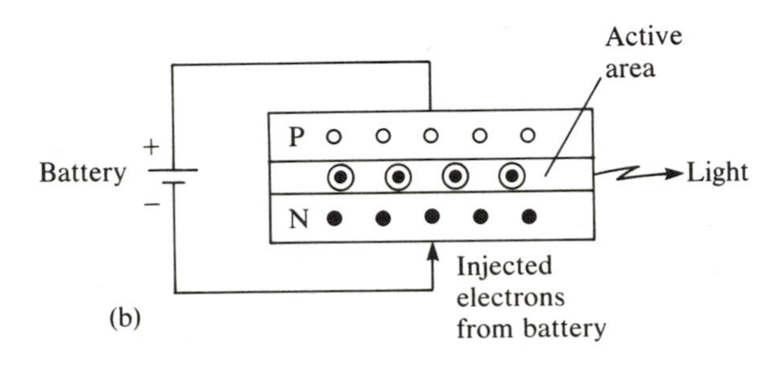

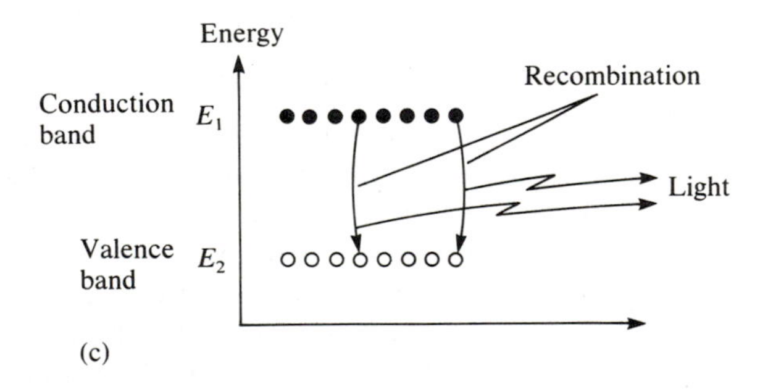

FIGURE 9–4 P-N structure. (a) P-N junction (no applied voltage). (b) P-N junction (forward biased). (c) Energy-level transition.

emitting diode (IRED), rather than LED, is used for sources operating in the infrared region.

9–2–2 Stimulated Emission

The light emission of lasers is a result of stimulated emission. Laser stands for *light amplification by stimulated emission of radiation.*

In principle, stimulated emission means that photons stimulate the emission of more photons. This is much like the avalanche effect in an avalanche (Zener) diode. The collision between photons and high-energy electrons induces an energy transition of the electron and radiates photons. These stimulated photons proceed to collide with electrons and produce more photons and so on.

You will see that it is important to be able to confine the photons to the active area to encourage this avalanche process. As was noted in the last section, to obtain spontaneous emission, it is necessary to have an electron population inversion (a high concentration of high-energy electrons that can recombine with holes [energy transition]). It is also necessary to choose a material for which these electron energy transitions produce radiation (highly radiative material). In addition, for stimulated emission, you need a source of photons large enough to produce the stimulated emission. This external source of photons, called an **energy pump,** is often provided by an external light source or suitable electrical current that produces a high concentration of photons. In the semiconductor injection laser, the pumping is provided by the injected electrical current. Other lasers, such as the gas lasers (CO_2, argon-ion, helium, and neon), liquid dye lasers, yttrium-aluminum-garnet (YAG), and ruby lasers require an external light pump.

9–3 LIGHT-EMITTING DIODES

The structure of a LED (or IRED) is essentially that of a semiconductor diode. The materials used, however, are such that the electron-hole recombination is highly radiative. The active area, where recombination occurs and radiation takes place, is usually very small, and the light output must propagate through a portion of the material.

Diodes are constructed in various ways. The **surface-emitting LED** (SLED) allows the active *surface* to radiate out. Other structures such as the **edge-emitting LED** (ELED) are designed to allow the *edge* of the active area to radiate out. In addition, various optical systems such as lenses and direct-mounted optical fibers are constructed as an integral part of the LED to improve output efficiency.

9–3–1 Simple Structure

Figure 9–5 shows a simple light-emitting structure. The P region is relatively thin, about 1 μm. Recombination takes place around the P-N junction and radiates in all directions. Most of the output light comes through the thin P layer. GaAs is the material used, rather than germanium or silicon, because GaAs is much more radiative. Some other materials used are gallium arsenide phosphor (GaAsP), gallium phosphor (GaP), aluminum gallium arsenide (AlGaAs), and indium gallium arsenide phosphor (InGaAsP). The light radiates downward, upward, and sideways. Because only a small portion of the internally generated light is output in a particular direction, this LED is very inefficient. This P-N structure is referred to as a homojunction. It consists of a single material (in this case, GaAs) and a single P-N interface.

To improve the performance of the LED, other P and/or N layers are added to the structure. Figure 9–6 shows a LED with a multilayer structure, a **heterostructure.** The two P and N regions differ from each other by the type

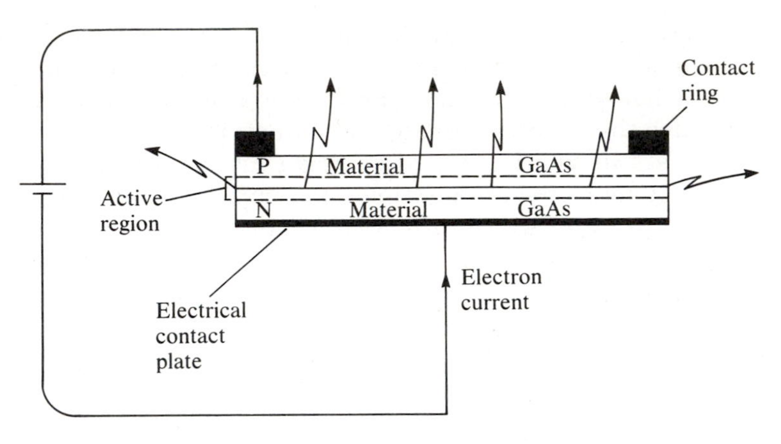

FIGURE 9–5 Simple light-emitting structure.

of material used and the doping (introducing desired impurities into the atomic structure) concentration. (This results, quite deliberately, in different refractive indices.) This structure increases the concentration of holes and electrons in the active region, and also confines the light emitted, so that most of the light is radiated upward. The lower P region serves as a reflector, while the N regions are transparent. The resulting beam is approximately 120° at its half-power points.

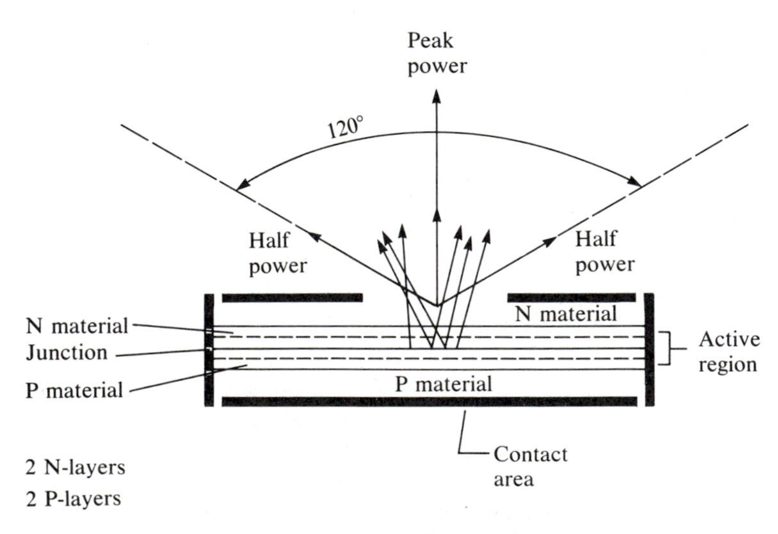

FIGURE 9–6 Double heterostructure—surface emitter.

9–3–2 Surface-Emitting LED

The structure shown in Figure 9–6 radiates from the junction surface, not from the edge, as in an edge-emitting diode (Section 9–3–3). A number of variations on this basic structure have been introduced to improve the radiation efficiency.

The Burrus type LED is shown in Figure 9–7. The active area is kept small by allowing electrical contact with only a small portion of the lower P surface. (The SiO_2 insulates the metallized layer from the P layer material.) The two lower P layers (GaAs and AlGaAs) serve, in part, as light reflectors. And, in part, they restrict the recombination to the area near the P-N junction. The circular well etched into the upper N layer focuses the light emitted by the active area onto the optical fiber. This type of LED is much more efficient than the homojunction or the simple heterostructure.

There are numerous ways to improve the coupling of light into an optical fiber. Section 11–2 examines some of these ways. Note that, by making the electric current flow through a small area, the current density is increased and the radiation efficiency is improved.

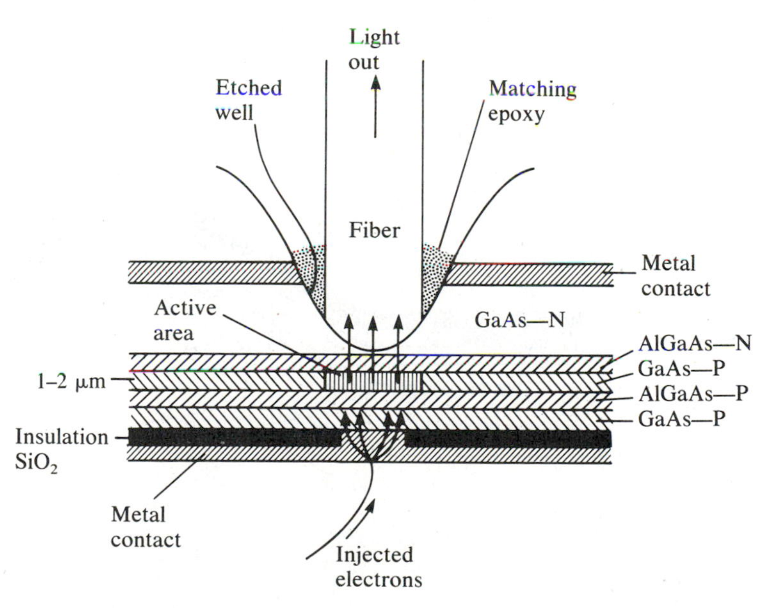

FIGURE 9–7 Burrus LED (etched well).
Source: C. A. Burrus and B. I. Miller, "Small Area Double-Heterostructure Aluminum Gallium Arsenide Electroluminescent Diode Sources for Optical Fiber Transmission Lines," *Optical Communications* 4:307–69 (1971).

9–3–3 Edge Emitters

Typically, the surface LED has a wide beam, and various methods are used to focus the beam onto an optical fiber (see Section 11–2). The ELED is designed to produce a more focused beam.

A typical edge-emitting structure is shown in Figure 9–8. The multiple layers serve two purposes. P layer 1 and N layer 1 concentrate the electrons and holes injected by the supply voltage in the active area. P and N layers 2 keep the light generated near the active region. In this way, you obtain a high electron-hole concentration, and you have more transitions; additionally, more light is emitted. Also, the light emitted at the edge of the LED is produced by the whole diode length L and not only by the active area near the edge. The light is guided by the two layer 2's toward the edge.

The edge-emitting structure produces a relatively narrow beam, about 30° (at the half-power points), in the perpendicular direction. It produces the same beam width as the SLED, 120° in the horizontal direction. This reduced beam width produces more efficient coupling to a low N.A. fiber. The SiO_2

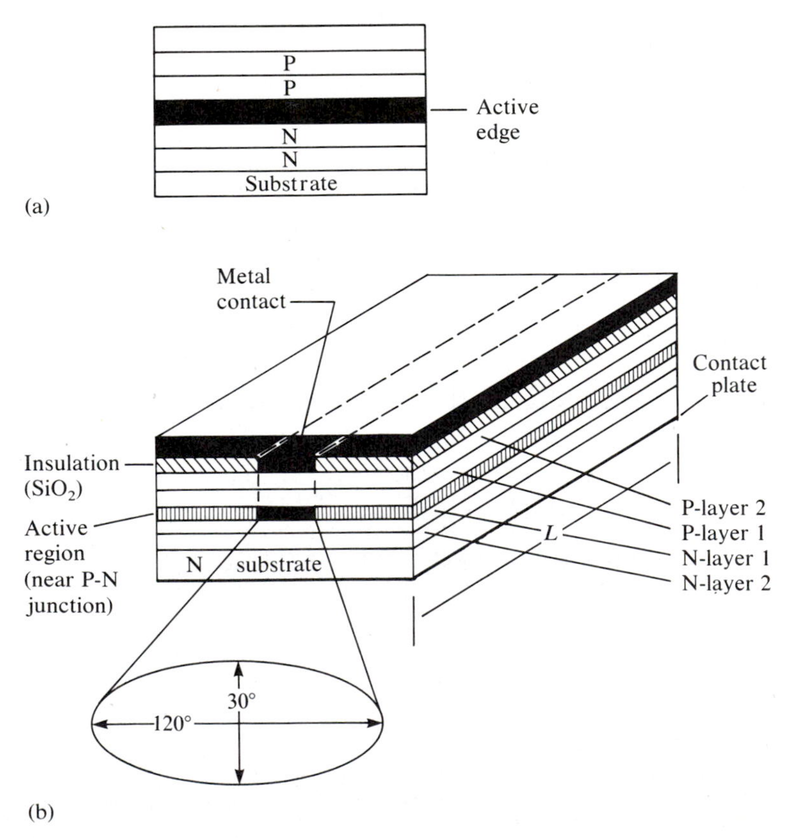

FIGURE 9–8 Edge-emitting diode. (a) Side view. (b) Detailed view.

insulation allows contact with the diode layers only along a narrow stripe. (The structure is called striped geometry.) This further concentrates the injected current to a small region and improves the efficiency of light generation.

9–3–4 Characteristics

Efficiency. One of the most important characteristics of a LED is its efficiency. Generally, the efficiency relates the electrical power (the input power) to the light power (the output power). You must, however, be careful in defining the output power. Quantum efficiency η_Q relates the input current to the light generated internally, not to the light that is finally emitted from the LED (see Section 9–1–1). In practice, η_Q varies between 0.5 and about 0.8 (50 and 80%). That means that only about 50% of the injected electrons produce photons. To simplify matters, η_Q is defined as

$$\eta_Q = P_{\text{int}}/P_{\text{el}} \qquad \textbf{(9–9)}$$

where P_{int} is the light generated internally and P_{el} is the electrical power dissipated. Another efficiency, η_{PT}, was defined in Section 9–1–1 in Equation 9–2.

It is important to realize that the light emitted by the LED P_{out} is substantially less than the light generated P_{int}, because of many factors. First, the light generated has to pass through semiconductor layers (reasonably transparent) before it reaches the outside. The interface between the LED material (refractive index of about 3.5–3.7) and air (refractive index 1) causes reflections back into the LED, resulting in substantial losses. (In Chapter 11, we will discuss the LED to fiber coupling loss. Only part of P_{out} can be made to propagate in the fiber [See Equation 9–3].) Because of these factors, only a small fraction of the generated light is emitted. This means that η_{PT} is relatively small (about 0.5–2%).

EXAMPLE 9–4

Find the emitted power P_{out} for a LED with

$$\eta_{\text{PT}} = 1.0\%$$
$$I_D = 50 \text{ mA} \quad \text{(diode current)}$$
$$V_F = 1.6 \text{ V} \quad \text{(diode voltage drop)}$$

Solution

The electrical power (the voltage across the diode times the current through it) P_{el} can be found by

$$P_{\text{el}} = I_D \times V_F = 50 \times 10^{-3} \times 1.6 = 80 \text{ mW}$$

$$P_{\text{out}} = (1/100) \times 80 = 0.8 \text{ mW} = 800\ \mu\text{W}$$

In Example 9–4, you found P_{out} to be 800 μW, which is a typical value for the average LED. Remember that the 800 μW generated is radiated in all directions; special means must be used to try to focus this energy into a fiber.

V-I and P-I Curves. The LED has a current-voltage relation similar to that of a regular diode. Figure 9–9 gives a typical *V-I* curve for a LED. The forward voltage drop of a LED varies from about 1.3 to 2 V. The forward current is, typically, limited to 50- to 100-mA, of direct current (dc) for the standard LEDs and goes up to a 1- to 3-A peak for pulsed operation. (The GaAlAs LED OPC216 made by TRW, Inc., can operate at a 3-A peak under pulsed conditions.)

From the discussion on efficiency, you know that higher input current (larger P_{el}) will result in larger P_{out}. Two typical I_F versus P_{out} curves are shown in Figure 9–10. Line 1 represents a high-power LED; line 2 represents a lower power LED. Here, I_F is the forward diode current, which is directly related to the number of injected electrons. Note that at I_F = 100 mA (the maximum I_F), P_{out} for graph 1 is 7 mW. Because its V_F is about 1.6 V, η_T turns out to be [7/(1.6 × 100) × 100] = 4.4%, which is a relatively high total efficiency.

The plot in Figure 9–10 is nearly a straight line, implying that the relation between current input and light power out is linear. This is true for many LEDs but not for all. The NDL4201A, made by the NEC Corporation, is much less linear, for example.

Line Width (Spectral Half Width and Spectral BW 50%). Most LEDs emit light with a relatively large line width (in comparison with that of the laser). The terms used to describe the range of wavelengths emitted vary from manufacturer to manufacturer. Some of these are **spectral response,**

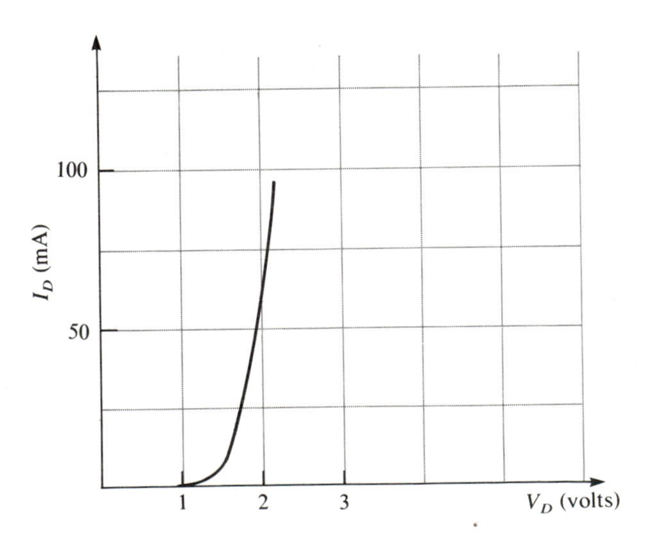

FIGURE 9–9 Voltage versus current plot of typical LED.

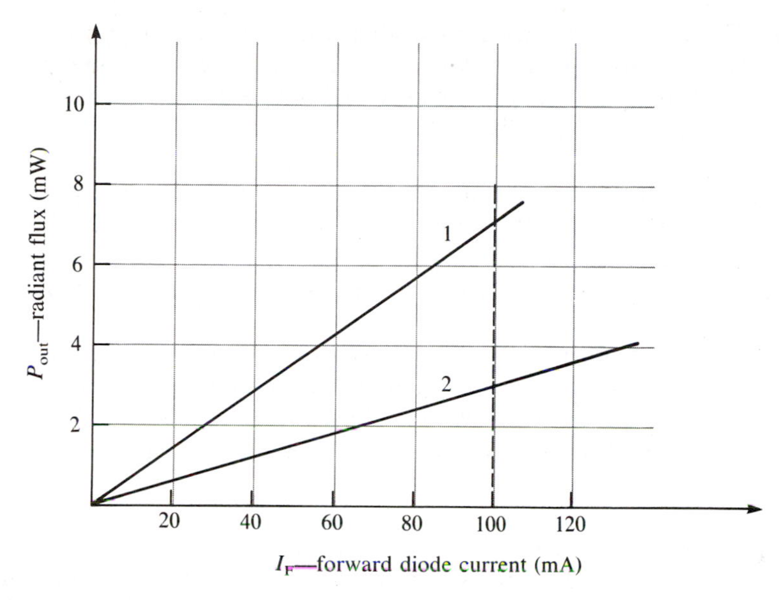

FIGURE 9–10 Radiant flux versus dc forward current for a typical LED.

spectral radiant flux, and **emission spectrum.** Figure 9–11 gives a spectral response plot for a typical LED. The line width is the range of wavelengths at half power, 50% in Figure 9–11. The line width for this LED is about 40 nm (between 920 and 960 nm). Remember that a large $\Delta\lambda$ results in a large dispersion and hence a lower modulation bandwidth and lower data rate.

FIGURE 9–11 Typical spectral response of LED.

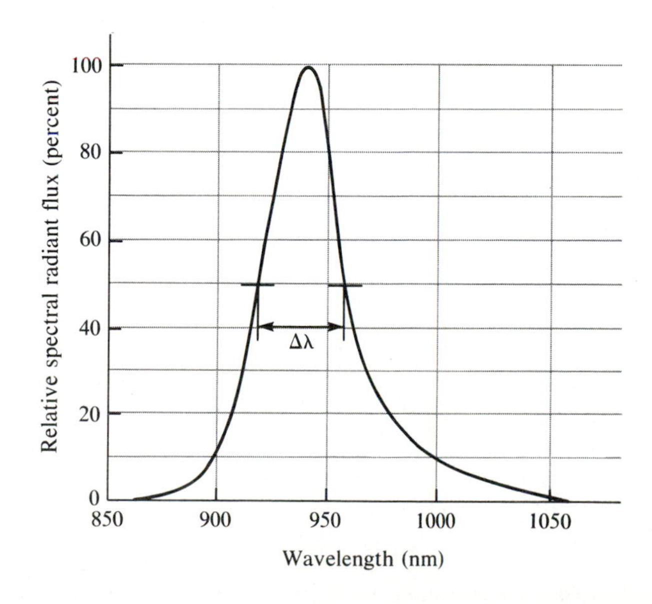

Angle of Radiation. In the general discussions of the LED, you noted that the LED (without any optical enhancement) radiates as a Lambertian source; that is, the intensity is related to the cosine of the angle of direction. Figure 9–12 demonstrates this distribution. At $\theta = 0$, the direction perpendicular to the emitting surface, the radiation is maximum. In relative terms, using normalization, call this maximum the unity intensity. At angle θ, the relative intensity is cos θ. At $\theta = 60°$, the relative intensity is 0.5. Note that, in Figure 9–12, the intensity scale is represented by the distance from point O, maximum at $\theta = 0$. The half-power width for the Lambertian source is 120° (+60° to −60°).

Most LEDs have radiation distributions narrower than the Lambertian source. This is accomplished by various optical means. A plot of radiant intensity versus angle for two LEDs is shown in Figure 9–13. Figure 9–13(a) has a half-power width of about 16° (+8° to −8°), while Figure 9–13(b) has a half-power width of 60° (+30° to −30°).

Response Time and Bandwidth. Response time, the speed at which you can turn the LED on and off, is usually given by the rise time and the fall time of the LED. These terms refer to the light radiation and not to the LED current. A rise time of 300 ns (t_r = 300 ns) at I_F = 50 mA (given for the General Electric LED model GF0E1A1) means that it takes that much time for the radiated light power to increase from 10 to 90% of full value.

In Figure 9–14, the LED current is assumed to rise and fall instantaneously. The radiated light takes time to rise and fall. The modulation rate

FIGURE 9–12 Intensity plot of Lambertian source.

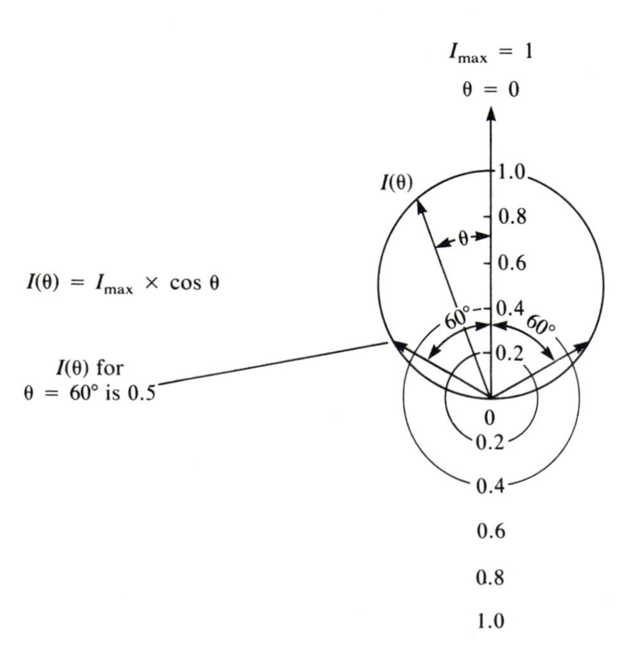

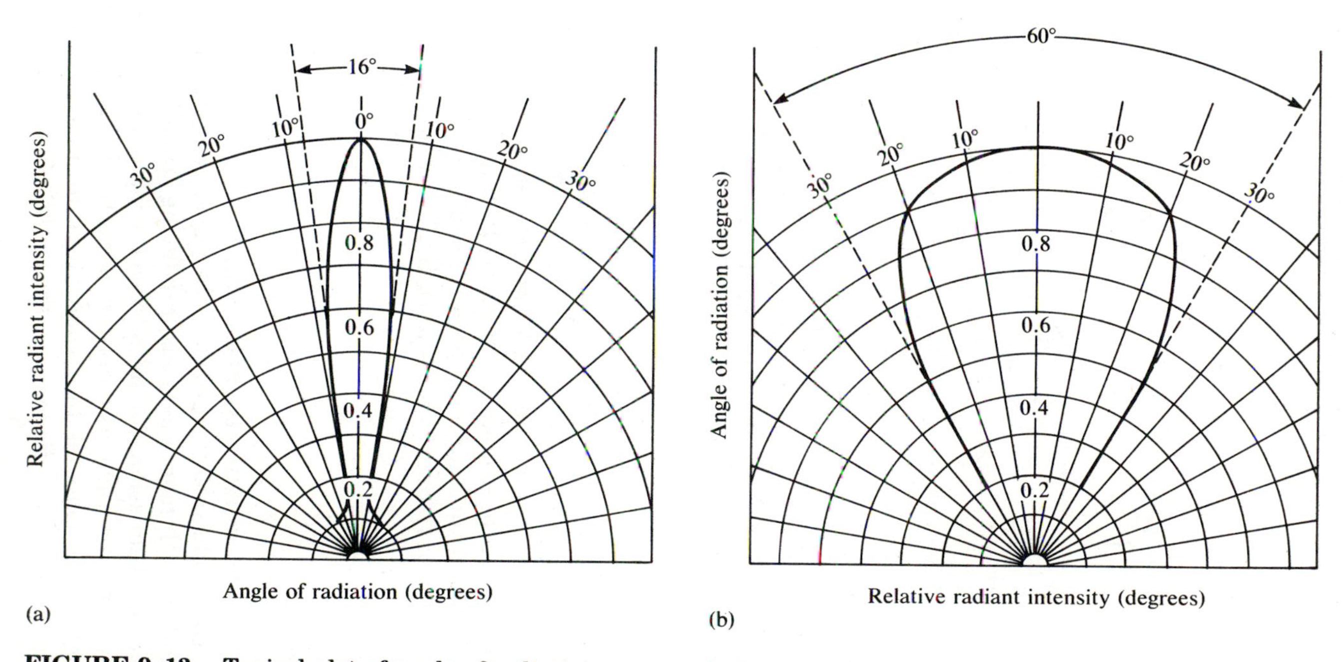

FIGURE 9–13 Typical plot of angle of radiation versus relative intensity. (a) Narrow angle lensed LED. (b) Wider angle.

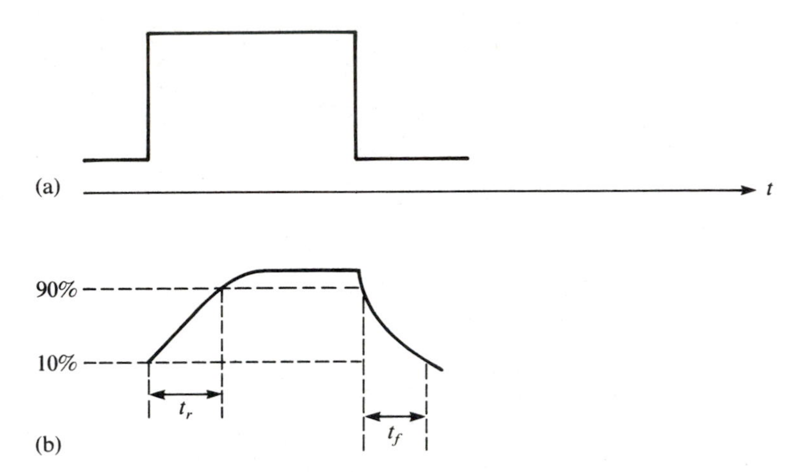

FIGURE 9–14 Rise and fall times. (a) LED current. (b) Radiated light intensity.

(the maximum data rate at which the LED can operate) must allow for both rise time and fall time. This means it is less than$1/(t_r + t_f)$. An estimate of the optical BW and computation of maximum data rate are given in Equation 9–6 and Example 9–2. This is the same expression used in the relation between rise time and BW for electronic circuits.

Figure 9–15 shows the complete data sheet of a LED made by the electron device division of NEC Corporation. Most items listed are either self-explanatory or have already been covered. The cutoff frequency referred to in the data sheet is the optical BW just mentioned. The test conditions given indicate that the 30-MHz (minimum) BW was measured when the LED was biased at 50 mA with a peak-to-peak signal of 10 mA. $P_{out} = -3$ dB simply means that at 30 MHz (minimum) the power dropped by 3 dB from its level at a midband frequency, a frequency substantially below the cutoff.

9–4 LASER

The laser operates by **stimulated emission;** that is, photons collide with electrons and cause electron transitions. In turn, the transitions produce photons that collide with more electrons, producing transitions, and so forth. To achieve stimulated emission, you need some initial radiation (external light source pump or some spontaneous emission, as in the LED) and a method for confining the generated photons to a small area, to increase the chances of photoelectron collision. In the diode laser, the initial photons are generated by the injected electrical current. The photons are confined to the active area (the area where there is a population inversion, hence, a large number of high-energy electrons) by mirrors at the ends of the diode.

9–4–1 Simple Structure (Homojunction)

Figure 9–16 gives a simplified diagram of the basic semiconductor laser structure. The photons generated by the injected current travel to the edge mirrors and are reflected back into the active area. Photoelectron collisions take place and produce more photons, which continue to bounce back and forth between the two edge mirrors. This process eventually increases the number of generated photons until **lasing** takes place. Note that the edge mirrors (at least one) must be partially transparent to allow the generated light to leave the structure. Although the details of the lasing process are somewhat more complex, it is sufficient to say that lasing will take place at particular wavelengths that are related to the length of the cavity L. For the student of microwaves, the radiation wavelength is such that **standing waves** (wave appears stationary) are present in the cavity. This means that L must be a multiple of $\lambda/2$. (The standing waves can be longitudinal, along the distance L, or transverse, along the width of the active area. These two types of waves lead to two different modes of radiation.)

As a result of the relation between L and $\lambda/2$, the simple laser structure produces radiation at a number of wavelengths. Figure 9–17 shows the emission spectrum of laser model LDM3–H made by Ortel Corporation. The width of each of the peaks shown is approximately 0.1 to 0.2 nm, and the spacing is about 0.2–0.3 nm. (This frequency or node spacing is given by $c/2L$.) This laser diode is a multimode laser. Each of the radiation peaks is referred to as a "mode." Another spectral response of the LDS3–H laser, also from Ortel Corporation, is shown in Figure 9–18. This is a single-mode laser. One radiation peak occurs, at about $\lambda = 840$ nm, and all other peaks are suppressed. The structure of such single-mode lasers is somewhat more complex than that shown in Figure 9–16.

9–4–2 Heterostructures

The heterostructure laser is a laser diode with more than single P and N layers. To illustrate the heterostructure, examine a GaAs/AlGaAs laser. Figure 9–19(a) shows the layered structure, and Figure 9–19(b) shows the refractive index distribution. The notations P^+ and N^+ and P^- and N^- indicate heavy doping and light doping, respectively. The P-N structure consists of the two double layers, $P^+ - P^-$ and $N^+ - N^-$.

LIGHT EMITTING DIODE
NDL4201A

850 nm OPTICAL FIBER COMMUNICATION AlGaAs LIGHT EMITTING DIODE

DESCRIPTION

NDL4201A is an AlGaAs double heterostructure light emitting diode designed for a light source of medium distance and medium transmission capacity data link.

PACKAGE DIMENSIONS

in millimeters

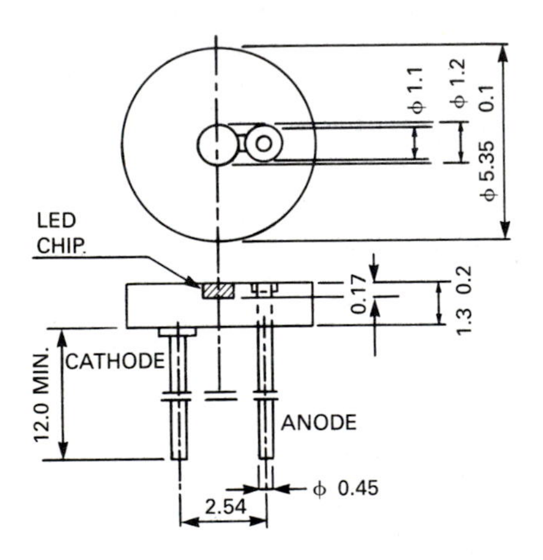

FEATURES

- Optical output power P_O = 1.0 mW TYP.
- Cutoff frequency f_c = 35 MHz TYP.

ABSOLUTE MAXIMUM RATINGS (T_a = 25°C)

Forward Current	I_F	80	mA
Reverse Voltage	V_R	2.0	V
Operating Case Temperature	T_C	−40 to +70	°C
Storage Temperature	T_{stg}	−40 to +90	°C

ELECTRO-OPTICAL CHARACTERISTICS (T_a = 25°C)

CHARACTERISTIC	SYMBOL	MIN.	TYP.	MAX.	UNIT	TEST CONDITIONS
Optical Output Power	P_O	0.6	1.0		mW	I_F = 50 mA
Forward Voltage	V_F		1.7	2.3	V	I_F = 50 mA
Peak Emission Wavelength	λp	840	850	870	nm	I_F = 50 mA
Spectral Half Width	$\Delta\lambda$		45	50	nm	I_F = 50 mA
Cutoff Frequency	f_c	30	35		MHz	I_F = 50 mA, I_S = 10 $mA_{p\text{-}p}$ P_O = −3 dB
Emitting Area Diameter	ϕ		35		μm	

FIGURE 9–15 Description of LED.
Source: NEC Corporation, Mountain View, Calif.

TYPICAL CHARACTERISTICS (T_a = 25°C)

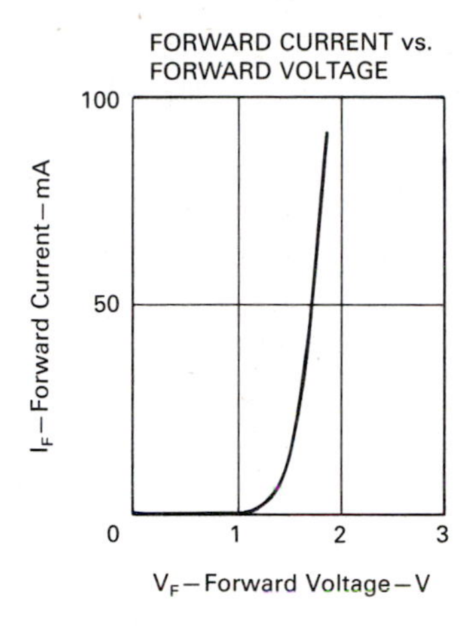

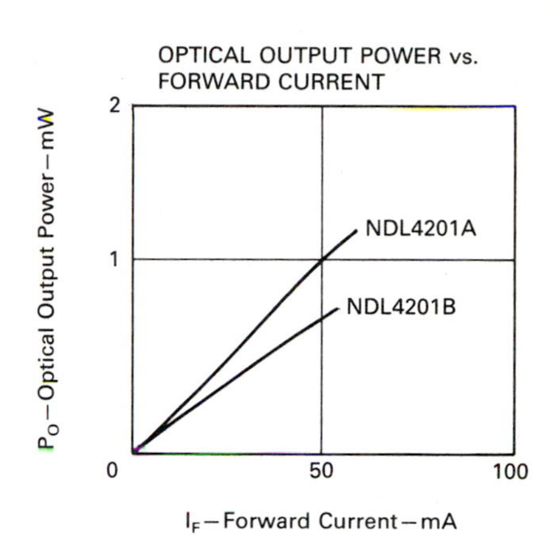

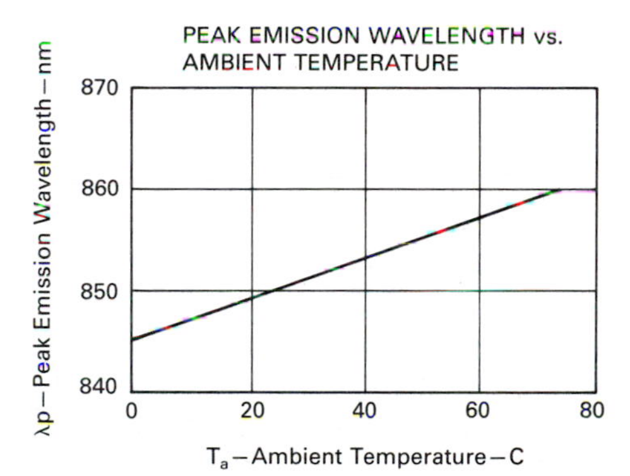

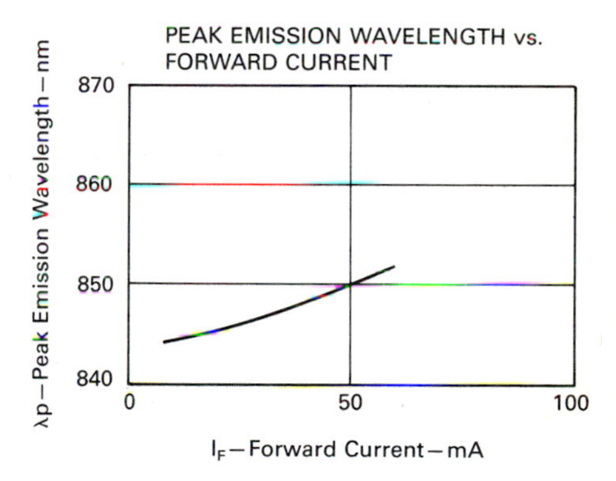

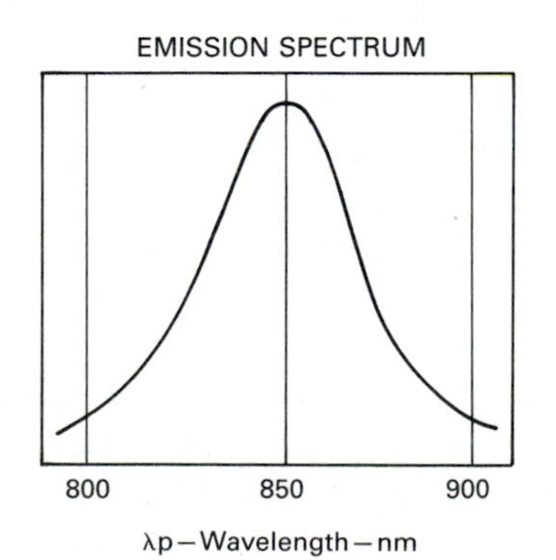

FIGURE 9–15 *(Continued)*

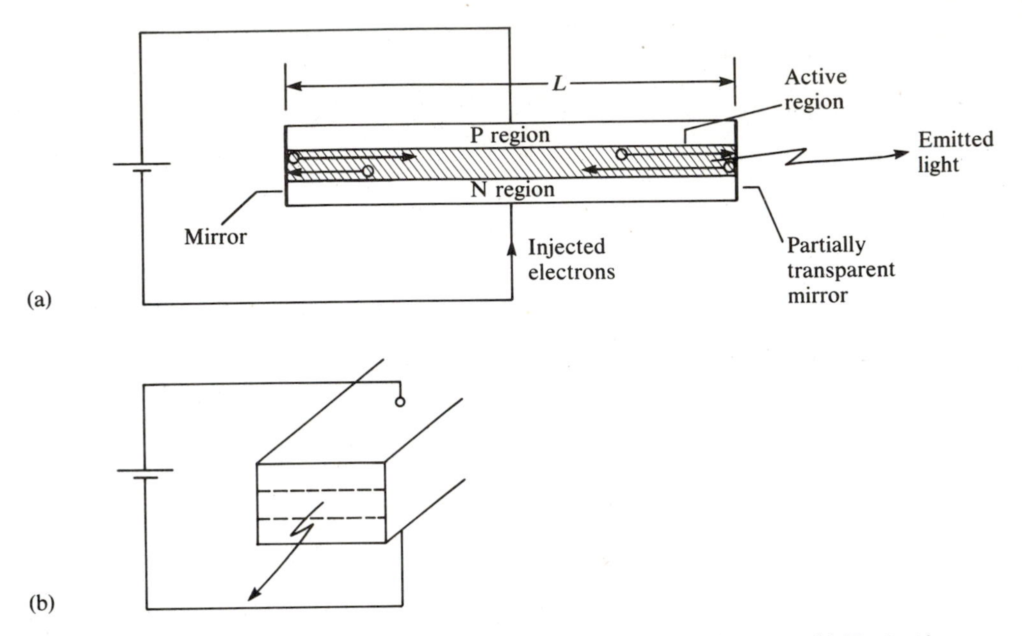

FIGURE 9–16 Basic semiconductor laser structure. (a) Side view. (b) Projection.

A thin layer (0.2 to 0.5 μm) of GaAs is placed at the junction, the active region. This substance is selected because the electron-hole recombinations are highly radiative. This increases radiation efficiency. The P^- and N^- layers are lightly doped regions that have an index of refraction n_2 less than n_1 of the active region ($n_1 - n_2 = 0.2$). These three layers, $n_2 - n_1 - n_2$, form a light waveguide much like the optical fiber, so that light generated is confined to the active region.

The light output is from the front and back surfaces (assuming partially reflecting mirrors on both sides) similar to the basic laser structure shown in Figure 9–16. The double-layered structure (double heterostructure) also functions to restrict the electron concentration to the junction area, further improving radiation efficiency.

In Figure 9–19, the injected current flows through the full cross-section of the structure. The light is then emitted from the total width of the active region.

The striped heterostructure is a modification of the structure shown in Figure 9–19, with the aim of concentrating the current in a small portion of the active region, as shown in Figure 9–20. Here, the electrical connection to the P side of the diode is restricted to a 5- to 15-μm stripe, so the active region is similarly reduced. Only about 5 to 15 μm of the total width of the GaAs

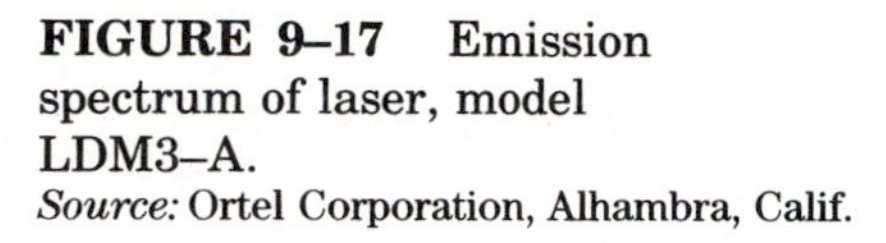

FIGURE 9–17 Emission spectrum of laser, model LDM3–A.
Source: Ortel Corporation, Alhambra, Calif.

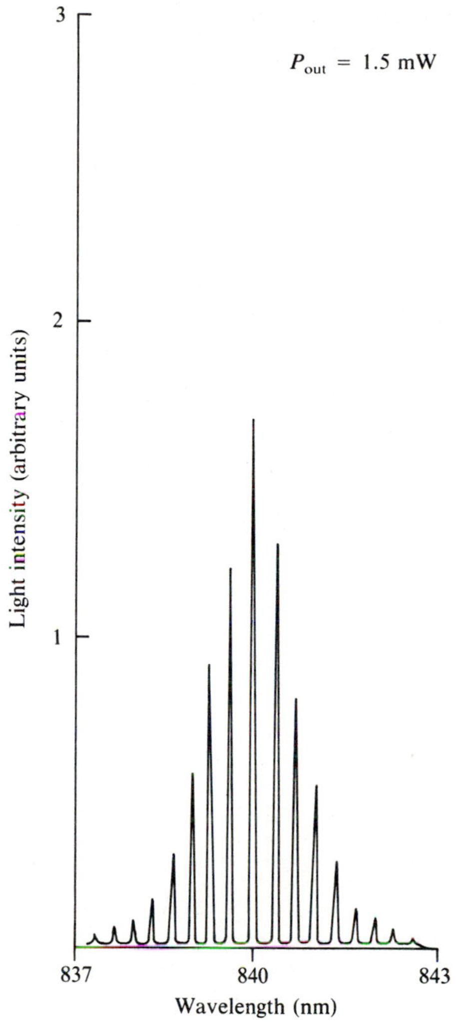

junction is active. Note that this structure increases current density (in amperes per square centimeters) in the diode, which improves efficiency. There is a higher concentration of electrons in a smaller region, increasing the number of transitions per unit area, hence, producing more photons and higher light intensity. With the additions of edge mirrors, producing a **Fabry Perot cavity,** high-efficiency lasing takes place. The few laser structures shown here are only a sample of the large variety of structures and materials used, and serve only to demonstrate the principles of the basic lasing process and of the electron and photon confinement techniques.

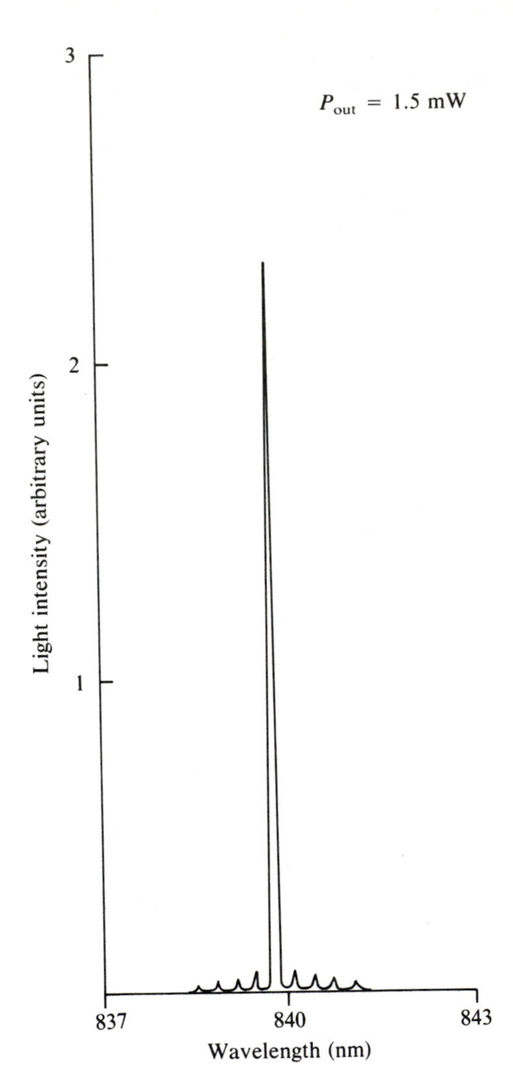

FIGURE 9–18 Emission spectrum of laser, model LDS3–H.
Source: Ortel Corporation, Alhambra, Calif.

9–4–3 Long Wavelengths

The lasers shown in Figures 9–16, 9–19, and 9–20 operate at wavelengths of about 0.8–0.9 μm because of the particular materials selected and their energy gap. (The precise proportions of Al, Ga, and As in the alloy determines its energy band gap and hence the wavelength of radiation.)

In the discussion of the choice wavelength (Section 4–4), you noted that $\lambda = 1.3\ \mu$m or $\lambda = 1.5\ \mu$m are best suited for communication applications. It is with this in mind that much research has been invested in developing suitable

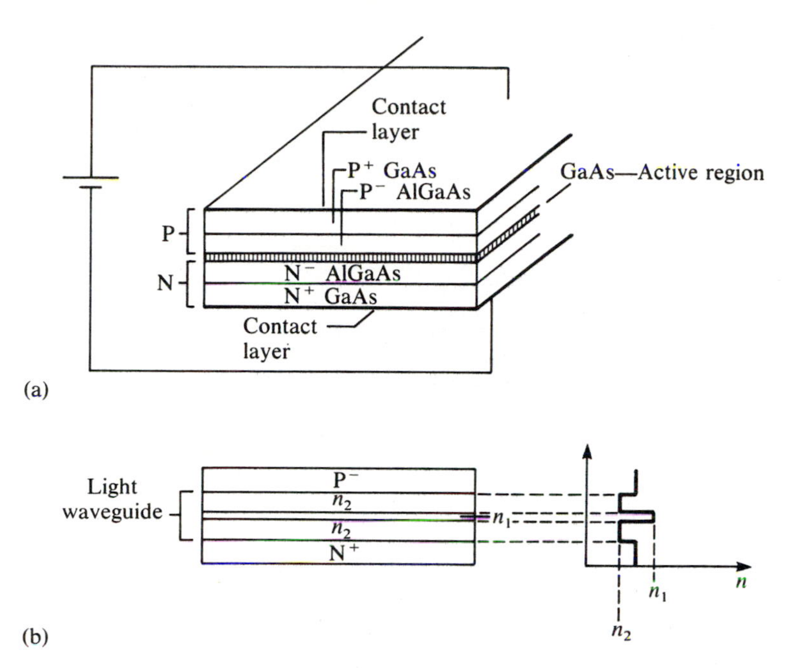

FIGURE 9–19 Laser heterostructure. (a) Schematic projection. (b) Refractive index profile.

long-wavelength lasers. (λ of 1.3 and 1.5 μm, are called long wavelengths, as opposed to λ of 0.82 μm, the wavelength used first.)

The problem of long-wavelength lasers is that of finding suitable materials with internal efficiency (quantum efficiency) high enough to use in a structure as shown in Figures 9–19 and 9–20. Two suitable alloys are InGaAsP/InP and GaAlAsSb/GaSb. (The elements not noted previously are indium [In], phosphorus [P], and antimony [Sb].) These alloys are suitable for operation at wavelengths of 1.0 to 1.7 μm. Here, as in the other alloys, the exact proportions of In, Ga, As, and P in one and Ga, Al, As, and Sb in the other, determine the exact wavelength of radiation. By controlling these ratios very carefully, lasers can be designed to operate at wavelengths anywhere in the range of 1.0–1.7 μm.

9–4–4 Characteristics

Laser diodes (LD, to be distinguished from LED) come in a large variety of packages. Some are mounted in dual-in-line integrated circuit (I/C) packages, while others are permanently attached to an optical fiber **(pigtail mounting).**

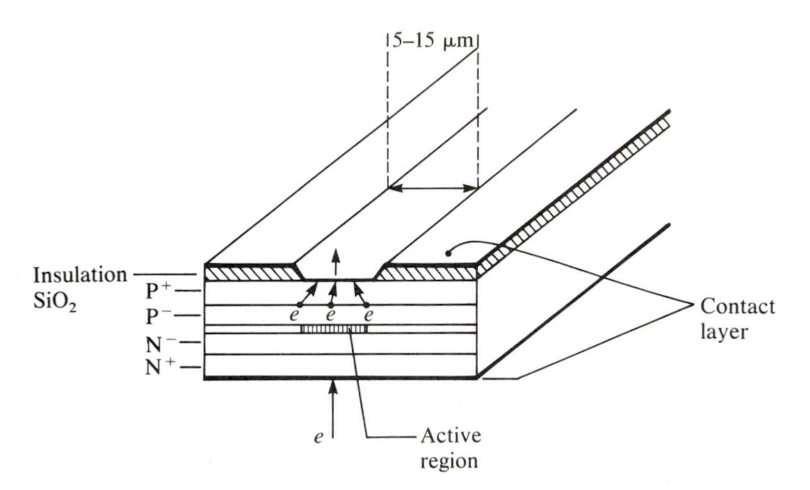

FIGURE 9–20 Striped heterostructure.

To simplify the discussion of LD characteristics, relate them to the data sheet shown in Figure 9–21.

V-I Curve. Remember that the laser diode is a diode and has all the typical characteristics of a diode. The forward voltage drop V_F is between 1.0 and 1.5 V. In the data sheet shown, the maximum V_F is given as 1.3 V (at I_F = 30 mA). The plot of forward voltage V_F versus the forward current I_F is that of a diode (Figure 9–21(d)).

Threshhold Current I_{th} and P_{out}. An important parameter for the LD is its **threshhold current** I_{th}. This is the minimum forward current required to produce lasing. It can be obtained from Figure 9–21(b), as shown. For currents below I_{th}, note that the LD behaves like a LED. For the NDL5004, the operation with current below 20 mA at (30°C) does not produce lasing. The output power for this current range is very small, like that of a LED.

When laser diodes were first developed, I_{th} was of the order of 0.5 A or larger. By confining the radiation and the electrical current to a very small area (making the active area very small), I_{th} has been reduced to its present range, typically 30 to 100 mA. This simplifies the device circuitry and reduces the power dissipation of the LD. Note that excessive power dissipation raises the internal temperature of the LD, shifting the P_{out} versus I_F curve (see Figure 9–21(b)). To combat this temperature effect, some cooling method is often used. Some high-power LD packages contain special electrical cooling elements to help stabilize the output power. (See model NDL500SP in the NEC Corporation catalog.) Similarly, feedback methods that use the **monitor photodiode** are used to stabilize the operating conditions of lasers.

1300 nm OPTICAL FIBER COMMUNICATION
InGaAsP DOUBLE HETEROSTRUCTURE LASER DIODE

DESCRIPTION

NDL5004 is a long wavelength laser diode especially designed for long distance high capacity transmission systems. The DCPBH (Double Channel Planar Buried Heterostructure) can achieve stable fundamental oscillation in wide temperature range.

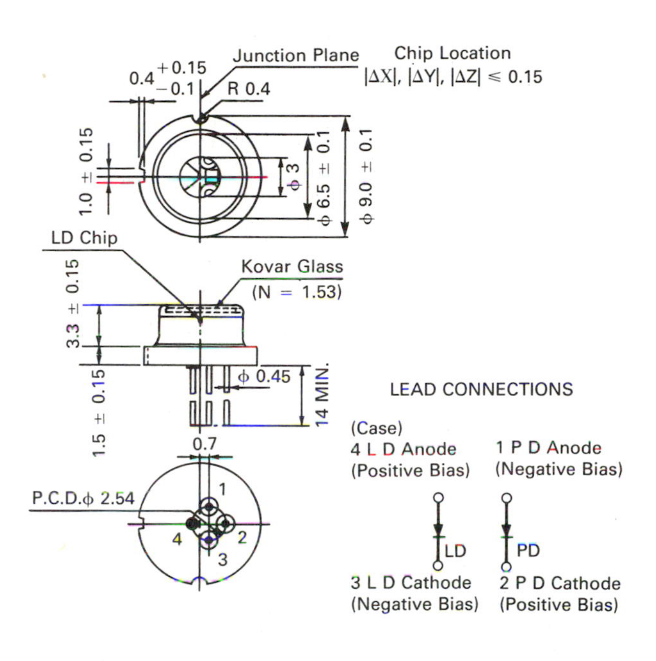

FEATURES

- High output power. P_O = 8 mW TYP. @I_F = I_{th} +30 mA
- Fundamental transverse mode.
- Wide operating temperature range.
- Long wavelength. λ_P = 1300 nm
- Low threshold current. I_{th} = 20 mA TYP.
- Narrow vertical angle and wide lateral beam angle $\phi_\perp \times \phi_{/\!/}$ = 35° × 28°

ABSOLUTE MAXIMUM RATINGS (T_a = 25°C)

Reverse Voltage	V_R	2.0	V
Optical Output Power	P_O	15	mW
Operating Case Temperature	T_C	−40 to +70	°C
Storage Temperature	T_{stg}	−55 to +125	°C

ELECTRO-OPTICAL CHARACTERISTICS (T_a = 25°C)

CHARACTERISTICS	SYMBOL	MIN.	TYP.	MAX.	UNIT	TEST CONDITIONS
Forward Voltage	V_F			1.3	V	I_F = 30 mA
Threshold Current	I_{th}		20	35	mA	
Optical Output Power	P_O	6.0	8.0		mW	I_F = I_{th} +30 mA
Peak Emission Wavelength	λ_P	1270	1300		nm	P_O = 6.0 mW
Half Power Spectral Width	$\Delta\lambda$			4.0	nm	P_O = 6.0 mW
Vertical Beam Angle	$\phi_\perp$		35		deg.	P_O = 6.0 mW, FAHM*
Lateral Beam Angle	$\phi_{/\!/}$		28	1.0	deg.	P_O = 6.0 mW, FAHM*
Rise Time	t_r		0.5	1.0	ns	10–90 %
Fall Time	t_f		0.7	1330	ns	90–10 %
Monitor Current of PD	I_m	300	500	1500	μA	V_R = 5 V, P_O = 6.0 mW
Dark Current of PD	I_D			3	μA	V_R = 5 V

* FAHM: Full Angle at Half Maximum

FIGURE 9–21 Laser diode NDL5004. (a) Specifications. (b) Typical characteristics. (c) Far-field patterns. (d) I_F versus V_F. (e) Pulse response. (f) Longitudinal mode emission spectrum.
Source: NEC Corporation, Mountain View, Calif.

ELECTRO-OPTICAL CHARACTERISTICS (T_a = 60°C)

CHARACTERISTIC	SYMBOL	MIN.	TYP.	MAX.	UNIT	TEST CONDITIONS
Forward Voltage	V_F			1.3	V	I_F = 30 mA
Threshold Current	I_{th}		40	60	mA	
Optical Output Power	P_O	5.0			mW	$I_F = I_{th}$ +30 mA
Peak Emission Wavelength	λ_P	1275	1315	1350	nm	P_O = 5.0 mW
Half Power Spectral Width				4.0	nm	P_O = 5.0 mW
Rise Time	t_r		0.5	1.0	ns	10–90 %
Fall Time	t_f		0.7	1.0	ns	90–10 %
Monitor Current of PD	I_m	200			μA	V_R = 5 V, P_O = 5.0 mW
Dark Current of PD	I_D		12	25	μA	V_R = 5 V

(a)

TYPICAL CHARACTERISTICS (T_a = 25°C)

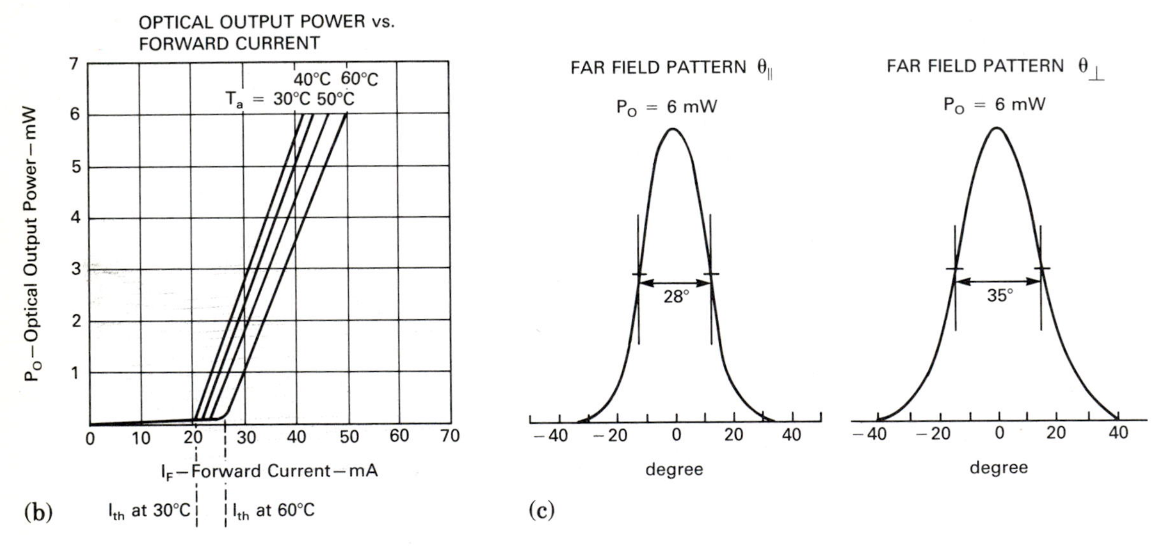

(b) (c)

FIGURE 9–21 *(Continued)* (a) Specifications. (b) Typical characteristics. (c) Far-field patterns. (d) I_F versus V_F. (e) Pulse response. (f) Longitudinal mode emission spectrum.
Source: NEC Corporation, Mountain View, Calif.

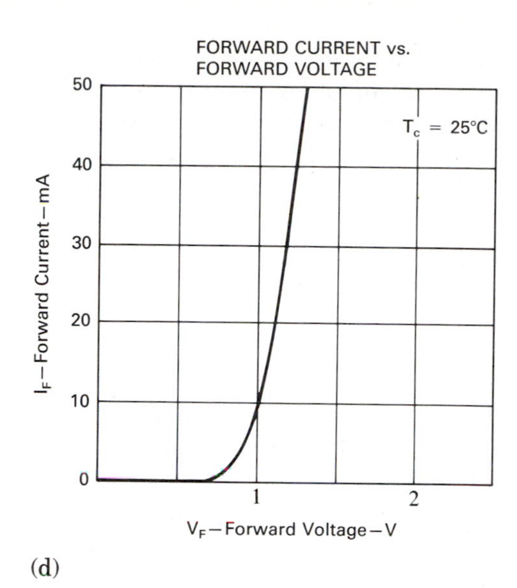

(d)

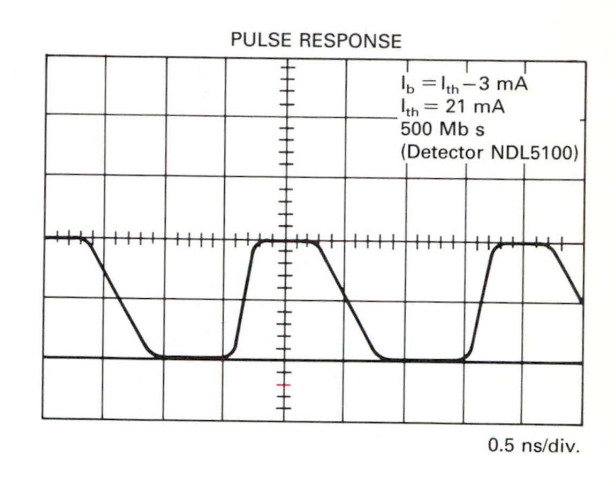

(e)

LONGITUDINAL MODE EMISSION SPECTRUM

P_O = 3 mW

P_O = 5 mW

P_O = 8 mW

T_c – Case temperature – C

50
40
30
20
10

1290 1300 1310

Wavelength (nm)

1295 1307

(f)

FIGURE 9–21 *(Continued)*

Figure 9–21(b) shows how sensitive I_{th} is to temperature. At 30°C, I_{th} is about 20 mA, while at 60°C, it is above 25 mA, a change of more than 25%. In addition, the power output P_{out} varies with temperature. The data sheet under electro-optical characteristics, at T_a = 25°C (Figure 9–21(a)), shows a minimum P_{out} of 6.0 mW at I_F of I_{th} + 30 mA. This test condition must be specified because P_{out} varies as I_F changes above I_{th}. At 60°C (next table in Figure 9–21(a)), the minimum P_{out} is only 5.0 mW; that is, for an increase in temperature from 30°C to 60°C, the power output dropped from 6.0 to 5.0 mW.

Wavelength λ_P. λ_P = 1,270 nm (the peak wavelength) and $\Delta\lambda$ = 4.0 nm (the half-spectral width [or the line width]), shown on the data sheet (Figure 9–21(a)), have been discussed before. Note that the peak wavelength is very sensitive to temperature and to P_{out} as demonstrated by Figure 9–21(f). Here, for P_{out} of 5 mW (center panel), λ_P is 1295 nm when T_c is 10°F, and λ_P is 1307 nm when T_c = 50°C. This change in wavelength is three times the line width.

Beam Angle. The **vertical beam angle** $\theta_\perp$ and the **lateral,** or **horizontal beam angle** $\theta_\parallel$ are given as 35° and 28°, respectively, and are graphed in Figure 9–21(c). The angles are defined as the **full angle at half maximum,** shown for clarity in Figure 9–22. (See test conditions and note under electro-optical characteristics in Figure 9–21(a)). The vertical beam angle $\theta_\perp$ is the half-maximum angle in a direction perpendicular to the active surface, while $\theta_\parallel$ is in a direction parallel to the active surface. (An explanation of half-maximum beam angle was given in Section 9–3–4, in connection with the LED.) The beam angle can only be measured at some distance from the emitting surface.

Rise and Fall Time. The rise time t_r and the fall time t_f given in the data sheet (Figure 9–21(a)) are, typically, 0.5 and 0.7 ns, respectively. This is translatable into a modulation BW of more than 0.5 GHz. Figure 9–21(e) shows a pulse response of the LD, as measured by a photodetector. This means that the t_r and the t_f include the effects of the photodetectors.

Monitor Photodiode. To understand the meanings of monitor current of photodiode (PD), and dark current of PD, you need to reexamine the complete LD package. As was noted before, the device is sensitive to temperature variations; P_{out} varies with temperature. It is advisable to keep P_{out} constant or at least constantly to monitor P_{out}. To accomplish this task, the manufacturer includes a PD (which converts light to current) in the LD package. The PD is exposed to radiation from the back side of the LD structure. This means that the end mirrors in the LD are only partially reflective. They allow light to go through on both ends: at one end, the output radiation; at the other, the light for the PD (Figure 9–23). The monitor current I_m given for the PD in the NDL5004 data sheet (Figure 9–21) is given for an output power P_{out} of 6.0 mW. The PD current is directly related to P_{out}. A more complete relation between

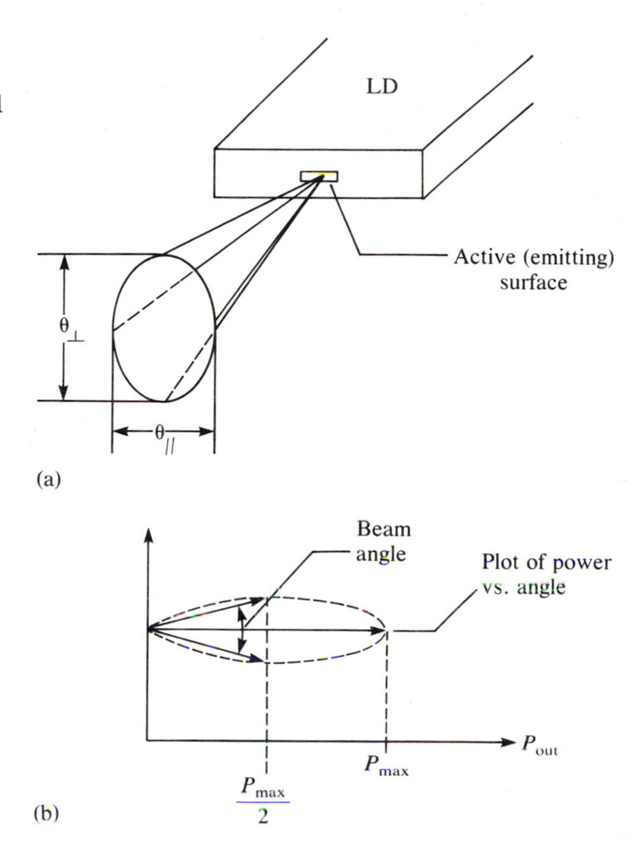

FIGURE 9–22 Beam angles. (a) $\theta_\perp$ and $\theta_\parallel$. (b) Definition of beam angle—full angle at half maximum.

P_{out} and I_m is given in Figure 9–24(d), for the NDL5004P. Here, however, I_m is related to P_f, the output power from the attached optical fiber. The NDL5004P is identical to the NDL5004 except for the fiber that is integrated into the package (see Figure 9–24(a).) Here, you can only speak of the power out from the fiber P_f. (In Figures 9–24(c) and 9–24(d), P_f replaces P_{out}.) P_{out} is greater than P_f for identical input conditions because not all the generated

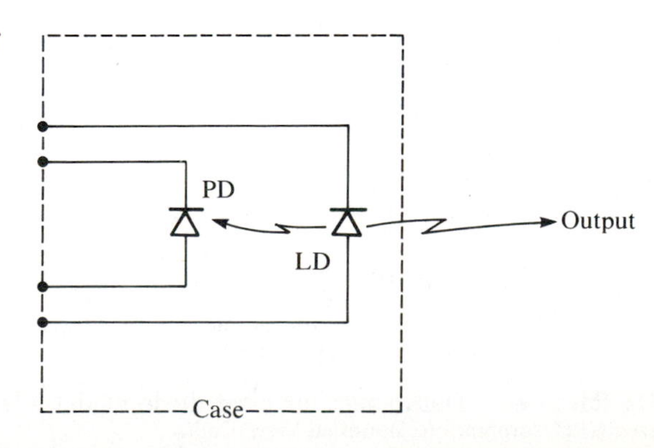

FIGURE 9–23 Schematic of laser diode-photodiode IC package.

LASER DIODE
NDL5004P

1300 NM OPTICAL FIBER COMMUNICATION
InGaAsP DOUBLE HETEROSTRUCTURE LASER DIODE MODULE

DESCRIPTION

NDL5004P is an InGaAsP Laser Diode Module especially designed for long distance high capacity transmission systems. The DC-PBH (Double Channel Planar Buried Heterostructure) can achieve stable fundamental oscillation in wide temperature range.

PACKAGE DIMENSIONS
in millimeters

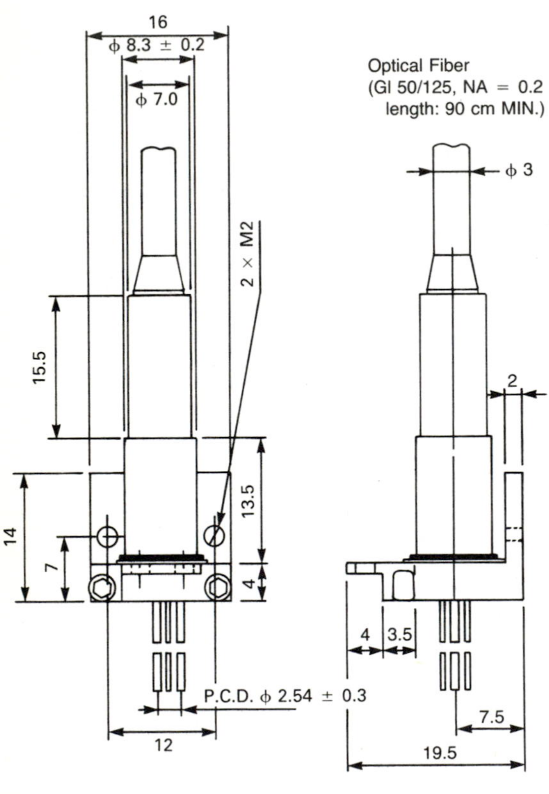

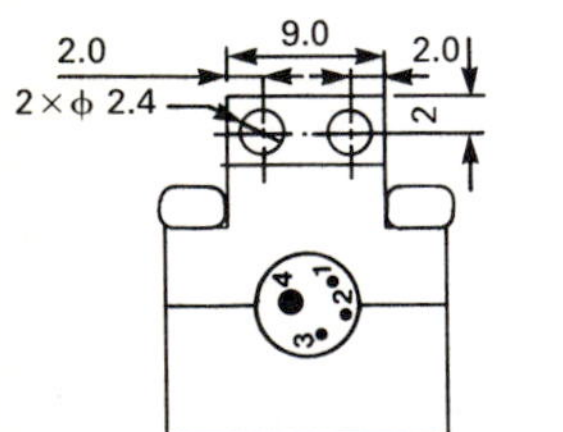

FEATURES

- High output power from fiber. P_f = 2.5 mW TYP.
- Long wavelength. λ_p = 1300 nm
- Low threshold current. I_{th} = 20 mA TYP.
- Fundamental transverse mode operation.
- High reliability.
- Hermetically Sealed.
- Monitor photodiode incorporated.

ABSOLUTE MAXIMUM RATINGS (T_a = 25°C)

Parameter	Symbol	Rating	Unit
Operating Case Temperature	T_C	−20 to +65	°C
Storage Temperature	T_{stg}	−40 to +70	°C
Optical Output Power From Fiber	P_f	4.0	mW
Reverse Voltage	V_R	2.0	V
Forward Current *1	I_F	50	mA
Reverse Voltage *1	V_R	20	V

*1 :Monitor diode

LEAD CONNECTIONS

4 Case and LD Anode (Positive Bias) — LD — 3 LD Cathode (Negative Bias)

1 PD Anode (Negative Bias) — PD — 2 PD Cathode (Positive Bias)

FIGURE 9–24 Data sheet for laser diode model NDL5004P.
Source: NEC Corporation, Mountain View, Calif.

TYPICAL CHARACTERISTICS (T_a = 25°C)

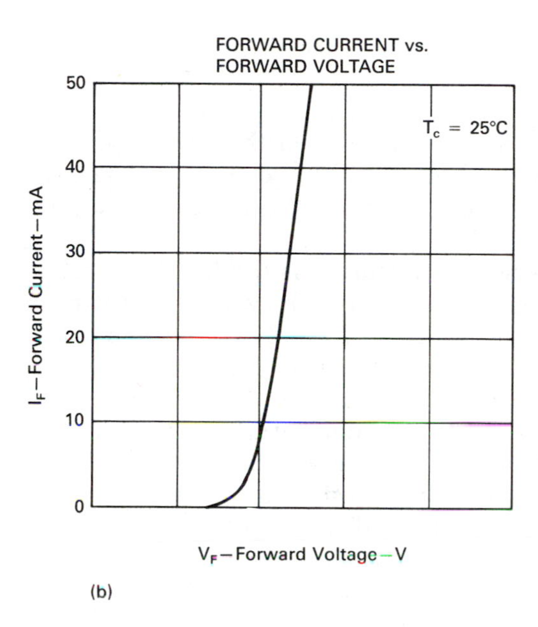

(b)

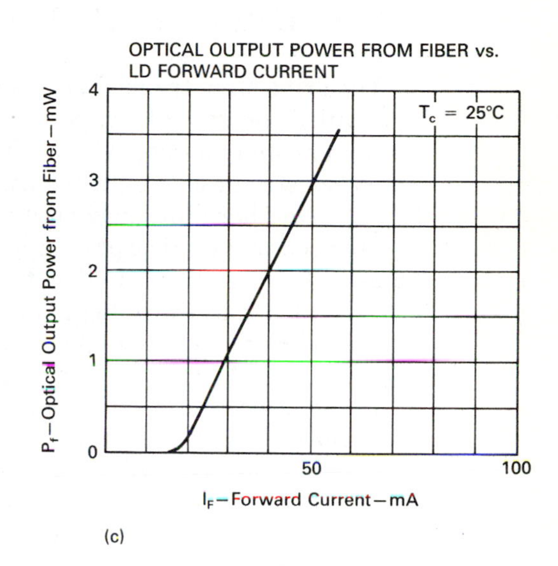

(c)

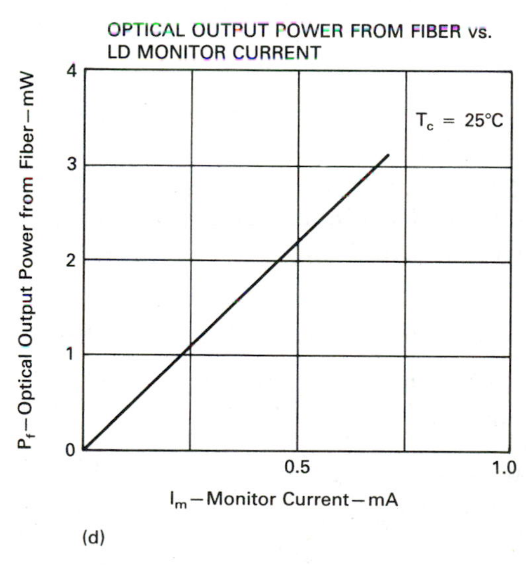

(d)

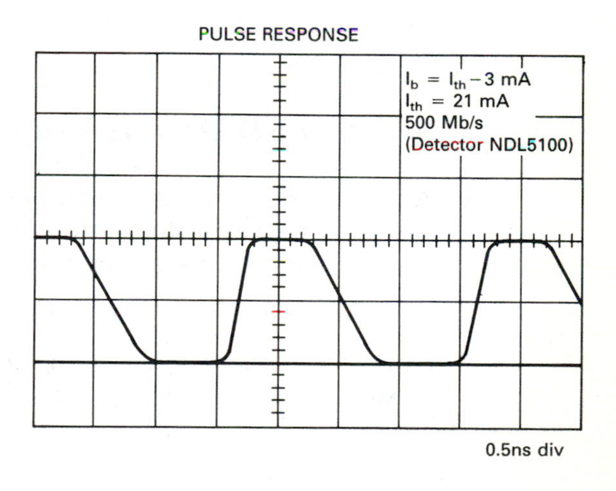

(e)

FIGURE 9–24 *(Continued)*

light is coupled to the fiber. P also depends on the particular fiber used. For the NDL5004P, the fiber used is a graded-index 50 to 125, N.A. = 0.2 fiber, as shown in Figure 9–24(a) under package dimensions. (The notation 50 to 125 refers to the ratio of the core diameter to the cladding diameter.)

9–5 DRIVE ELECTRONICS

The drive circuitry for LEDs is much simpler than that for the laser. The LED requires no special temperature-compensating circuitry to stabilize the bias current and voltage. The LED requires no special cooling, neither heat sink nor electronic cooling elements.

Another difference between the two light sources is the fact that lasers are used in digital or pulse applications almost exclusively, while LEDs can be used for digital or analog transmission. This is a result of the nonlinear current versus the optical power P_{out} relation of the laser. Figure 9–25 shows typical current–optical power plots for a LED and a laser.

In the LED, there is a wide linear region, suitable for analog modulation. In the laser, the linear region is too small to allow analog operation. Any shift in the bias current will result in distortion (clipping) similar to that of a poorly biased transistor. Bias point shifts have little effect in pulse applications, where it is only necessary to turn the LD on and off. Both the LED and LD can be operated with no bias current in digital applications. For analog transmission, the LED must be biased as shown in Figure 9–25(a), permitting operation in the linear region with positive and negative input signals.

9–5–1 LED Drivers—Digital

A simple driver circuit is shown in Figure 9–26. An input pulse of 3 V or more saturates the transistor so the LED current I_F equals I_{Csat}. To calculate I_F, we find I_{Csat} (the collector saturation current):

$$I_F = I_{Csat} = (V_{CC} - V_F)/R_1 \tag{9–10}$$

where V_F is the forward voltage drop in the diode. R_B should be selected to produce a base current I_B that will saturate the transistor.

$$I_B \geqq I_{Csat}/\beta \tag{9–11}$$

$$I_B = (V_{in\text{-}peak} - V_{BE} - V_F)/R_B \tag{9–12}$$

EXAMPLE 9–5

Given $I_F = 50$ mA, $\beta = 100$, $V_F = 1.5$ V, $V_{CC} = 8$ V, $V_{in\text{-}peak} = 4$ V, and a silicon transistor, find (1) R_1 and (2) R_B.

Solution

1. $I_F = I_{Csat} = 50$ mA
   $R_1 = (V_{CC} - V_F)/I_F = (8 - 1.5)/(50 \times 10^3) = 130\ \Omega$
2. $I_B = I_{Csat}/100 = 0.5$ mA
   $R_B = (4 - 0.7 - 1.5)/(0.5 \times 10^{-3}) = 3.6\ \text{k}\Omega$

An IC driver circuit is shown in Figure 9–27. It uses a standard peripheral driver, 75450 or 75451, which consists of a Nand gate and a transistor. When the "enable" terminal is high (about +5 V), transmission is

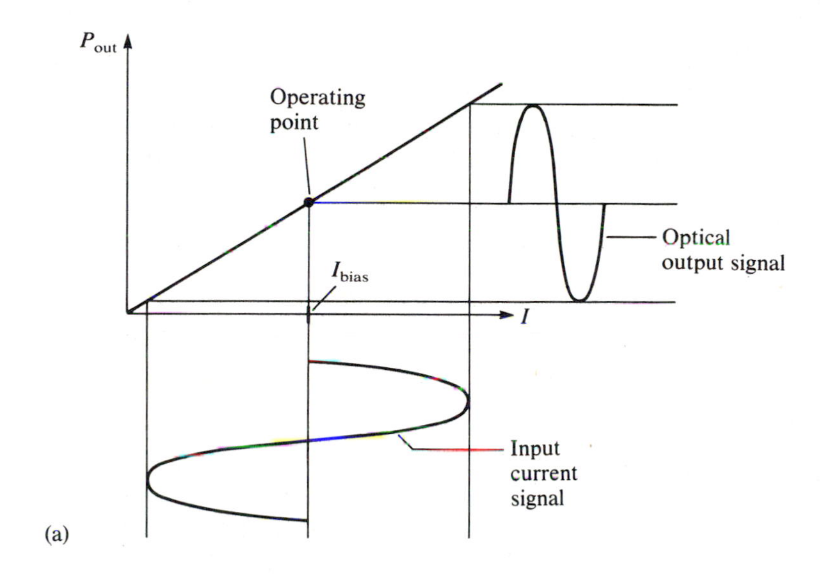

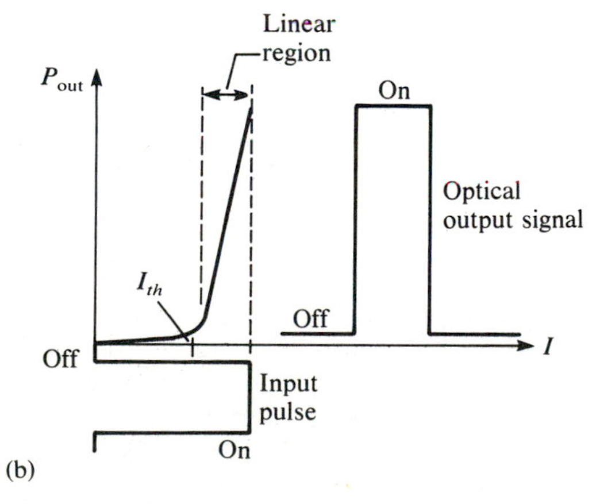

FIGURE 9–25 Input-output relations. (a) light-emitting diode. (b) "On" and "Off" switching of laser diode.

FIGURE 9–26 Transistor driver.

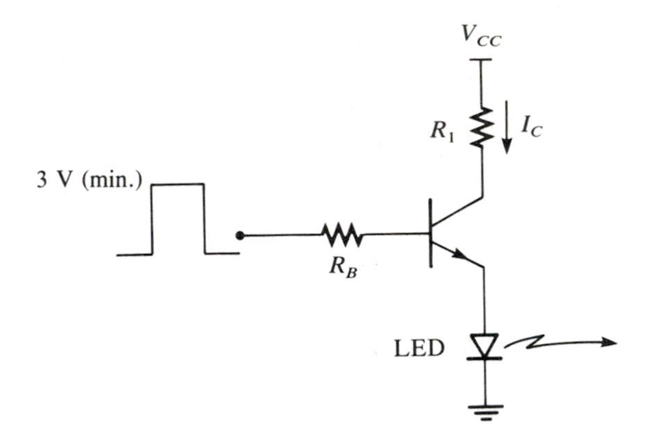

enabled. That is, the "data in" signal controls the LED. Data in high (+5 V) means that the LED is off, while data in low (0 V) means the LED is on. Logic 0 is transmitted by the presence of light—LED on, while logic 1 is represented by no light—LED off. The resistor R is selected to get the desired LED forward current I_F.

$$R_1 = (V_{CC} - V_F)/I_F \tag{9–13}$$

V_F is the forward drop of the LED. The collector-emitter saturation voltage was neglected in Equation 9–13. C_1 is often included to reduce spikes and overshoots. Its value is about 1000 pF.

EXAMPLE 9–6

The NEC NDL4201A LED is used in the circuit of Figure 9–27, with V_{CC} = 5 V. The "on" current of the LED is to be 40 mA, half its maximum allowable current. Find R_1.

FIGURE 9–27 Integrated circuit driver 1.

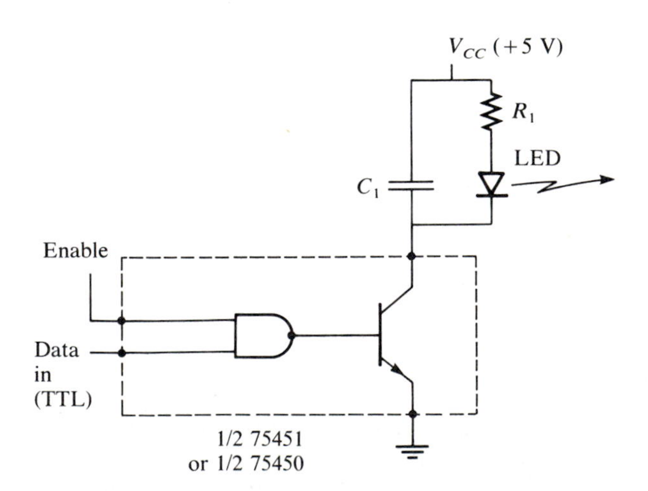

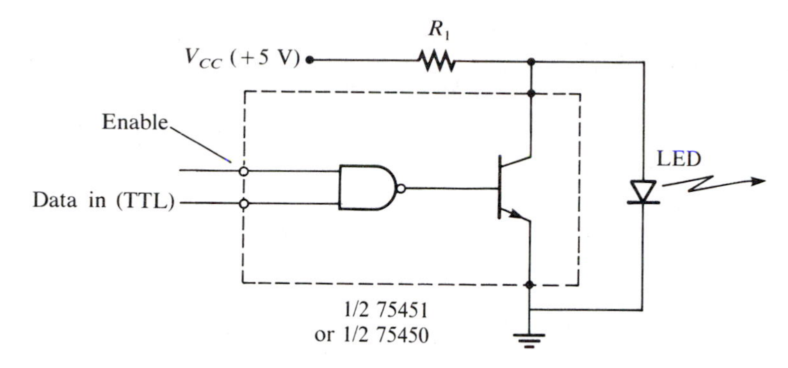

FIGURE 9–28 Integrated circuit driver 2.

Solution

From the specifications of the NDL4201A, you find that the typical $V_F = 1.7$ V. Hence,

$$R_1 = (5 - 1.7)/(40 \times 10^{-3}) = 82.5\ \Omega$$

Another pulse driver is shown in Figure 9–28. Here, the LED is on for data in high and off for data in low (provided enable terminal is high). R_1 is chosen as is shown in Equation 9–13. For this circuit, however, the transistor is off when the LED is on.

A common problem in digital circuits is the introduction of a dc component by the signal itself. Assume that the LED (or laser) must be biased at a particular dc level V_{bias}. The average voltage of the typical return to zero or nonreturn to zero (NRZ) digital signal is not zero; that is, it has a dc component. (For example, computing the average of a "1010" NRZ pulse train with logic levels of 5 and 0 V, you find a dc level of 2.5 V.) To compound the problem, the dc component depends on the data. This means that the bias point, the total dc applied to the LED (or laser), will vary with the data. A simple solution to this problem is the use of special codes, such as Manchester (biphase) codes, where the dc component of the data pulse train does not depend on the data. Other techniques use complex circuitry to reduce the effect of the dc component of the pulse train.

9–5–2 LED Drivers–Analog

Analog operation of a LED requires a dc bias current as that shown in Figure 9–25(a). The input signal then causes the LED current to vary above and below the biasing current.

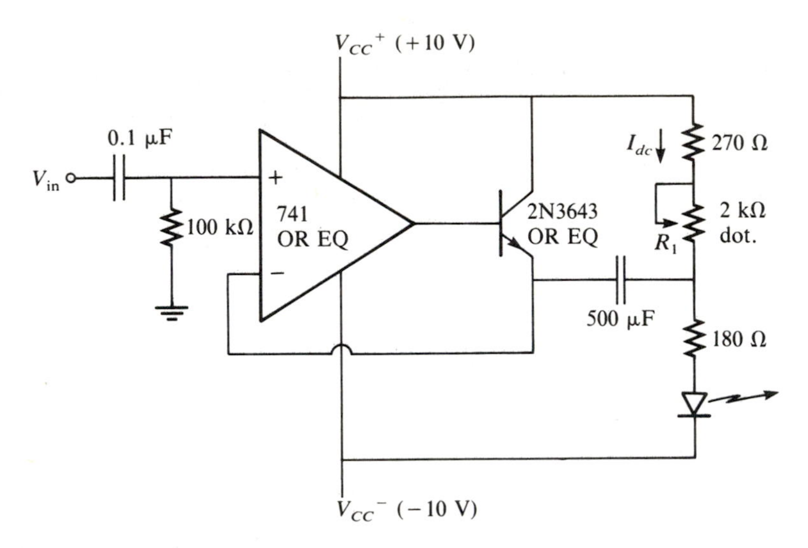

FIGURE 9–29 Analog driver.

A circuit using an operational amplifier is shown in Figure 9–29. The resistor values were selected to set up a bias current between about 7.5 mA and 40 mA, depending on the setting of R_1. An input signal of ± 2 V ($V_{\text{in } P\text{-}P} = 4$ V) will result in approximately a ± 10-mA change in diode current.

In another circuit, shown in Figure 9–30, the LED is placed in the feedback path of the operational amplifier. Here, potentiometer R_2 is used to set up the bias current by adjusting V^+ appropriately.

$$I_{\text{bias}} = V^+/R_1 \tag{9–14}$$

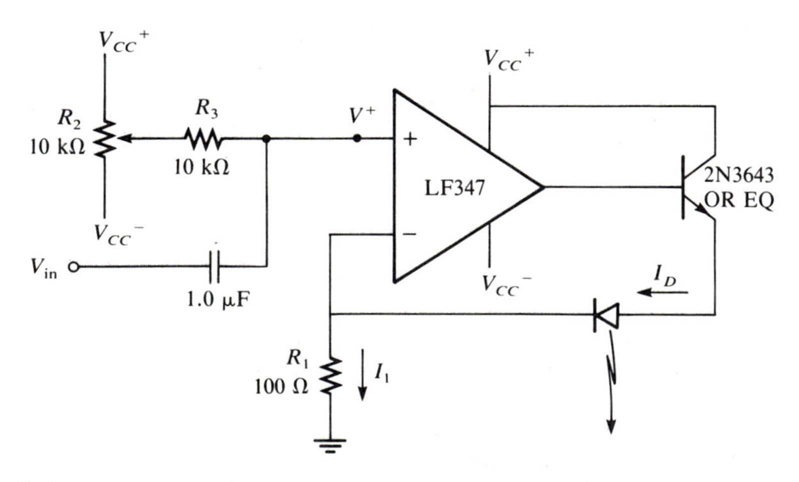

FIGURE 9–30 Analog driver with LED in feedback path.

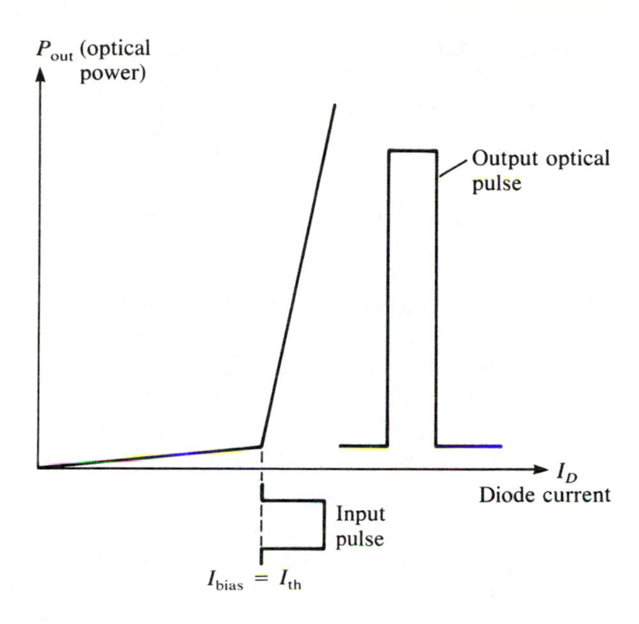

FIGURE 9–31 Laser diode bias and modulation.

The signal current through the diode is

$$I_{sig} = V_{in}/R_1 \tag{9–15}$$

9–5–3 Laser Diode Drivers

Most LDs drive circuits are designed to stabilize the LD current to minimize the effects of temperature. Often, the LD is biased at the threshhold current I_{th} and the pulse signal turns the laser fully on and off. Figure 9–31 shows the input pulse and the optical output. The monitor photodiode is used to stabilize the output, that is, to make sure that the optical power output (when the LD in "on") remains unaffected by temperature. This is accomplished in a feedback circuit, as shown in Figure 9–32. The optical output controls the drive

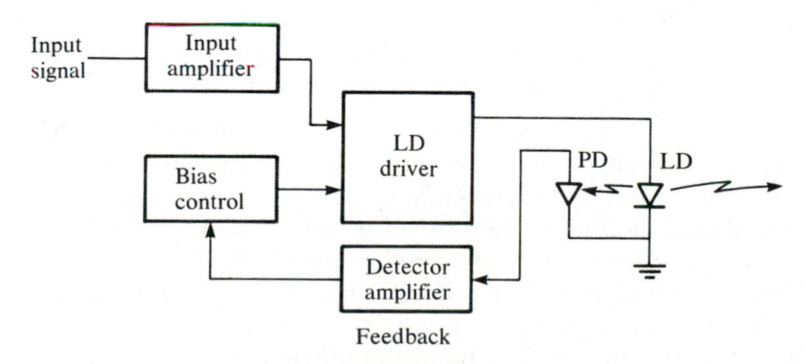

FIGURE 9–32 Laser temperature stabilization.

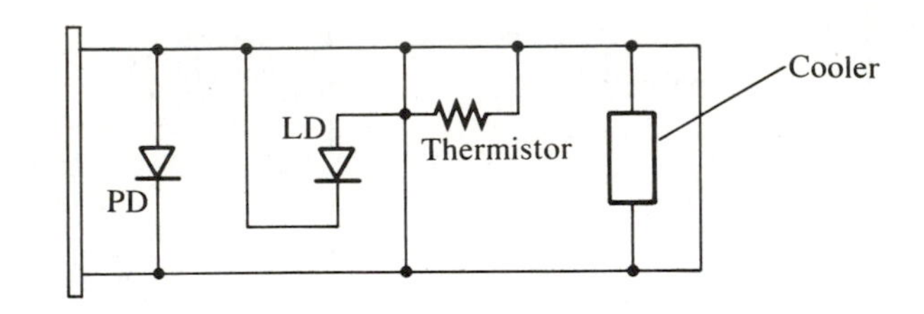

FIGURE 9–33 The NDL5007P laser diode integrated circuit package.

circuitry through the photodiode and detector amplifier, in a way that will keep the output constant. For example, an increase in P_{out} is detected by the photodiode, which causes the LD current to decrease, thus decreasing P_{out}.

Another stabilization method often used involves a thermistor, a device whose resistance decreases with increases in temperature. The LD NDL5007P, manufactured by NEC Corporation (Nippon Electric Company, Japan), includes a thermistor as well as a photodiode in the LD package (see Figure 9–33). The LD current is directly or indirectly controlled by the thermistor. For example, by connecting the thermistor in series with the LD, the diode current can be made to track I_{th}. An increase in temperature tends to increase I_{th}. Because the thermistor resistance decreases with temperature, it causes the diode current to increase, tracking the change in I_{th}.

SUMMARY AND GLOSSARY

As in other chapters, this section is largely provided for reviewing purposes.

Beam angle. The angle of the radiating beam as determined by the half-power points. (See "full angle at half maximum," "horizontal beam angle," "lateral beam angle," and "vertical beam angle.")

> **Horizontal beam angle.** The beam angle parallel to the emitting surface.
> **Lateral beam angle.** See "horizontal beam angle."
> **Vertical beam angle.** The beam angle in a direction perpendicular to the emitting surface.

Edge emitter. Light emitter with radiation coming from the edge of the active region (sometimes abbreviated as ELED).

Efficiency. Generally defined as the ratio of output power to input power. (See "coupling efficiency," "external efficiency," "internal efficiency," "power efficiency," and "quantum efficiency.")

> **Coupling efficiency.** The ratio of power inside (coupled to) the fiber to the power of the source connected to the fiber.
> **External efficiency.** The ratio of radiated power to the input electrical power for a light source.
> **Internal efficiency.** See "quantum efficiency."
> **Power efficiency.** See "external efficiency."

Quantum efficiency. The number of photons generated divided by the number of electrons injected into the light source.

Emission. Emission of photons (optical energy). (See "emission spectrum," "spontaneous emission," and "stimulated emission.")

Emission spectrum. The range of wavelengths between the half-power points.

Spontaneous emission. Photons emitted as a direct result of injected electrons, with no secondary stimulation.

Stimulated emission. Mode of radiation in which generated photons collide with electrons, stimulating more emission of photons (the multiplication effect).

Energy pump. To sustain stimulated emission, this external energy source is needed.

Fabry Perot cavity. A cavity with highly reflective parallel plates at both ends, causing a light beam to bounce back and forth, producing standing waves. The cavity is said to resonate when standing waves are produced.

Full angle at half maximum. In reference to radiation patterns, the width (in degrees) between the two points at which radiated power is half its maximum intensity.

Heterostructure. A light-emitting diode or laser structure with multiple P and N layers.

Infrared-emitting diode (IRED). A light-emitting diode that operates in the infrared region.

Intensity modulation. Varying of the intensity of optical radiation to represent an input signal.

Lasing. Operation in a sustained stimulated emission mode.

Monitor photodiode. A photodiode integrated into a laser package used to monitor output power.

Optical transducer. A device that converts electrical energy to optical radiated energy.

Pigtail mounting. A light-emitting diode or laser with a fiber permanently attached to it.

Population inversion. The condition in which there is a high concentration of electrons (and holes) in a region, where recombination is likely.

Recombination. When an electron and a hole recombine, that is, when an electron occupies a hole. In terms of energy levels, the transition of an electron from a higher to a lower level (conduction band to valence band).

Spectral radiant flux. See "emission spectrum."

Spectral response. See "emission spectrum."

Standing waves. When a transmitted wave is reflected at the far end of a cavity, the wave in the cavity sometimes appears to be standing still. This situation occurs only when the reflected wave has the appropriate phase. It depends on the length of the cavity measured in wavelengths.

Surface emitter. Emission emanates from the active area surface (SLED).

Threshhold current. The minimum laser current at which lasing takes place.

FORMULAS

$$\eta_Q = \text{Photons generated/Electrons injected} \tag{9–1}$$

Quantum efficiency.

$$\eta_{PT} = P_{out}/P_{el} \tag{9–2}$$

External efficiency. Optical power output over electrical power input.

$$\eta_c = P_F/P_{out} \tag{9–3}$$

Coupling efficiency. Power in fiber over power output of the source connected to the fiber.

$$\eta_T = \eta_{PT} \times \eta_c = P_F/P_{el} \tag{9–4}$$

Total efficiency including external and coupling efficiencies.

$$\eta_c = \sin^2 \theta_a = (\text{N.A.})^2 \tag{9–5}$$

Coupling efficiency for situation shown in Figure 9–1(b).

$$\text{BW}_{opt} = 0.35/t_r \tag{9–6}$$

Optical bandwidth related to rise time.

$$E_p = E_2 - E_1 = h \times f \tag{9–7}$$

The relation between the optical frequency f and the energy of the photon E_p.

$$\lambda = (h \times c)/(E_2 - E_1) = (h \times c)/E_g \quad \textbf{(9–8)}$$

The wavelength of the emitted optical energy related to the energy gap E_g and the speed of light c.

$$\lambda = 1.24/E_g \quad \textbf{(9–8a)}$$

Equation 9–8 with all constants combined.

$$\eta_Q = P_{int}/P_{el} \quad \textbf{(9–9)}$$

Redefining quantum efficiency as the internally generated optical power (not the power radiated) over the electrical input power.

$$I_F = I_{Csat} = (V_{CC} - V_F)/R_1 \quad \textbf{(9–10)}$$

Forward LED current in circuit of Figure 9–26.

$$I_B \geqq I_{Csat}/\beta \quad \textbf{(9–11)}$$

To cause saturation, the base current must be greater than or equal to the collector saturation current over the amplification factor, β.

$$I_B = (V_{in\text{-}peak} - V_{BE} - V_F)/R_B \quad \textbf{(9–12)}$$

Computation of I_B when LED is on (for circuit of Figure 9–26).

$$R_1 = (V_{CC} - V_F)/I_F \quad \textbf{(9–13)}$$

Computation of R_1 in Figure 9–27.

$$I_{bias} = V^+/R_1 \quad \textbf{(9–14)}$$

The bias current for the circuit of Figure 9–30.

$$I_{sig} = V_{in}/R_1 \quad \textbf{(9–15)}$$

The signal current through the diode for the circuit of Figure 9–30.

QUESTIONS

1. What are the important characteristics of a communication light source?
2. What is the difference between quantum efficiency and external efficiency?
3. What is the significance of coupling efficiency?
4. What is meant by Lambertian intensity distribution?
5. If you have two fibers, one with N.A. of 0.2 and the other with N.A. of 0.35, which will have a higher coupling efficiency (using the same source)?
6. What is one of the main factors that determines the wavelength of radiation of a LED?
7. Is linearity of a LED important for digital or analog transmission? Explain.
8. To check whether a LED or laser is emitting radiation, do you simply look at the radiating surface? Explain.
9. What is a photon?
10. How does the radiation process in LEDs and lasers involve electron-hole recombinations?
11. What is spontaneous emission?
12. Why is electron population inversion important in optical radiation?
13. What is the difference between a homojunction and a heterojunction?
14. What is the difference between spontaneous and stimulated emission?
15. What is one of the main differences in performance between edge emitters and surface emitters?
16. The quantum efficiency is typically 50–80%; however, the external efficiency is only 1 to 2%. Why?
17. Is the LED used in the forward biased mode?
18. Why is a narrow line width important in communication applications?
19. Which of the following two LEDs (with the same P_{out}) would you prefer for communication applications? One with an angle of radiation of 45° or one with an angle of radiation of 16°?
20. What does the "laser" stand for?
21. Is a laser simply a high current LED? Explain.
22. What is the laser cavity called? What is its function?
23. How is optical confinement achieved in lasers? Why is it necessary?
24. What is the purpose of the stripe structure?
25. On what does the laser wavelength depend?
26. Does the laser diode operate in the forward biased mode? Explain.
27. What is the significance of the threshhold current I_{th}?
28. How do temperature changes affect the operation of a laser? Relate it to I_{th}, λ_P, and P_{out}.
29. How would you define vertical and horizontal beam angles? Explain each.
30. For what is the monitor photodiode used?
31. Can a LED be used for both digital and analog transmission? Explain.
32. Can a laser be used for both digital and analog transmission? Explain.
33. When is it necessary to bias the LED?

34. What does temperature compensation for lasers mean?
35. Is the line width of a LD narrower or wider than that of a LED?

PROBLEMS

1. A photodiode receives −55 dBm of optical power. How much is that in milliwatts and in microwatts?
2. A LED couples −10 dBm of optical power into a fiber. Total transmission loss is 35 dB. Calculate the power at the receiver.
3. A LED-fiber combination has $\eta_Q = 50\%$, $\eta_{PT} = 1\%$, and $\eta_c = 5\%$. The LED input power is 100 mW. Find the optical power in the fiber.
4. A fiber with N.A. of 0.2 is coupled to a LED (Lambertian source) that dissipates a total of 200 mW. The power in the fiber must be at least 100 μW. Find the minimum η_{PT}.
5. A fiber with a total acceptance angle of 48° is illuminated by a LED (Lambertian pattern) that covers the fiber edge. The LED total output is 0.8 mW. Find the optical power in the fiber.
6. Given that optical rise and fall times are 15 ns, find the optical BW.
7. If the energy of a photon is 1.4 eV, find the frequency of radiation and the wavelength of radiation.
8. An alloy with a band gap of 1.3 eV is used in a LED. Find λ_P (wavelength of peak emission).
9. For the circuit of Figure 9–26, $V_{CC} = 10$ V, $V_F = 2$ V, "on" LED current = 40 mA, and $\beta = 100$. Find R_1 and R_B, if $V_{\text{in-peak}} = 5$ V.
10. In Figure 9–27, $R_1 = 80\ \Omega$, $V_F = 1.5$ V, and $V_{CC} = 6$ V. Find the LED "on" current.
11. In Figure 9–28, find the current through the LED if $V_F = 1.4$ V, $R_1 = 120\ \Omega$, and $V_{CC} = 10$ V.
12. In Figure 9–29, $R_1 = 1\ \text{k}\Omega$, $V_F = 2$ V, and $V_{\text{in}} = 4\ \text{V}_{P\text{-}P}$. Find the bias current through the LED and the *P-P* signal current through the LED. (Ignore capacitive reactance.)
13. In Figure 9–30, $R_1 = 100\ \Omega$, $V^+ = +5$ V dc, and $V_{\text{in}} = 4\ \text{V}_{P\text{-}P}$. Find the LED bias current and the LED *P-P* signal current.

10
Optical Detectors

CHAPTER OBJECTIVES

This chapter covers the principles and functional characteristics of a number of popular photodetectors. You should gain an appreciation of how the structure of a photodiode is related to its characteristics and how the functional characteristics affect the performance of a photoreceiver.

A variety of photodetector circuits are discussed, and some design and analysis details are included. These discussions will allow you to compute signal voltages from input optical power.

The end of this chapter covers noise and detectability and presents numerous computational examples. This will give you an understanding of the relations among minimum detectable signal, internal noise, and system bandwidth (BW).

10–1 INTRODUCTION

The **photodetector,** particularly the photodiode, can be viewed as the inverse of a light-emitting diode (LED). Here, the input is optical and the output is electrical. Indeed, the principles of operation of the photodiode rely on the same physical phenomena as the LED. The photodetector absorbs photons and emits electrons; that is, it produces an electrical current.

To perform well as an optical transducer, particularly in communication applications, the photodetector should have the following features:

1. It should be highly sensitive. Electrical currents that are as large as possible should be produced in response to incident light. Because the photodetector is wavelength selective (it responds to a limited range of wavelengths), the sensitivity should be high in the operating wavelength region.
2. The **response time** should be fast. The detector should respond (produce electrical current) to the shortest possible light pulse. This will allow operation at high data rates and will result in an efficient communication system.
3. For analog transmission, **photodetector linearity** is required to minimize signal distortion. The term "linearity" refers to the relation between the incident optical energy and the generated electrical current.
4. The internal noise generated by the detector should be minimal to allow the smallest possible optical input to be detected.

Other important characteristics include reliability, stability, insensitivity to environmental conditions, and cost.

10–2 PRINCIPLES OF PHOTODETECTION

For efficient photodetection of incident light, it is essential that most (if not all) of the photons be converted to electrons (electrical current). Each photon may cause the emission of only one electron. The term "emission" refers to the generation of a free electron: an electron that is mobile and will become an electrical current when a voltage is applied.

A free electron (and a mobile hole) is generated by transferring an electron from the valence band in the atomic structure to the conduction band. What is left behind in the valence band is a mobile hole. The process of creating an electron-hole pair is called electron-hole **photogeneration.**

This process is shown schematically in Figure 10–1. The photon is absorbed by the atom and causes an electron to move from the valence band (corresponding to the valence energy level) to the conduction band. The change in energy of the electron is E_g; that is, the photon must have an energy level of at least E_g. Because the energy of the photon is related to its frequency (or wavelength), you find that the **energy gap** E_g determines the **spectral response** of the photodetector. E_p must be larger than E_g (or equal to it if no other losses are involved):

$$E_p \geq E_g \qquad h \times f \geq E_g \tag{10–1}$$

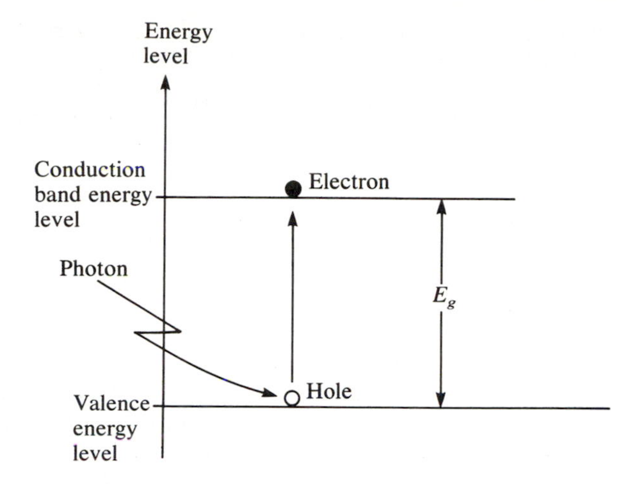

FIGURE 10–1 Electron-hole generation.

The photodetector will not respond to frequencies below

$$f_c = E_g/h \tag{10–2}$$

nor wavelengths longer than

$$\lambda_c = (c \times h)/E_g \tag{10–3}$$

(Units used are meters, seconds, and joules.) Equations 10–2 and 10–3 give the **cutoff frequency** f_c and **cutoff wavelength** λ_c, respectively, of a photodiode with a given E_g.

EXAMPLE 10–1

Germanium has an energy gap of 0.81 eV. Find the cutoff wavelength of a germanium photodiode.

Solution

$$\lambda_c = (h \times c)/E_g$$
$$h = 6.626 \times 10^{-34} \text{ J-s}$$
$$E_g = 0.81 \times 1.602 \times 10^{-19} \text{ J}$$
$$\lambda_c = (6.626 \times 10^{-34} \times 300 \times 10^6)/(0.81 \times 1.602 \times 10^{-19})$$
$$= 1.531 \times 10^{-6} \text{ m} = 1.531\ \mu\text{m}$$

Example 10–1 indicates that germanium detectors can be used only with wavelengths shorter than 1.531 μm, a range that covers all the wavelengths used in communications (0.8, 1.3, and 1.5 μm). In contrast to germanium, silicon has a band gap of about 4.1 eV (direct band gap) and is usable for ultraviolet (about 0.3 μm) or shorter wavelengths.

Combining the various constants in Equation 10–3, you obtain

$$\lambda_c = 1.24/E_g \tag{10–4}$$

where λ_c is in μm, and E_g is in eV. (Compare this to Equation 9–8*a* for the LED.) Example 10–1 can be solved by

$$\lambda_c = 1.24/0.81 = 1.53\ \mu\text{m}$$

The efficiency of the photodetector is often given in terms of quantum efficiency η_Q, similar to the quantum efficiency of the LED. η_Q is defined as

$$\eta_Q = \frac{\text{Number of mobile electrons generated}}{\text{Number of incident photons}} = N_e/N_p \quad \textbf{(10–5)}$$

The most you can expect is an η_Q of 1, which means that every absorbed photon produces a mobile electron. η_Q is, typically, about 70–80%.

A more practical term relating to the operation of the photodetector is the **responsivity** R (sometimes called the **radiant sensitivity**), defined as the ratio of **photocurrent** (current generated by absorption of photons) to the incident photon power.

$$R = I_p/P_i \quad \textbf{(10–6)}$$

R is usually given in amperes per watt (A/W), or μA per μW. I_p is the photocurrent and P_i is the incident optical power.

It is useful to relate R to η_Q. By definition,

$$I_p = (N_e \times e)/\text{s} \quad \textbf{(10–7)}$$

where e is electron charge (1.602×10^{-18} C) and N_e is the number of mobile electrons generated.

$$P_i = (N_p \times E_p)/\text{s} \quad \textbf{(10–8)}$$

E_p is the energy of the photons, and N_p is the number of incident photons. Substituting Equations 10–7 and 10–8 for variables in Equation 10–6, you get

$$R = (N_e \times e)/(N_p \times E_p) = (\eta_Q \times e)/E_p \quad \textbf{(10–9)}$$

(In the expression, $\eta_Q \times e/E_p$, E_p is in joules. It can be replaced by $R = \eta_Q/E_p$, with E_p in eV.)

$$R = \eta_Q \times e/(h \times f) = \eta_Q \times (e \times \lambda)/(h \times c) \quad \textbf{(10–10)}$$

η_Q is a fraction (a ratio, not a percent), λ is in meters. R is in amperes per watts, c is in meters per second, and e is in coulombs. Both quantum efficiency and responsivity are usually given for the particular wavelength at which responsivity peaks. Equation 10–10 becomes Equation 10–10*a* with R in A/W and λ in micrometers.

$$R = (\eta_Q \times \lambda)/1.24 \quad \textbf{(10–10a)}$$

EXAMPLE 10–2

A photodetector has quantum efficiency of 70% (0.7) and is operating with λ of 0.82 μm. Find the responsivity.

Solution

$$R = (0.7 \times 0.82)/1.24 = 0.463 \text{ A/W}$$

Equations 10–10 and 10–10*a* imply that responsivity increases with wavelength. This is a strictly theoretical prediction. It ignores a number of practical considerations. The incident light has to penetrate the P region (or the N region) to reach the depletion, or intrinsic, region. The photodiode is encapsulated in glass or plastic, which must be penetrated by the light. Both the P (or N) region and the encapsulating material have a filtering effect. Transparency to optical power varies with wavelength. The typical plot of responsivity as a function of wavelength looks like a resonance response that tapers off on both sides of the peak response (Figure 10–2).

10–3 PHOTODETECTORS

10–3–1 P-N Photodiodes

One of the most popular photodetectors is the photodiode. It is a diode operated in reverse bias mode, with the depletion region exposed to the optical energy. Remember that an unbiased diode has a relatively narrow depletion region, a

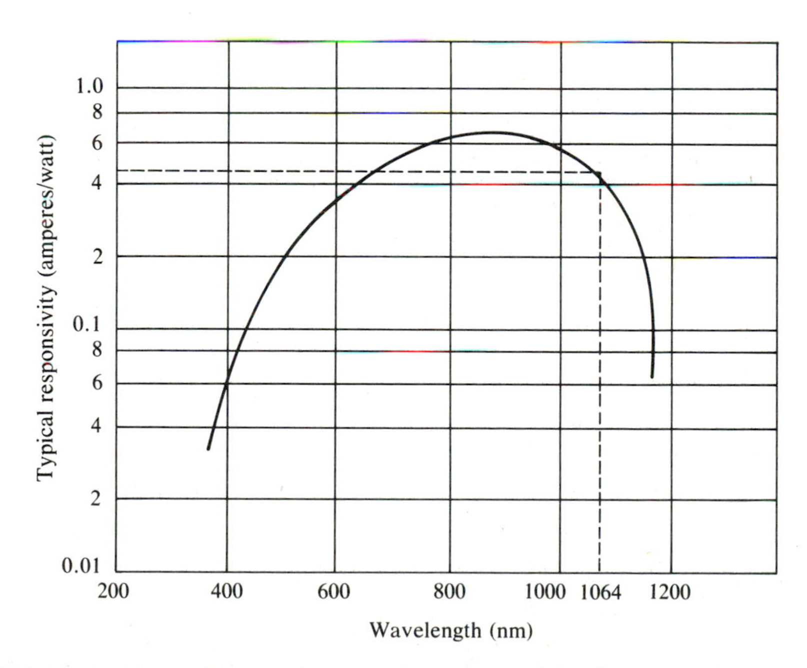

FIGURE 10–2 Typical spectral responsivity for model AP4010.
Source: Antel Optronics, Inc.

region where there are very few mobile charges (electrons or holes). With increased reverse bias, the depletion region gets wider. The photons incident on the depletion region produce electron-hole pairs (mobile charges), which can then be moved by the applied voltage across the junction. Figure 10–3 shows this process.

Within the depletion region, the electrons and holes travel because of the applied voltage. Note that because the depletion region is a high-resistance region, most of the voltage E appears across that region and accelerates the generated mobile electrons and holes. Some of the photons may be absorbed in the P and N regions outside the depletion region. The corresponding mobile charges, however, move by diffusion at a very low velocity, and most of them recombine along the way, contributing very little to the resulting photocurrent.

Consequently, it is desirable to enlarge the depletion region as much as possible so that more photons can be absorbed and yield higher photocurrent. To accomplish an effective increase in the sensitive area, the positive-intrinsic-negative **(PIN)** diode was developed. It is important to note that the photocurrent (the current generated by the photons) has the same polarity as the diode reverse current (the leakage current). It is, therefore, important to keep the leakage current **(dark current)** as small as possible.

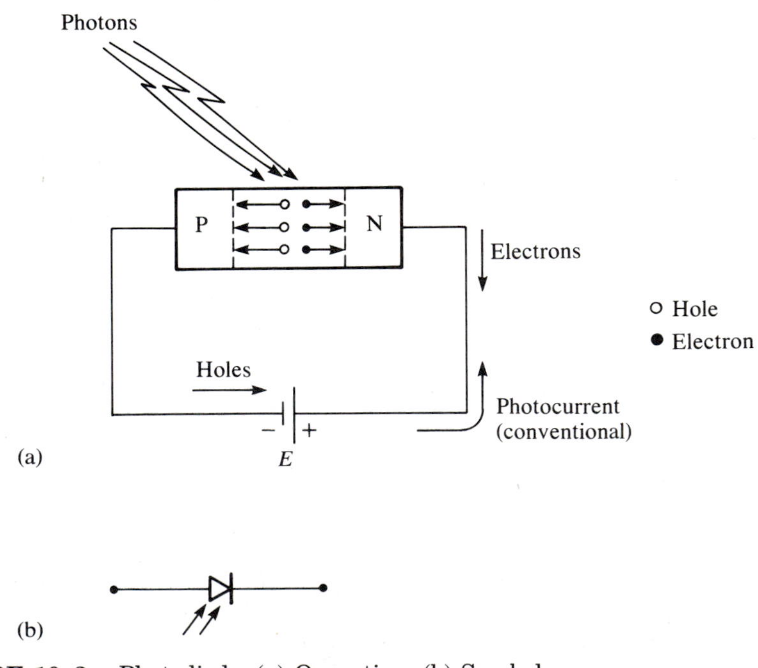

FIGURE 10–3 Photodiode. (a) Operation. (b) Symbol.

10–3–2 PIN Photodiodes

The structure of the PIN photodiode consists of P-N regions separated by an intrinsic material region (Figure 10–4). This material is not specially doped. It is a semiconductor, silicon for the silicon diode, germanium for the germanium diode, and so on.

The intrinsic material region has few mobile charges, typical of all semiconductors. It is like the depletion region. In other words, the PIN diode is predepleted. There is no need for a reverse bias to produce a large depletion region. This allows the PIN to be operated with zero reverse bias (**photovoltaic mode**) or reverse bias (**photoconductive mode**).

Because the photosensitive area is increased by the introduction of the intrinsic region, the responsivity is also increased. In addition, the response time is reduced in comparison with the P-N diode. First, there is no photogeneration in the P or N regions (the regions where charges move very slowly) to cause a slow response. Second, because the depletion region is larger, the interelectrode capacity is smaller, leading to a faster response.

The two major advantages of the PIN over most other photodetectors are speed and low-voltage operation. The P-N diode requires about 20–40 V or higher in the photoconductive mode, whereas the PIN can operate with a 8- to 10-V reverse bias in the photoconductive mode or 0 V in the photovoltaic mode. In both cases, increasing the reverse bias tends to produce a faster response. The time required for the electrons to travel from P to N is reduced by increasing the reverse bias.

10–3–3 Avalanche Photodiodes

The **avalanche photodiode** (APD) has the basic characteristics of the avalanche diode. Photogenerated electrons are accelerated by the relatively large reverse voltage and collide with other atoms, generating new free electrons that continue to collide and produce new conduction electrons. A single photogenerated electron may be multiplied to well over 100 free electrons by the **avalanche multiplication** effect. This yields a highly

FIGURE 10–4 The PIN diode.

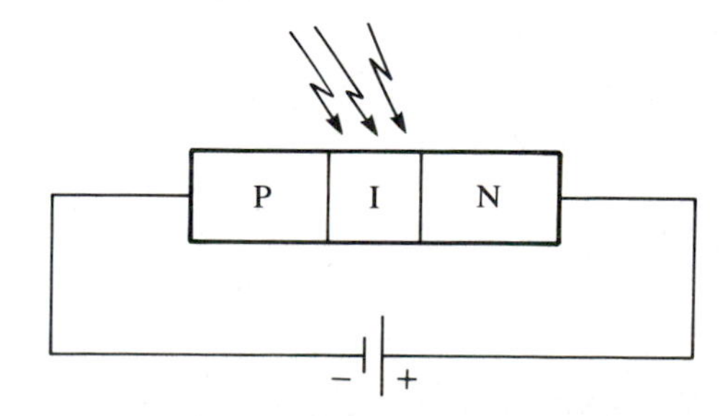

sensitive photodetector with responsivity of about 25–100 A/W, in comparison with 0.5–0.8 A/W of the P-N photodiode.

To produce the avalanche effect, high reverse bias voltages are required. The APD usually operates with 40- to 400-V reverse bias. With these high voltages, the response time is very short. The electrons attain a very high velocity as they travel across the P-N structure. (Typically, the turn-off time of the APD is larger than the turn-on time, thus reducing the speed of operation.)

The APD has a higher internal noise level (caused by random electrical current) than the P-N or PIN diode. It is also more temperature sensitive than the P-N or PIN diodes. The internal noise is somewhat improved in the *reach-through* APD, where voltage that develops across the internal active region is very carefully controlled.

10–3–4 Other Photodectors

Two photodetectors, both with active gain, are the **phototransistor** and the **PIN-FET** (FET = field-effect transistor). Both can operate at extremely low light levels.

The phototransistor can be viewed as a transistor whose current is controlled by the incident light, as though a photodiode were placed in the base circuit. The symbol of the phototransistor is shown in Figure 10–5(a). Similarly, the photodarlington is a darlington circuit with the input base current replaced by optical input power (Figure 10–5(b)). The phototransistor is relatively slow and suitable for low-speed (t_r about 10 μs) applications.

FIGURE 10–5
Photodetector symbols.
(a) Phototransistor.
(b) Photodarlington.

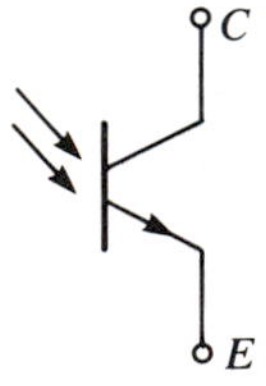

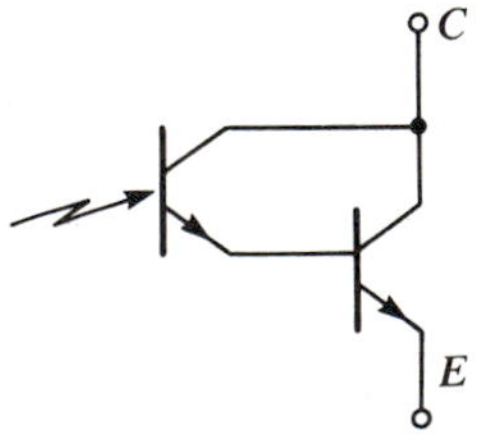

The PIN-FET is an integrated circuit, a PIN detector and a FET amplifier. To improve sensitivity and to reduce the effects of internal noise, a FET amplifier with a PIN diode at its input is packaged in a single integrated circuit. Integrated PIN-FET circuits have been developed with a high BW of over 100 MHz (the Plesscor Optronics model AR–13–100).

A plot of I_C versus V_{CE} of a phototransistor with input light power as a parameter is shown in Figure 10–6, for the GE transistor models GFOD1A1 and GFOD1A2. Note that the plot is similar to that of a regular transistor, with P_{in} replacing I_B. P_{in}, the input optical power, has the same effect as I_B in the standard transistor. (Note that in Figure 10–6, both the I_C axis and the V_{CE} axis use logarithmic, nonlinear scales.) The I_C scale is in relative amperes, not absolute values of milliamperes or amperes. The values $I_C = 1$ for $V_{CE} =$ 5 V and $P_{in} = 10\ \mu$W were selected arbitrarily. All other I_C values are shown in proportion to this unity value.

A more specific relation between optical power input and I_C is shown in Figure 10–7, for the Optoelectronic Division of TRW, Inc., models OP600–OP604. Here the P_{in} is expressed indirectly as irradiance in milliwatts of optical power per square centimeter. To obtain P_{in} approximately, it is necessary to multiply irradiance by the active area of the particular device. For example, the incident power for a device with an active area of 2500 μm^2 exposed to 4 mW/cm^2 is 100 nW ($4 \times 2500 \times 10^{-8} \times$ mW). Note that 1 $(\mu\text{m})^2 = 10^{-8}$ cm^2.

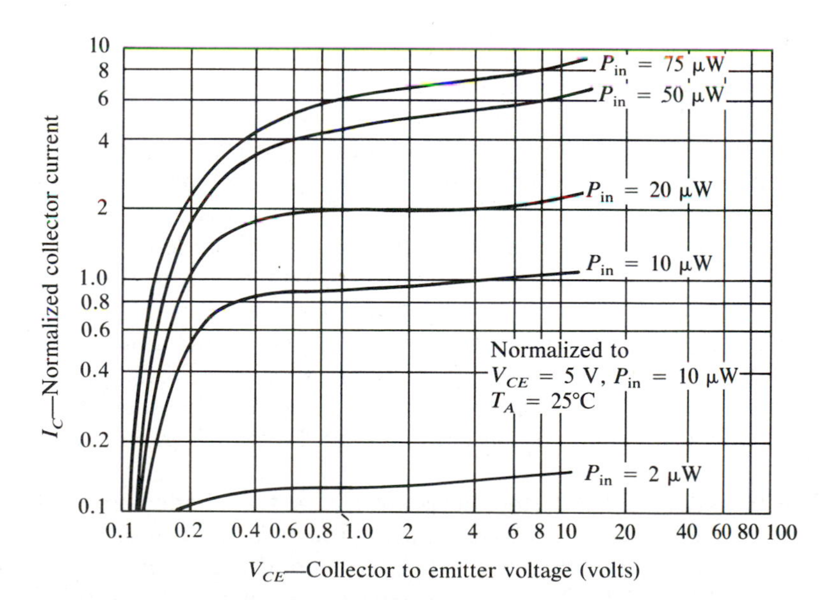

FIGURE 10–6 Phototransistor I_C versus V_{CE} plot. (Collector characteristics for GE models GFOD1A1, and GFOD1A2.)
Source: General Electric Semiconductor Products Department.

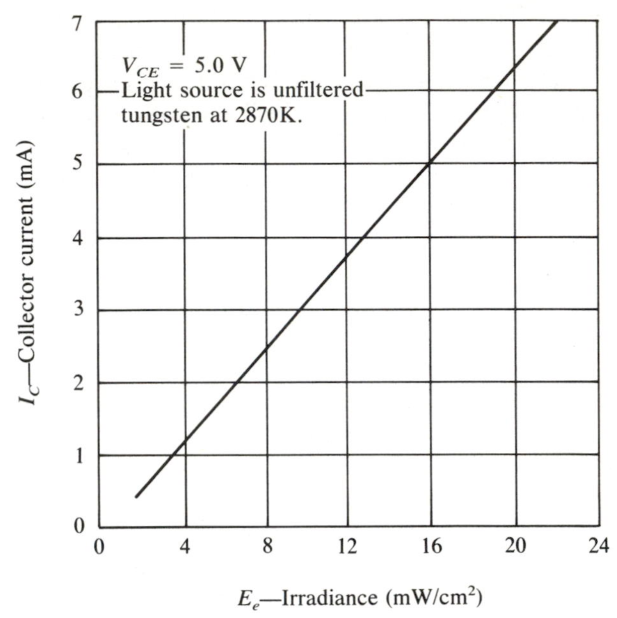

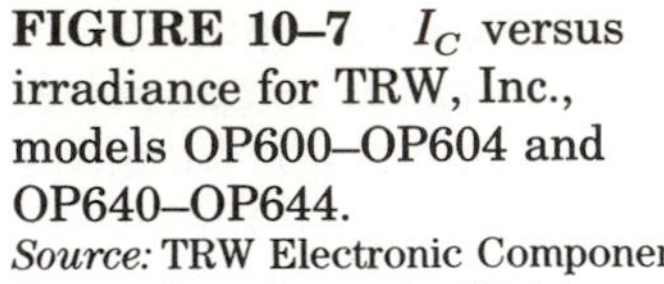

FIGURE 10–7 I_C versus irradiance for TRW, Inc., models OP600–OP604 and OP640–OP644.
Source: TRW Electronic Components Group, Optoelectronics Division

10–4 CHARACTERISTICS OF PHOTODETECTORS

Unfortunately, there is no uniformity in the descriptions of the functional characteristics of the photodetectors. A number of parameters are, however, fairly standard. Terms such as responsivity (sometimes radiant sensitivity is used), dark current, peak wavelength (or detailed spectral response), rise time, **noise-equivalent power** (NEP), and active area are common to most data sheets.

To explore the functional behavior of the photodetectors, look at a number of PIN silicon photodiodes, using data sheets from the Hamamatsu Corporation; these data sheets are relatively complete.

Various types of packages are shown in Figure 10–8. Some have lenses, some have very large areas, some have square sensitive areas, and some have round sensitive areas. The **window** (the opening in a photodetector exposing the active area to incident light) material can be simple glass, lens glass, or resin.

The parameters are listed horizontally in the shaded area. This text examines that list in as much detail as is appropriate for each term. In addition to the model number, the data sheets briefly list one or two important features. For example, the S1190–01 is a "high-speed response, lense window" diode. Additionally, the data sheets provide the following information.

1. The *outline* drawings given by the circled numbers are shown at the bottom of Figure 10–8. (Only outlines 1 through 6 are shown.) The types of material used, K, L, F, and R, are listed under note A.

2. The effective area can be calculated from data in the *size* column. For example, the 0.8-mm diameter of the S1188–02 gives an area of 0.5 mm^2.
3. *Range* gives the wavelength range for which the diode has a meaningful responsivity (radiant sensitivity). The spectral response graph shows that the S1188 series has a range of 400–1060 nm. The range is determined here, somewhat arbitrarily, by a drop of responsivity to 0.1 A/W. The peak wavelength for the S1188 series is given as 900 ± 50 nm. Again, the spectral response plot agrees with this value.
4. *Radiant sensitivity,* or responsivity, is given for a number of different wavelengths, the peak wavelength, 633 nm (produced by a helium-neon laser) and 930 nm (produced by a GaAs LED).
5. *Short-circuit current* is the current produced by a particular light source (100 lux of optical power) while the diode has no bias ($V_D = 0$). (See Figure 10–9.)
6. *Dark current* I_D is the current in the diode with a given reverse bias with no incident light. This current contributes to the internal noise and must be considered in circuit design as diode leakage current.
7. *Rise time* t_r is given for a 50-Ω load ($R_L = 50\ \Omega$). Typically, t_r increases almost linearly with R_L. If R_L doubles so does t_r. For high-speed operations, this implies that very low load resistors should be used. However, this also reduces the amplitude of the signal voltage.
8. *Cutoff frequency* f_c is the 3-dB frequency. This is the frequency at which the photocurrent decreases by 3 dB from its low-frequency response, the best (highest) response.
9. The *noise equivalent power,* (NEP) is defined as the optical power required to produce a load current equal to the noise current. For reverse biased diodes and no signal input (no signal light), the dominant contributor to noise current is the diode dark current. As it turns out, the noise current is proportional to, among other things, the square root of the BW used, so that NEP is always given in watts per square root of BW ($W/\sqrt{BW}$) or watts per square root of hertz ($W/\sqrt{Hz}$). You will examine NEP again in Section 10–6, where this text also covers detectability.

An important parameter not listed in the data table in Figure 10–8 is described by the directivity plot. Similar to the radiant angle of the LED or laser, there is a limited range of incident angles for which a current is produced in the photodiode. The lensed photodiode, the S1190–01 of Figure 10–8, has an **angular response** (defined as the angle at which the diode sensitivity is 50% of its maximum) of ±12° (or 24°). The S2506–01, an example of an unlensed diode, has an angular response of ±70°. (To find the value, follow the 50% circle until it intersects the diode plot and read the degrees on the corresponding radial line.) The narrow-angle diodes are usually designed for coupling to optical fibers that operate with very shallow angles, while the large-angle diodes are commonly used in instrumentation or control systems.

PIN Silicon Photodiodes

Type No.	Features	Ⓐ Outlines / Window Materials	Package	Photosensitive Surface: Size (mm)	Photosensitive Surface: Effective Area (mm^2)	Spectral Response: Range (nm)	Spectral Response: Peak Wave-length (nm)	Characteristics (at 25°C): Radiant Sensitivity Typ. (A/W): Peak Wave-length	Radiant Sensitivity Typ. (A/W): 633nm He-Ne Laser	Radiant Sensitivity Typ. (A/W): 930nm GaAs LED	Short Circuit Current I_{sh}, 100 lux: Min. (μA)	Short Circuit Current I_{sh}, 100 lux: Typ. (μA)
S1188-02	Ultra-high speed response	④ /K	TO-18 (3 pin)	0.8 dia.	0.5	400 ~ 1060	900±50	0.6	0.4	0.6	0.4	0.5
S1188-06				0.4 dia.	0.12						0.1	0.14
S2216-01	Ultra-high speed response, low bias type	④ /K	TO-18 (3 pin)	0.8 dia.	0.5	400 ~ 1060	900±50	0.6	0.4	0.6	0.4	0.5
S2216-02				0.4 dia.	0.12						0.1	0.14
S1190	High-speed response	① /K	TO-18	1.1 × 1.1	1.2	400 ~ 1100	920±50	0.6	0.4	0.6	0.8	1.1
S1190-01	High-speed response, lens window	③ /L									6	8
S1190-03	High-speed response	④ /K	TO-18 (3 pin)								0.8	1.1
S1190-04	Low capacitance, lens window	③ /L	TO-18				940±50				6	8
S1223	2.4 × 2.8mm sensitive area	⑥ /K	TO-5	2.4 × 2.8	6.6	400 ~ 1100	920±50	0.6	0.4	0.6	4	5.5
S1223-01	3.7 × 3.7mm sensitive area			3.7 × 3.7	13.6						8	11
S2506	Visible light cutoff molded resin type	㉗ /R	Molded resin	3 × 3	9	700 ~ 1100	950±50	0.55	–	0.5	4.0	5.5
S2506-01	High immunity to ambient fluorescent lighting					840 ~ 1100	980±50	0.5		0.4	2.2	3.0
S2336	4 element type	㉘ /R	Molded resin	See ㉘ on P.37.		400 ~ 1060	900±50	0.6	0.4	0.6	–	–
S2337	6 element type											

Ⓐ See pages 34 to 37 below for outlines.
Window materials are
K: Borosilicate glass
L: Lens type borosilicate glass
F: Visible light cutoff filter
R: Resin molding

• **Spectral Response**

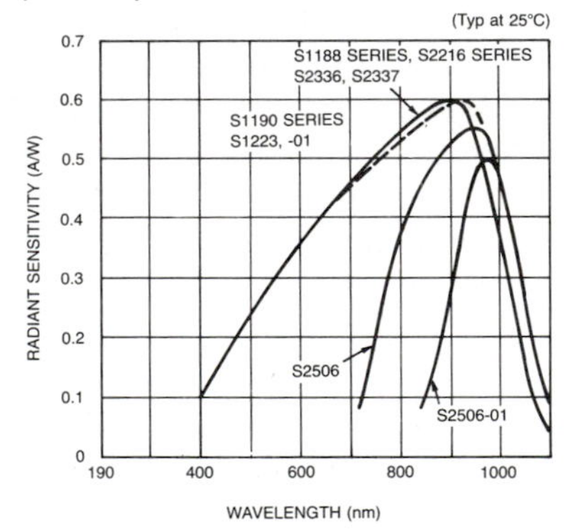

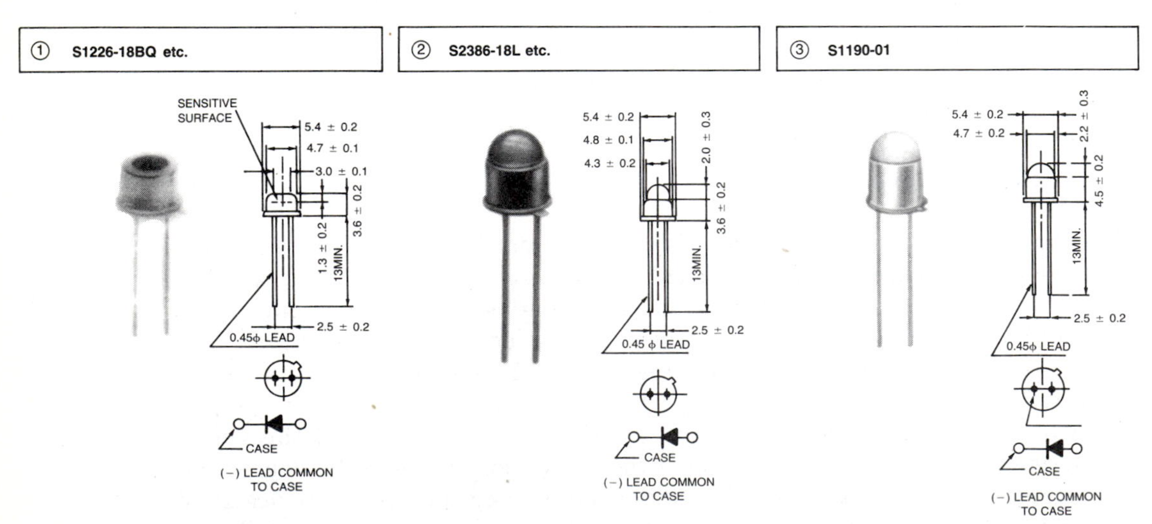

FIGURE 10–8 Specifications for PIN photodiodes.
Source: Hamamatsu Corporation, Bridgewater, N.J.

Characteristics (at 25°C)							Maximum Ratings					
Dark Current I_d Max. (nA)	Temperature Dependence of Dark Current Typ. (Times/°C)	Junction Capacitance C_j Typ. (pF)	Rise Time t_r R_L = 50 Ω Typ. (ns)	Cutoff Frequency f_c Typ. (MHz)	NEP Typ. ($W/Hz^{1/2}$)	D* Typ. ($cmHz^{1/2}/W$)	Reverse Voltage $V_R$max. (V)	Current 1 max. (mAp)	Power Dissipation P max. (mW)	Temperature Range Operating (°C)	Temperature Range Storage (°C)	Type No.
2 (V_R = 20V)	1.15	4 (V_R = 20V)	3 (V_R = 20V)	80 (V_R=20V)	5×10^{-14} (V_R = 20V)	2×10^{15} (V_R = 20V)	30	0.5	50	−20 ~ +80	−55 ~ +100	S1188-02
1 (V_R = 20V)		2 (V_R = 20V)	0.8 (V_R = 20V)	100 (V_R=20V)	4×10^{-14} (V_R = 20V)							S1188-06
1 (V_R = 5V)	1.15	2.5 (V_R = 5V)	1 (V_R = 5V)	40 (V_R = 5V)	6×10^{-15} (V_R = 5V)	2×10^{13} (V_R = 5V)	30	0.5	50	−20 ~ +80	−55 ~ +100	S2216-01
0.5 (V_R = 5V)		1.5 (V_R = 5V)		50 (V_R = 5V)	4×10^{-15} (V_R = 5V)							S2216-02
2 (V_R = 10V)	1.15	8 (V_R = 10V)	3 (V_R = 10V)	30 (V_R=10V)	6×10^{-14} (V_R = 10V)	2×10^{12} (V_R = 10V)	20	0.5	50	−20 ~ +80	−55 ~ +100	S1190
												S1190-01
												S1190-03
3 (V_R = 10V)		3 (V_R = 20V)	2 (V_R = 10V)	20 (V_R=10V)	1.3×10^{-14} (V_R = 10V)	8×10^{12} (V_R = 10V)	30					S1190-04
10 (V_R = 20V)	1.15	13 (V_R = 20V)	5 (V_R = 20V)	30 (V_R=20V)	7×10^{-14} (V_R = 20V)	8×10^{12} (V_R = 20V)	30	0.5	100	−20 ~ +80	−55 ~ +100	S1223
20 (V_R = 20V)		25 (V_R = 20V)	10 (V_R = 20V)	20 (V_R=20V)	1×10^{-13} (V_R = 20V)							S1223-01
10 (V_R = 12V)	1.15	16 (V_R = 12V)	50 (V_R = 12V, R_L = 1kΩ)	25 (V_R=12V)	1×10^{-13} (V_R = 12V)	3×10^{13} (V_R = 12V)	35	0.5	150	−20 ~ +80	−40 ~ +100	S2506
											−40 ~ +80	S2506-01
0.2 (V_R = 10V)	1.15	1 (V_R = 10V)	1 (V_R = 10V)	—	—	—	20	0.5	10	−20 ~ +80	−40 ~ +100	S2336
												S2337

• **Dark Current vs. Reverse Voltage**

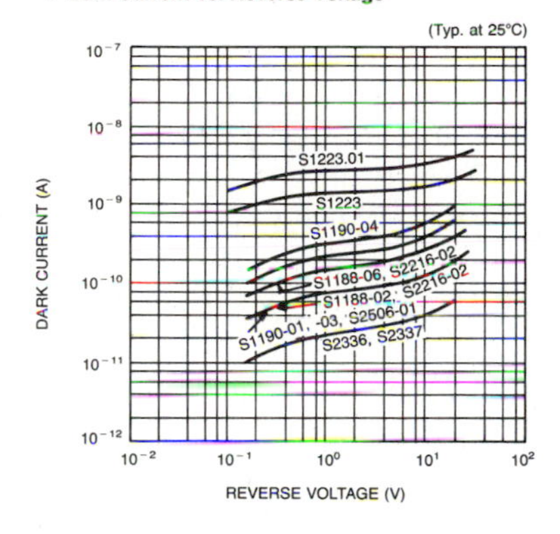

• **Directivity**

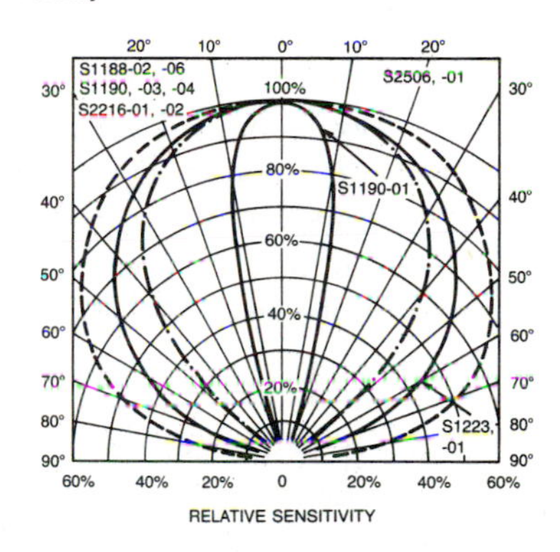

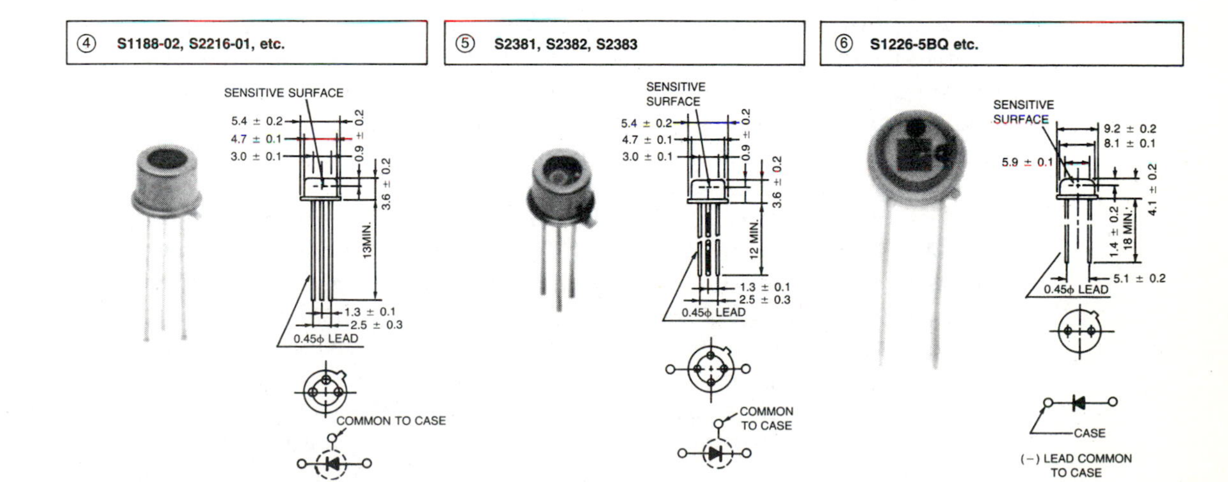

FIGURE 10–8 *(Continued)*

FIGURE 10–9 Photodiode circuits. (a) Short-circuit current. (b) Diode with load resistor R_L.

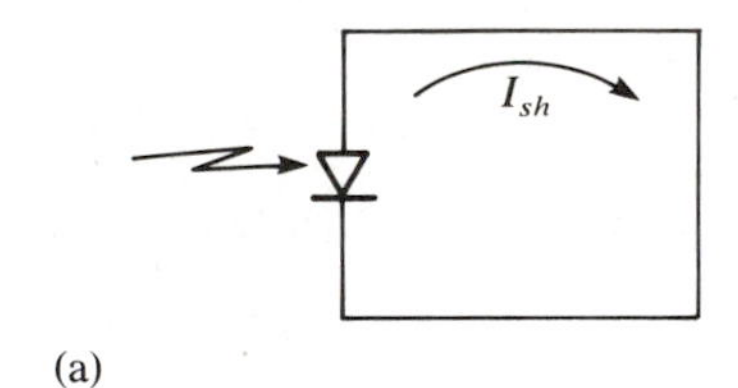

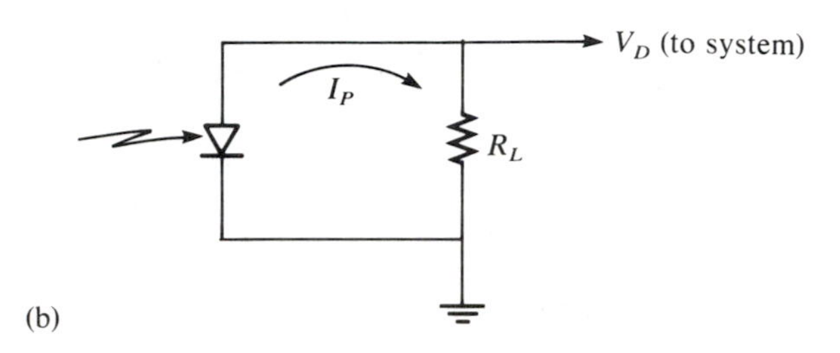

Some manufacturers include the quantum efficiency of the diode at a particular wavelength. The data sheet shown in Figure 10–8 is for PIN diodes. The data sheet for a P-N diode is similar. All the parameters listed are relevant to a P-N diode.

Note that the photodiode, P-N or PIN, when forward biased, yields the typical diode forward characteristics. However, the reverse current is a function of the incident light, as shown in Figure 10–10. This reverse current is the photocurrent I_p.

The responsivity of the APD is often given in two parts: the responsivity, assuming no avalanche effect (which is typically low, about 0.4 A/W), and a multiplication factor M, which represents the avalanche effect. Consequently, an APD with a responsivity of 0.4 A/W and M of 100 has an effective responsivity of 40 A/W. Because the factor M increases as the applied voltage increases, the effective responsivity increases. At the breakdown voltage, the diode current (both the photocurrent and the dark current) increases very sharply, indicating that the device must be operated below the breakdown voltage. (See the sharp bend on the left side of Figure 10–10.)

The specifications of the phototransistor are similar to those of a transistor. Instead of input current I_B, you have optical input power, and responsivity replaces β (or h_{fe}). That is, I_C can be obtained by multiplying P_{in} by the responsivity R (Figure 10–11). Note that the typical responsivity of the phototransistor is more than 100 times that of the P-N or PIN photodiodes, which is a result of transistor action. Similarly, the responsivity of the photodarlington is more than 10,000 times that of the P-N diode.

A qualitative comparison among the various photodetectors may be helpful in outlining possible applications, Table 10–1 provides such a comparison. For this comparison, assume reverse bias P-N or PIN operation

FIGURE 10–10 Typical photodiode *V-I* characteristics.
Source: Hamamatsu Corporation, Bridgewater, N.J.

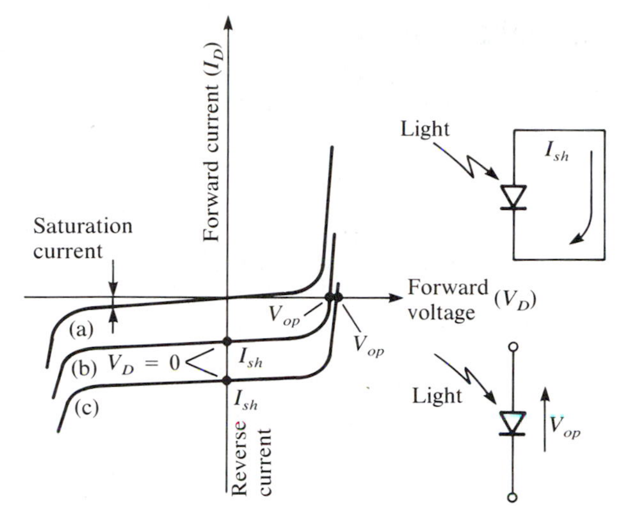

(a) No incident light (dark current)

(b) Some incident light

(c) Higher-intensity incident light

(photoconductive mode). Although the characteristics of most photodetectors vary with the specific operating conditions, the following comparison will ignore such variations. It is intended to serve as a rough guideline for device selection.

To operate the APD efficiently; that is, to obtain high responsivity, it must be biased near the avalanche knee (a sharp bend in the curve) but not beyond (to the left of) it. Because of sensitivity to temperature, it may be necessary to have a feedback system to stabilize the bias point. The NEP value should be below the expected optical signal strength. Remember that the actual noise power depends on the square root of the BW. (The NEP is specified for 1 Hz.) It is usually specified for a given wavelength. The rise time is highly dependent on the load resistance. High values of R_L result in longer rise times.

10–5 DETECTOR CIRCUITS—OPTICAL RECEIVERS

Photodiodes can be operated in two basic modes: photovoltaic mode (where no bias is applied to the diode) and *photoconductive mode* (where the diode operates with reverse bias). In the photovoltaic mode, operating speed is low and responsivity is also relatively low. The circuit is, however, somewhat less complex. It requires no biasing voltage. Because of their low speed, phototransistors are more suitable for control functions such as alarm systems and photorelays.

40.11 4/81 Preliminary

Fiber Optic Detectors GFOD1B1 — GFOD1B2

Silicon Photo-Darlington Detectors for Fiber Optic Systems

The General Electric GFOD1B1 and GFOD1B2 are silicon photo-darlington detectors which detect and convert light signals from optical fibers into electrical signals. They are packaged in a housing designed to optimize fiber coupling efficiency, reliability, and cost. They mate directly with AMP OPTIMATE™ fiber optic connectors for easy interconnection and use. Mounting is compatible with SAE and metric fasteners of both through hole and self-tapping types.

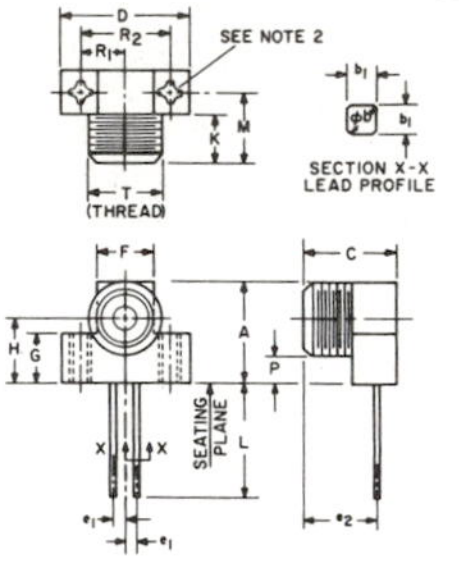

SYM.	MILLIMETERS		INCHES		NOTES
	MIN.	MAX.	MIN.	MAX.	
A	10.67	11.17	.420	.440	
ϕb	.61	.66	.024	.026	1
b1	.50	NOM.	.020	NOM.	1
C	9.88	10.26	.389	.404	
D	13.47	13.97	.530	.550	
e_1	1.27	NOM.	.050	NOM.	
e_2	7.93	8.07	.312	.318	
F	5.87	6.12	.231	.241	
G	5.08	5.58	.200	.220	
H	6.84	7.08	.269	.279	
K	5.11	5.25	.201	.207	
L	12.22	—	.481	—	
M	7.73	7.97	.304	.314	
P	3.00	REF.	.118	REF.	
R_1	4.70	4.82	.185	.190	
R_2	9.40	9.65	.370	.380	
T			5/16-32 NEF 2A		

NOTES:
1. Two Leads
2. Mounting Holes see attached drawing or M2x0.4 or Self-Tapping Screws

absolute maximum ratings

(25°C unless otherwise specified)

Voltages		
Collector to Emitter Voltage	V_{CEO}	30 V
Emitter to Collector Voltage	V_{ECO}	5 V
Current		
Collector Current (continuous)	I_C	100 mA
Dissipation		
Power Dissipation (TA = 25°C)*	P_T	150 mW
Temperatures		
Operating Temperature	T_{OP}	-55°C to +85°C
Storage Temperature	T_{STG}	-55°C to +100°C
Lead Soldering Temperature	T_L	5 seconds at 260°C

*Derate 2.5 mW/°C above 25°C ambient.

electrical characteristics (25°C unless otherwise specified)

	SYMBOL	MIN.	TYP.	MAX.	UNITS
Collector to Emitter Voltage (I_C = 10 mA, P_{in} = 0)	$V_{(BR)CEO}$	30	—	—	V
Emitter to Collector Voltage (I_E = 100 μA, P_{in} = 0)	$V_{(BR)ECO}$	5	—	—	V
Collector Dark Current (V_{CE} = 10V, P_{in} = 0)	I_{CEO}	—	—	100	nA

optical characteristics (25°C unless otherwise specified)

		SYMBOL	MIN.	TYP.	MAX.	UNITS
Responsivity (Note 1)	GFOD1B1	R	1000	—	—	μA/μW
(V_{CE} = 1.5V, P_{IN} = 2μW, λ_P = 940 nm)	GFOD1B2	R	500	—	—	μA/μW
Turn on Time (See Note 1)						
(V_{CC} = 5V, I_F = 10 mA, R_L = 750Ω		t_{on}	—	45	—	μs
(V_{CC} = 1.5V, I_F = 10 mA, R_L = 0)		t_{on}	—	10	—	μs
Turn Off Time (See Note 1)						
(V_{CC} = 5V, I_F = 10 mA, R_L = 750Ω)		t_{off}	—	250	—	μs
(V_{CC} = 1.5V, I_F = 10 mA, R_L = 0)		t_{off}	—	25	—	μs

Note 1: Radiation source used is a GFOE1A1 Fiber Optic Emitter coupled via 1 meter of CROFON® 1040 fiber terminated per AMP Incorporated instruction sheet IS 2878-2.

FIGURE 10–11 Photodarlington specifications for General Electric models GFOD1B1 and GFOD1B2.
Source: General Electric Semiconductor Products Department.

TABLE 10–1
Photodetector Comparisons

Photo-detector	Peak Wavelength (nm)	Responsivity (A/W)	Dark Current Approximate (nA)	Rise time (ns) (50 Ω load)	Operating Voltage	NEP[1] (W/Hz)	Notes
Silicon P-N	550–850	0.4–0.7	1–5	5–10	20–40	10^{-13}–10^{-14}	
Silicon PIN	850–950	0.6–0.8	3–300	1–5	5–40	10^{-13}–10^{-14}	High speed, highly linear
InGaAs PIN	300–1,500	0.8	10–30	1–3	5–40	10^{-13}–10^{-14}	Long wavelength, highly linear
Silicon APD	650–900	74–100	1	0.5–2	60–120	10^{-14}	High R, needs bias stability
Photo-transistor	800–900	70	25–100	5–15μs	10	No data	High R, stable operation

[1]NEP = noise-equivalent power.

10–5–1 P-N and PIN Circuits

The simplest photodetector circuit (usually not used as shown) consists of a photodiode and a load resistor, as shown in Figure 10–12. In Figure 10–12(a), the direction of the generated photocurrent is opposite the direction of the diode polarity. It makes the diode behave like a battery or, more precisely like a current source with the anode as its positive terminal. This is the direction of the reverse diode current. The diode becomes forward biased by its own generated potential. This mode of operation is slow. It has a low internal noise because the dark current I_d, which contributes to the noise, is very low. I_d increases with reverse bias. In Figure 10–12(b), the diode is reverse biased, increasing I_d and I_p, the photocurrent.

The two modes of operation are demonstrated again in Figure 10–13. Here, the photocurrent serves as the input current to an operational amplifier. In both circuits, the voltage across the diode is constant because point A is virtual ground (zero potential). In Figure 10–13(a), the diode operates with zero bias voltage. Its photocurrent is the short-circuit photocurrent with zero load resistance. This is a more stable operation than the circuit in Figure 10–12(a). It is substantially slower than the circuit of Figure 10–13(b). V_{out} for both is given as

$$V_{out} = -I_p \times R_f \tag{10–11}$$

FIGURE 10–12 Photodiode circuits. (a) Photovoltaic. (b) Photoconductive.

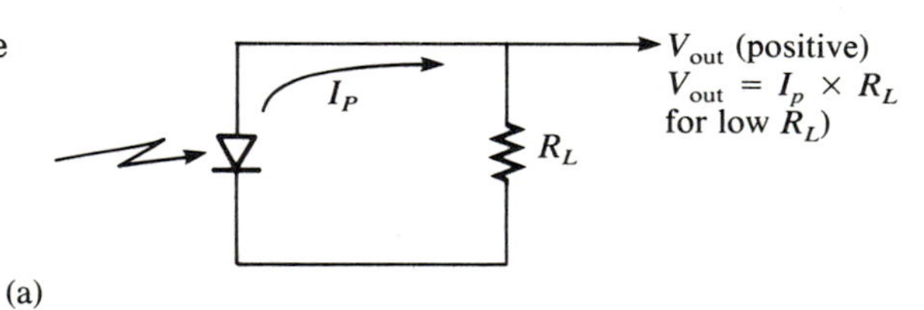

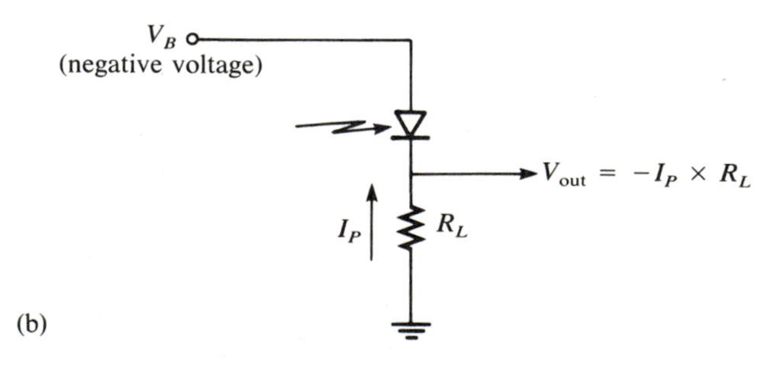

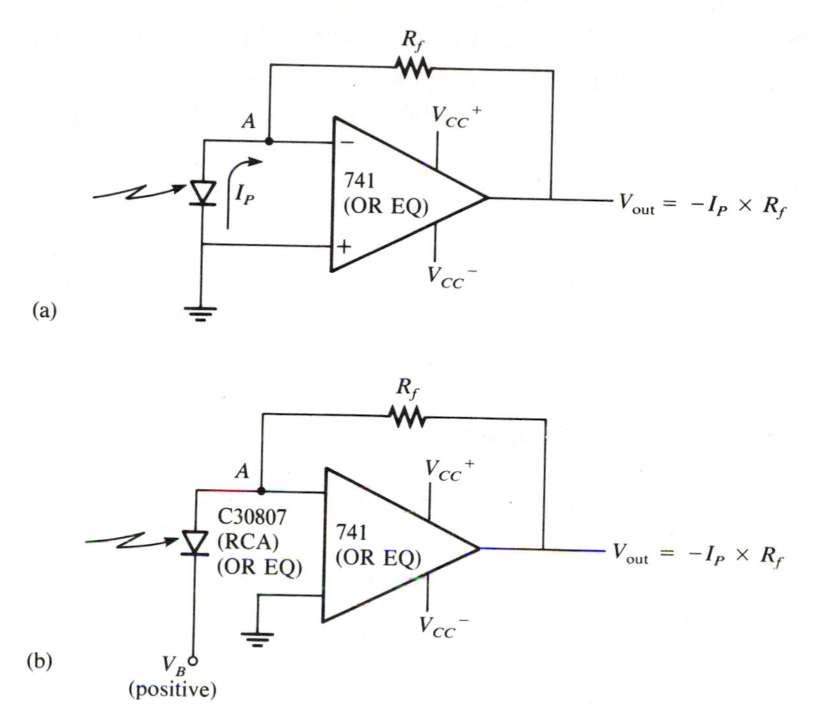

FIGURE 10–13 Photodetectors with operational amplifiers. (a) Photovoltaic (or photoamperic). (b) Photoconductive.

To compute V_{out}, you must know I_p and R_f. I_p can be obtained from the responsivity of the diode if the incident power is known. Because the definition of R is the current produced per incident unit power (A/W), you get the current I_p as

$$I_p = R \times P_{in} \tag{10–12}$$

EXAMPLE 10–3

Find V_{out} in Figure 10–13(b), given

$$R = 0.6 \text{ A/W}$$
$$I_d = 10 \text{ nA} \quad \text{for the C30807 diode}$$
$$R_f = 100 \text{ k}\Omega$$
$$P_{in} = 0.5\ \mu\text{W}$$

Solution
First, find I_p:

$$I_p = R \times P_{in} = 0.6 \times 0.5 \times 10^{-6} = 0.3\ \mu\text{A}$$

I_d must be added to I_p to obtain the total current in the operational amplifier:

$$I_d + I_p = 0.3 + 0.01 = 0.31\ \mu A$$
$$V_p = -0.31 \times 10^{-6} \times 100 \times 10^3 = -0.031\ V = -31\ mV$$

Note that the signal voltage is -30 mV, whereas the voltage from I_d is -1 mV.

Some diodes have a **guard ring,** an extra diode structure surrounding the active area. When the guard ring diode is heavily reverse biased, I_d can be reduced to near zero. (Reverse bias must be 50 V or more.)

10–5–2 PIN-FET Circuits

A FET receiver of alternating current signals is shown in Figure 10–14. To calculate V_{out}, first calculate V_{in}.

$$V_{in} = I_p \times R_L \parallel R_G \quad \textbf{(10–13)}$$

(You include the 1.0 MΩ, effectively in parallel with R_L. I_p is calculated in Example 10–3.)

$$V_{out} = -g_{mQ} \times R_D \times V_{in} \quad \textbf{(10–14)}$$

(g_{mQ}, the operating transconductance, must be obtained by analyzing the FET common source circuit.)

EXAMPLE 10–4

A photodiode with R of 0.6 A/W is used in the circuit of Figure 10–14, with the values as shown. The g_{mQ} for the FET is found to be 2.5 ms. The input light signal varies between 0.5 and 1.5 μW, for a peak optical signal of 1.0 μW. Find $V_{out\ p\text{-}p}$ (the peak-to-peak output voltage).

Solution

$$I_{p\ p\text{-}p} = 1.0 \times 10^{-6} \times 0.6 = 0.6\ \mu A_{p\text{-}p}$$
$$V_{in\ p\text{-}p} = 0.6 \times 10^{-6} \times 500\ k\Omega = 0.3\ V_{p\text{-}p}$$
$$V_{out\ p\text{-}p} = -g_{mQ} \times R_D \times V_{in\ p\text{-}p}$$
$$= -2.5 \times 10^{-3} \times 2.2 \times 10^3 \times 0.3 = -1.65\ V_{p\text{-}p}$$

The circuit of Figure 10–14 is suitable for ac analog operation. A digital optical signal will be distorted because the circuit cannot respond to a direct current (dc) input. (A digital input typically contains a dc component.) A simple digital detector circuit is shown in Figure 10–15. When the optical input is 0 (low logic level), the FET must be cut off. That is, V_{gs}, which is the voltage across the diode, must be more negative than V_P, the pinch-off voltage.

FIGURE 10–14 AC photoamplifier FET.

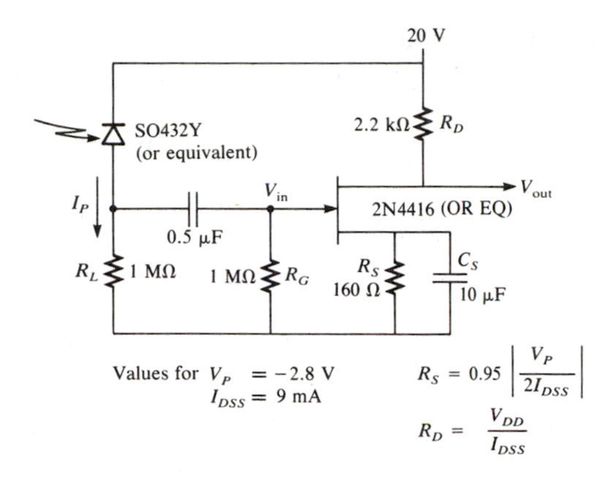

This depends directly on R_G and I_d, the dark current, because the photocurrent I_P is zero.

$$V_{gs} = -5 + (I_d + I_P) \times R_G \tag{10–15}$$

Under these conditions, V_{out} is 0 V (Q_2 is cut off). For logic 1 input, the input optical level must change V_{gs} so that the FET will conduct. The FET drain current related to V_{gs} must be large enough to turn the transistor on (saturation). V_{out} becomes approximately +5 V.

10–6 DETECTABILITY, NOISE, AND BANDWIDTH

For a light signal to be detected for further processing, it should be of the same magnitude as the internal noise of the detector, or larger. This means that **detectability** is determined by the inherent noise of the detector. This text deliberately ignores external noise because it does not depend on circuit parameters. Because the signal to the detector is expressed in terms of input

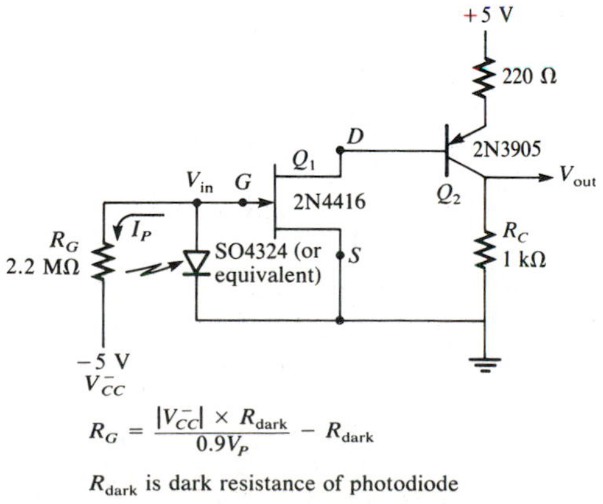

FIGURE 10–15 Digital photocircuit (FET input).

$$R_G = \frac{|V_{CC}^-| \times R_{dark}}{0.9V_P} - R_{dark}$$

R_{dark} is dark resistance of photodiode

power and noise is internal random currents, the NEP is used as the interface between noise current and input power. (See the definition of NEP in Section 10–4.)

The internal noise is expressed in terms of two internal currents. The **thermal noise** current (Johnson noise) is caused by random currents in the photodetector resistance. The **shot noise** is caused by the diode dc current. The total noise current i_n is given by

$$i_n = (i_t^2 + i_s^2)^{1/2} \quad \textbf{(10–16)}$$

where i_t is the thermal noise current and i_s is the shot noise current. The value for i_t at 27°C (300 K), room temperature, is given approximately as

$$i_t = (1.3 \times 10^{-10}) \times (\mathrm{BW}/R_{\mathrm{SH}})^{1/2} \quad \textbf{(10–17)}$$

where R_{SH} is the equivalent resistance of the diode. The i_t is proportional to the square root of the absolute temperature. It is also proportional to the square root of BW (the BW of the detector-receiver) and inversely proportional to the square root of R_{SH}, which is usually tens or hundreds of megohms.

This noise term is predominant in photovoltaic applications with no or small reverse bias. (See Figure 10–13(a).) For reverse biased diodes, with no incident light, the shot noise from I_d is the larger component. The i_s can be expressed by

$$i_s = (5.7 \times 10^{-10}) \times (I_d \times \mathrm{BW})^{1/2} \quad \textbf{(10–18)}$$

where i_s is proportional to $I_d^{1/2}$ and $\mathrm{BW}^{1/2}$.

It is useful, at this point, to refer briefly to NEP. All noise currents are related to $\mathrm{BW}^{1/2}$ (the square root of the BW), and the NEP is related directly to the noise current. As a result, NEP is related to $\mathrm{BW}^{1/2}$, and is given in terms of $\mathrm{W/Hz}^{1/2}$.

EXAMPLE 10–5

A reverse biased photodiode operates over a BW of 80 kHz. It has I_d of 50 nA, and internal resistance R_{SH} of 200 MΩ. Find (1) i_t at 27°C, (2) i_s, and (3) i_n.

Solution

1. $i_t = 1.3 \times 10^{-10} \times [(80 \times 10^3)/(200 \times 10^6)]^{1/2}$
   $= 2.6$ pA
2. $i_s = 5.7 \times 10^{-10} \times (50 \times 10^{-9} \times 80 \times 10^3)^{1/2}$
   $= 36.1$ pA
3. $i_n = (5.6^2 + 36.1^2)^{1/2}$
   $= 36.2$ pA

Example 10–5 shows how small i_t is relative to i_s when reverse bias is applied.

To find the NEP, use the relation

$$\text{NEP (at a given BW)} = i_n/R \tag{10–19}$$

where i_n is the total noise current and R is the diode responsivity. Equation 10–19 yields the NEP in terms of W/Hz$^{1/2}$, if i_n was obtained for unit bandwidth (BW = 1). The equation yields the total NEP for the particular BW involved if i_n was obtained based on the given BW and not per unit BW. The relation between NEP at a given BW, and NEP per unit BW, is

$$\text{NEP} = \text{NEP (at given BW)/Hz}^{1/2} \tag{10–20}$$

EXAMPLE 10–6

1. Find NEP for the problem of Example 10–5, given that BW is 80 kHz and R is 0.8 A/W.
2. Find NEP in W/Hz$^{1/2}$.

Solution

1. $\text{NEP (80 kHz)} = (36.2 \times 10^{-12})/0.8$
$= 45.25 \times 10^{-12}$ W
$= 45.25$ pW
2. $\text{NEP} = 45.25/\text{BW}^{1/2} = 45.25/(80{,}000)^{1/2}$
$= 0.16$ pW/Hz$^{1/2}$
$= 1.6 \times 10^{-13}$ W/Hz$^{1/2}$

This last value for NEP is close to the specified NEP for the S2506 diode (Figure 10–8). NEP is typically measured for a given BW and then translated into W/Hz$^{1/2}$, using Equation 10–20, as shown in Example 10–6.

The noise currents so far have been calculated with no incident light. That is, dark current only was used. With incident light and a specific load resistance, as in Figure 10–12, the dominant source of noise is the shot noise due to the total diode current $I_d + I_p$ (the dark current plus the photocurrent). I_P is not the signal current. It is the average photocurrent. Since I_d is usually much smaller than I_p, the shot noise is largely due to I_P and

$$i_n = i_s = (5.7 \times 10^{-10}) \times (I_p \times \text{BW})^{1/2} \tag{10–21}$$

The effects of I_d are neglected.

EXAMPLE 10–7

1. Find the NEP, given BW = 100 kHz, R = 0.6 A/W, and P_{in} varies between 0.2 and 0.4 μW, with an average of 0.3 μW.
2. Find NEP in W/Hz$^{1/2}$ for the given P_{in}.

Solution

1. First find I_p and i_{in}:

$$I_p = (0.3 \times 10^{-6})/0.6 = 0.5\ \mu A$$
$$i_n \approx i_s$$
$$= 5.7 \times 10^{-10} \times (0.5 \times 10^{-6} \times 100 \times 10^3)^{1/2}$$
$$= 127\ pA$$

NEP can be obtained from Equation 10–19:

$$NEP = (127 \times 10^{-12})/0.6 = 212 \times 10^{-12}\ W = 212\ pW$$

2. In terms of $W/Hz^{1/2}$,

$$NEP = 212 \times 10^{-12}/(100 \times 10^3)^{1/2} = 6.7 \times 10^{-13}\ W/Hz^{1/2}$$

Note that, with incident power (that is, with an optical signal present), the noise current and the NEP are substantially larger than they are without input light.

SUMMARY AND GLOSSARY

Some terms used in this chapter have been discussed in Chapter 9 in a slightly different context. The glossary given here should help you understand the distinctions between terms relating to light sources and those relating to light detectors (even when the same notation is used); and it should serve as a review of the material in this chapter.

Remember, the definitions given here are narrow in scope. They are relevant only to the material covered in this chapter.

Angular response. The relationship between the generated photocurrent and the angle of incidence. The angle is measured with respect to a line perpendicular to the sensitive area.

Avalanche multiplication. The increase in emitted electrons (generated electron-hole pairs) due to the avalanche phenomenon in the diode, given as a multiplicative factor. (A photogenerated free electron collides with an atom, producing another free electron, and so on.)

Avalanche photodiode. A photodiode that operates with a large enough reverse bias to cause operation in the avalanche region, where avalanche multiplication takes place.

Cutoff frequency. The lowest optical frequency that will produce a photocurrent. Frequencies below f_c do not have enough energy to produce a photocurrent.

Cutoff wavelength. Similar to cutoff frequency but expressed in wavelengths. Wavelengths longer than λ_c will not produce photocurrent.

Dark current. The photodiode leakage current, with no incident light.

Detectability. Input incident optical power that can be detected as an input signal.

Energy gap. The energy difference between a conduction electron and a valence electron.

Guard ring. An extra diode structure (also reverse biased) surrounding the photodiode-active area. It reduces the dark current.

Noise. See "shot noise" and "thermal noise."

Noise-equivalent power. The input power that would generate a photocurrent equal to the internal noise current.

Photoconductive mode. A reverse biased photodiode where the conduction of the diode varies with incident light.

Photocurrent. Current that results solely from incident optical power.

Photodetector. A device that converts optical energy to electrical energy.

Photodetector linearity. The deviation from a straight line in the relation between photocurrent and incident optical power.

Photogeneration. The generation by optical energy of electron-hole pairs, the electron having moved to the conduction band, leaving a mobile hole in the valence band.

Phototransistor. A transistor whose base current is produced by incident light.

Photovoltaic mode. A photodiode operating with zero bias, in which current (voltage) is generated as a result of incident light.

PIN. A photodiode with three regions: a P region, an intrinsic region, and an N region.

PIN-FET. An integrated package containing a positive-intrinsic-negative (PIN) diode at the input of a field-effect transistor (FET).

P-N. A photodiode with positive and negative regions.

Radiant sensitivity. Photocurrent divided by incident optical power.

Response time. The time it takes the photocurrent to build up to 63% of its maximum value in response to an optical step input.

Responsivity. See "radiant sensitivity."

Shot noise. Random current (noise) related to diode current, such as dark current or photocurrent.

Spectral response. The relation between generated photocurrent and the incident optical wavelength for a particular detector.

Thermal noise. Random motion of electrons (random current) related to the temperature of the device.

Window. The opening in a photodetector exposing the active area to incident light.

FORMULAS

$$E_p \geq E_g$$
$$h \times f \geq E_g \qquad \textbf{(10–1)}$$

The photon energy must be larger than the energy gap to produce photocurrent.

$$f_c = E_g/h \qquad \textbf{(10–2)}$$

The optical cutoff frequency.

$$\lambda_c = (c \times h)/E_g \qquad \textbf{(10–3)}$$

The cutoff wavelength.

$$\lambda_c = 1.24/E_g \qquad \textbf{(10–4)}$$

Equation 10–3 simplified.

$$\eta_Q = N_e/N_p \qquad \textbf{(10–5)}$$

Definition of quantum efficiency. The number of generated electrons divided by the number of incident photons.

$$R = I_p/P_i \qquad \textbf{(10–6)}$$

Definition of responsivity R. The photocurrent divided by the input optical power.

$$I_p = (N_e \times e)/\text{s} \qquad \textbf{(10–7)}$$

Definition of photocurrent. The number of generated electrons N_e times the electron charge e per second.

$$P_i = (N_p \times E_p)/\text{s} \qquad \textbf{(10–8)}$$

Incident power. The number of incident photons times the photon energy per second.

$$R = (\eta_Q \times e)/E_p \tag{10–9}$$

The expression for R in terms of η_Q. The e is the charge of the electron; E_p is in volts.

$$R = \eta_Q \times (e \times \lambda)/(h \times c) \tag{10–10}$$

R in terms of e and λ. λ is in meters, h is in Joule-seconds, and c is in meters per second.

$$R = (\eta_Q \times \lambda)/1.24 \tag{10–10a}$$

R in terms of η and λ (λ in μm).

$$V_{\text{out}} = -I_p \times R_f \tag{10–11}$$

Output voltage for Figure 10–13(a) and 10–13(b).

$$I_p = R \times P_{\text{in}} \tag{10–12}$$

The photocurrent with zero load resistance for Figures 10–13(a) and 10–13(b).

$$V_{\text{in}} = I_p \times R_L \tag{10–13}$$

Calculated input voltage for the FET circuit of Figure 10–14.

$$V_{\text{out}} = -g_{mQ} \times R_D \times V_{\text{in}} \tag{10–14}$$

Calculation of V_{out} for circuit of Figure 10–14.

$$V_{gs} = -5 + (I_d + I_P) \times R_G \tag{10–15}$$

Computation of V_{gs} in circuit of Figure 10–15.

$$i_n = (i_t^2 + i_s^2)^{1/2} \tag{10–16}$$

Total noise current: i_t is thermal noise current and i_s is shot noise current.

$$i_t = (1.3 \times 10^{-10}) \times (\text{BW}/R_{\text{SH}})^{1/2} \quad \textbf{(10–17)}$$

Thermal noise current. (R_{SH} is diode resistance, and the constant 1.3×10^{-10} is given in $\sqrt{W/Hz}$.

$$i_s = (5.7 \times 10^{-10}) \times (I_d \times \text{BW})^{1/2} \quad \textbf{(10–18)}$$

Shot noise current as a function of BW and dark current I_d, and the constant 5.7×10^{-10} is given by $\sqrt{A/Hz}$.

$$\text{NEP} = i_n/R \quad \textbf{(10–19)}$$

The noise equivalent power for a given bandwidth is the ratio of the noise current to the responsivity.

$$\text{NEP} = \text{NEP (at given BW)}/\text{Hz}^{1/2} \quad \textbf{(10–20)}$$

Relation between NEP at a given BW and NEP per unit BW.

$$i_n = i_s = (5.7 \times 10^{-10}) \times (I_p \times \text{BW})^{1/2} \quad \textbf{(10–21)}$$

Noise current with photocurrent I_p present.

QUESTIONS

1. What are the units of the input signal and the output signal in a photodetector?
2. What is meant by electron-hole photogeneration?
3. How is the band gap energy of the photodetector related to the wavelength and frequency of the detectable optical signal?
4. Define quantum efficiency of a photodetector.
5. What is meant by responsivity?
6. How is responsivity related to the optical wavelength?
7. Can a P-N photodetector operate with forward bias? Explain.
8. What was the purpose of introducing an intrinsic region (I) to the P-N structure, converting it to a PIN structure?
9. What is the operating principle of the APD?
10. How is an APD different from a P-N or a PIN?
11. Compare the responsivity of the APD, PIN, and phototransistor.
12. What is meant by noise equivalent power (NEP)?
13. If the active area of a photodiode is increased, do you expect its responsivity to increase or decrease? Explain.

14. What is the significance of the directivity of a photodetector, particularly in fiber optics?
15. What is meant by photovoltaic mode of operation? Explain.
16. What is meant by photoconductive mode of operation? Explain.
17. Can a photodetector used in analog reception be allowed to operate in saturation? Explain.
18. Can a photodiode be allowed to saturate in digital applications?
19. Why is the diode in Figure 10–13(a) said to operate in the photovoltaic mode? Explain.
20. If total NEP for a given BW and a given optical signal level is 10 pW, would you expect a 2-pW input signal to be detectable? Explain.
21. If the total NEP for a BW of 100 kHz is 2×10^{-14} W, what do you expect it to be for double the BW?
22. Do you expect a diode with I_d of 10 nA or one with I_d of 50 nA to have a lower NEP? (Assume reverse bias operation.)
23. Does the noise current increase or decrease with the application of optical power? Explain.
24. What is the dominant cause of noise in a photodiode exposed to substantial optical input?

PROBLEMS

1. A photodiode is made from indium-phosphor semiconductor material with a band gap of 1.35 V. What is the longest wavelength the diode can detect?
2. Calculate the lower frequency limit of a photodiode where E_g is 1.3 eV.
3. The responsivity of a photodiode is 0.7 A/W. Calculate the photocurrent resulting from an incident light of 0.5 μW.
4. Calculate the photon energy of a detector with $R = 0.6$ A/W (responsivity) and $\eta_Q = 70\%$ (quantum efficiency).
5. A photodiode with η_Q of 75% is operating at 0.85-μm wavelength. Calculate the responsivity.
6. Calculate η_Q for a detector with R of 0.6 A/W operating at λ of 1.5 μm.
7. Calculate V_{out} in Figure 10–12a for $R = 0.7$ A/W (responsivity), $R_L = 50$ kΩ (load), and $P_i = 800$ nW (incident power). Ignore the effect of dark current. Show the polarity of V_{out}.
8. Repeat Problem 7 for the circuit in Figure 10–12(b).
9. In Figure 10–13(a), $R = 0.8$ A/W, $R_f = 100$ kΩ, and $I_d = 50$ nA. Calculate V_{out} with P_i of 200 nw:
   a. Ignoring I_d
   b. Including I_d

   Would the circuit operate with the polarity of the photodiode reversed?
10. Calculate the incident power required to get V_{out} of -25 mV in Figure 10–13(b), if R_f is 1 MΩ. Responsivity is 0.6 μA/μW (0.6 A/W).
11. A GaAs photodiode is used in Figure 10–13(b), with R_f of 200 kΩ,

operating at λ of 1.5 μm, η_Q of 75%, and E_g of 1.43 eV. Calculate R and V_{out}, if P_{in} is 400 nW.

12. In Figure 10–14, the incident power varies from 300 to 600 nW, R is 0.6 A/W, and g_{mQ} is 4 ms. Assume C_s is a perfect short. Calculate the corresponding V_{out}. Remember that V_{out} for this FET is $-v_{gs} \times g_{mQ} \times R_D$. (*Hint:* Calculate I_p from R, then get v_{gs} from $(R_L \parallel R_G) \times I_p$.)

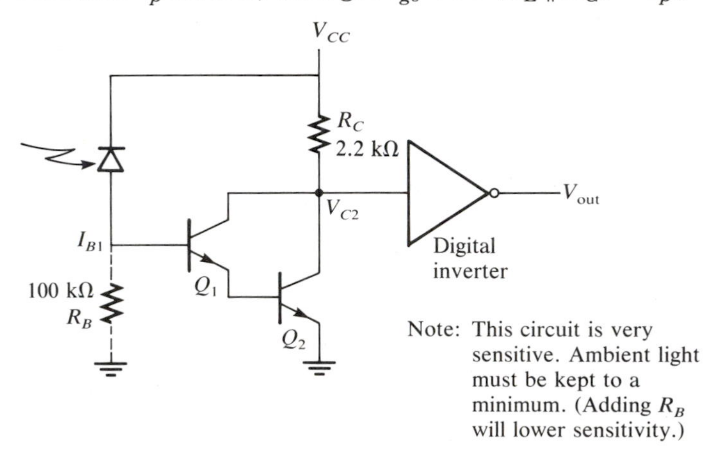

FIGURE 10–16 Figure for Problems 13 and 14: darlington digital photodetector.

13. In Figure 10–16, the responsivity of the photodiode $R = 0.6$ A/W, $\beta_1 = 100$, $\beta_2 = 100$, and $V_{CC} = 5$ V. For proper digital operation, Q_2 must be cut off (or nearly so) to produce logic 0 (about 0 V) and saturated to produce logic 1 (about 5 V). Show that an input optical signal power of 0.38 μW or higher will saturate Q_2. Ignore the loading effects of the inverter. (*Hint:* Calculate I_P for P_{in} and R. Then obtain I_C. Note that $I_P = I_{B1}$ and $I_{C2} = \beta_1 \times \beta_2 \times I_{B1}$.)
14. For the parameters of Problem 13, show that if P_{in} is less than 0.226 μW, V_{C2} is greater than 2 V (which produces logic 0 [approximately 0 V] at the output of the inverter).
15. Calculate the thermal noise current i_t for a BW of 100 kHz and R_{SH} of 100 MΩ (at 27°C).
16. Calculate the shot noise current for a BW of 200 kHz and I_d of 50 nA.
17. Find the total noise current (thermal and shot noise) for a photodetector operating at a BW of 100 kHz, with R_{SH} of 200 MΩ and I_d of 100 nA (at 27°C).
18. If the diode in Problem 17 has a responsivity of 0.65 A/W, find the NEP for the 100-kHz BW (total NEP) and in W/Hz$^{1/2}$.
19. A photodiode has an NEP of 1.5×10^{-14} W/Hz$^{1/2}$ and R of 0.6 A/W. Find the total noise current.
20. A photodiode operates with I_p of ± 15 μA, I_p average is 15 μA, I_d of 50 nA, R_{SH} of 200 MΩ, and R of 0.6 A/W. Calculate NEP in W/Hz$^{1/2}$.
21. Calculate the total NEP in Problem 20 for a BW of 200 kHz.
22. A photodiode has R of 0.7 A/W, I_d of 50 nA, and NEP of 1.5×10^{-12} W/Hz$^{1/2}$. Calculate total noise current and R_{SH}.

11
Fiber Optic Communication System

CHAPTER OBJECTIVES

This chapter covers the design choices for a fiber optic communication system. The coverage will give you the necessary tools to

- Choose among various modulation and multiplexing techniques,
- Analyze power budget and power margins,
- Analyze rise time and bandwidth (BW) characteristics,
- Evaluate effects of noise and the relation of noise to system bandwidth, data rate, and **bit error rate** (BER, the probability of an error occurring for a number of detections),
- Determine if repeaters are needed, and if so at what points along the link, and
- Use the eye pattern to evaluate system performance.

11–1 BASIC SYSTEM COMPONENTS

A generalized block diagram of a typical fiber optic communication system is shown in Figure 11–1. This general description is appropriate for analog as well as digital systems. The differences between these become apparent as more details are filled in.

Figure 11–2 is a somewhat more detailed block diagram, showing particular choices that either may or must be made in the design of a communication system. For analog inputs, you can use direct analog multi-

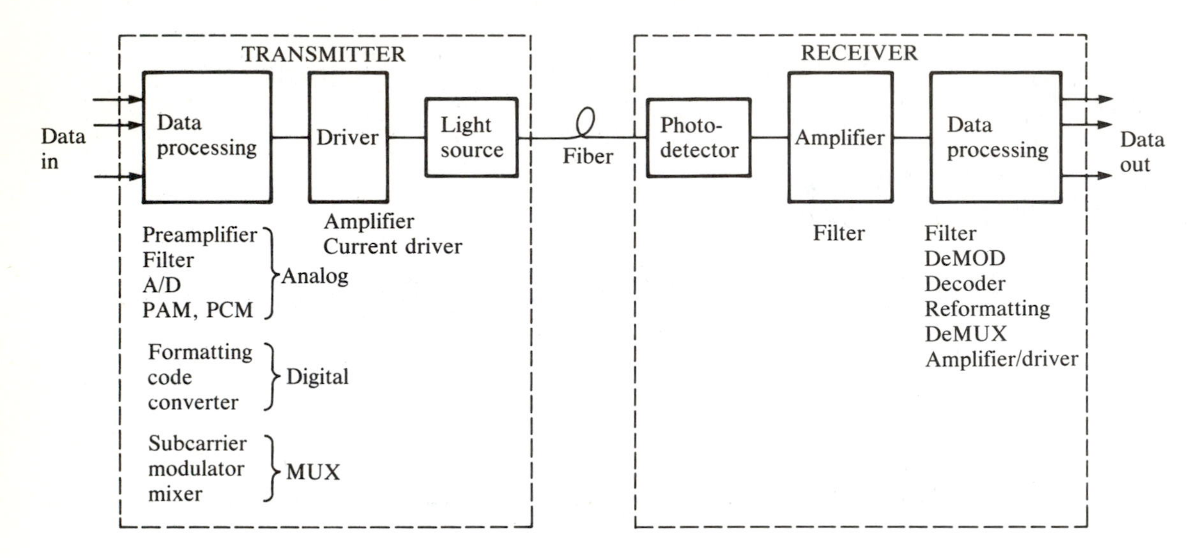

FIGURE 11–1 Block diagram of communication system.

plexing (if there is more than one channel), frequency division multiplexing (FDM), or time division multiplexing (TDM). You could also choose to convert the analog input to digital signals using pulse code modulation (PCM), pulse amplitude modulation, pulse width modulation, or pulse position modulation. Digital input may have to be code formatted; that is, it may be necessary (or at least desirable) to convert the incoming digital data into an nonreturn to zero (NRZ), Manchester, or other code format. Clearly, the multiplexer (MUX) must be compatible with the data to be transmitted: digital MUX for digital data and analog MUX for analog data.

Note that if FDM is used, you must carefully select the subcarrier frequencies, and you must also choose among different modulation schemes: frequency modulation (FM), amplitude modulation (AM), or phase modulation. (In PM, the phase of a carrier is modulated by the signal to be transmitted. PM has characteristics similar to FM.) For TDM, the switching rate must be determined, and provision must be made to synchronize the receiver demultiplexer (DeMUX) with the transmitting MUX. Note that most transmission systems are designed with some limits on the choice of techniques. For example, it is rare to find a system that uses both analog and digital TDMs. Nevertheless, the various techniques can be mixed and matched to suit particular needs. In addition, you can choose to use a light-emitting diode (LED) or a laser as the light source. A variety of fibers, such as single-mode, multimode, step-index, or graded-index fibers may be used. Some leeway in the size of the fiber is also available. (Some standardization is emerging in specific communication areas such as local area networks, computer networks, and telephone systems.)

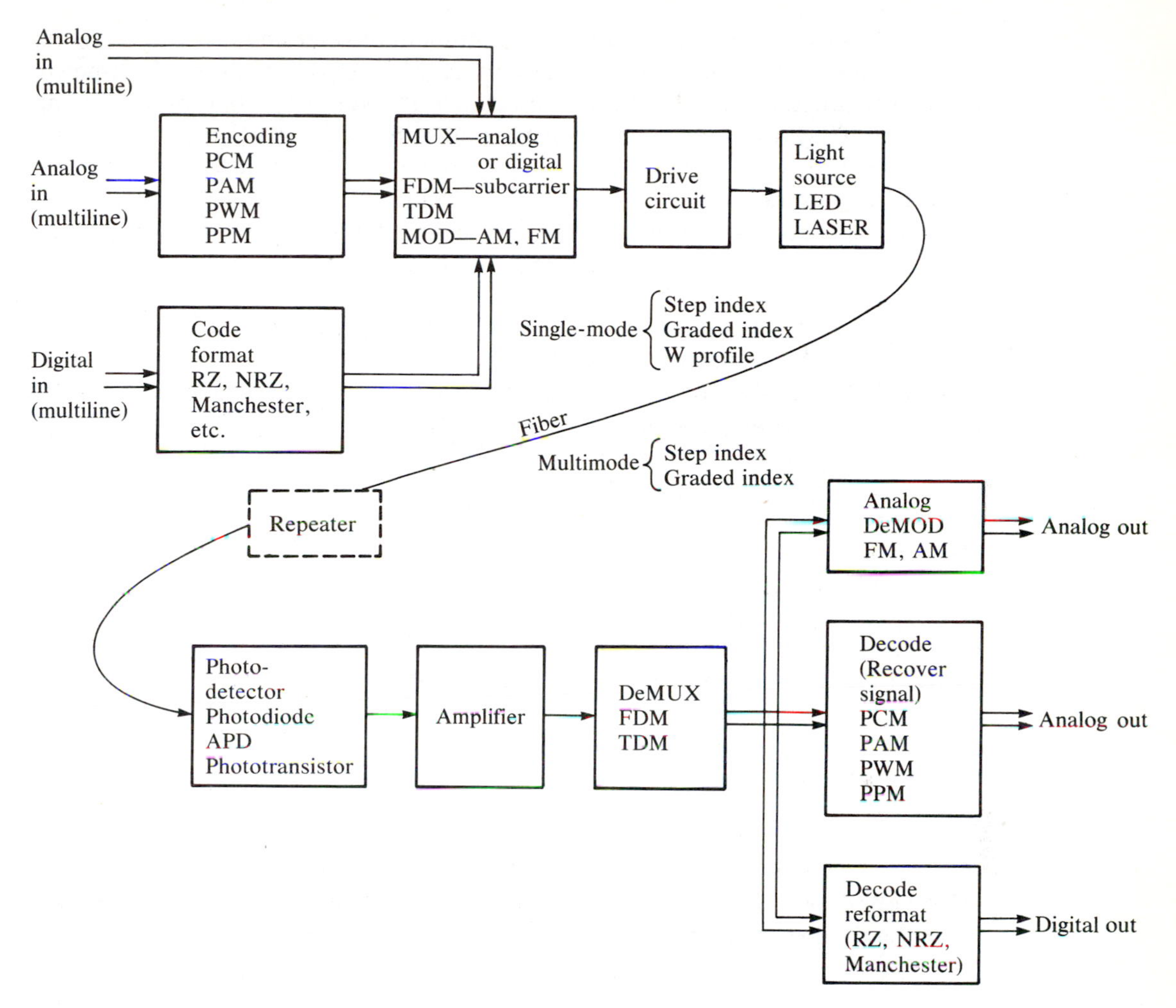

FIGURE 11–2 Communication system showing some design choices.

On the receiver side, some choices must be compatible with their counterparts on the transmitter side. The demultiplexing must recover the original channels from the multiplexed data stream, FDM, TDM, and so on. The photodetector can be chosen somewhat independently of the transmitted signal. A photodiode may be used for both LED- and laser-transmitted signals.

In addition to showing a transmitter and receiver, Figure 11–2 shows a **repeater** (dashed box). The repeater, essentially used to reamplify and reshape the signal, may or may not be necessary, depending on the distance and the type of transmitter used.

It is most important to realize that component selection, as well as choice of technique, determines the overall performance of the system. High-power laser transmitters, for example, can transmit over long distances without repeaters. They are relatively expensive and complex to use. A phototransistor at the receiving end is typically slow and will result in low capacity or in a low data rate channel and, probably, in a relatively low-cost receiver. The decisions concerning the use of splices or connectors, where necessary, or the type of couplers (*T* or star) that may be required are functions of the particular application.

A complete presentation of all design criteria and available choices will not be given here. Such a presentation would require a separate textbook. However, the design decisions are affected by two essential criteria: total power loss and the operating data rate.

The choice of fiber is directly related to the required unrepeatered (without a repeater) length, which in part determines the acceptable fiber loss per kilometer and the fiber dispersion. The total power loss is also affected by the type of connectors or splices used and the efficiency of the coupling between fiber and light source and light detector.

Typically, single-mode fibers offer the highest data rate and the lowest loss per kilometer at the highest cost. Graded-index fiber is rated second in data rate and power loss. Step-index is rated last but is the least expensive and is the easiest with which to work.

The choice of fiber is closely tied to the light source selection. A single-mode fiber is unsuitable for operation with a LED source. The LED beam angle is relatively wide, while the acceptance angle of the single-mode fiber is very small. Coupling of LED to single-mode fibers is extremely inefficient. A loss of 20 dB is to be expected. It also would be unwise to use a relatively expensive and fast laser with a multimode fiber that will severely limit the data rate and virtually eliminate the high data rate advantage of the laser.

11–2 COUPLING TO AND FROM THE FIBER

The power loss associated with the coupling from source to fiber is affected by the mismatches of two major elements: the area and the numerical aperture (N.A.). The active area of the source must always be smaller than the core area of the fiber. The loss is directly proportional to the area mismatch:

$$\text{Loss}_{\text{area}} = 10 \times \log(A_{\text{core}}/A_s) \tag{11–1}$$

For circular areas,

$$\begin{aligned}\text{Loss}_{\text{area}} &= 10 \times \log(D_{\text{core}}/D_s)^2 \\ &= 20 \times \log(D_{\text{core}}/D_s)\end{aligned} \tag{11–1a}$$

where A_{core} is the fiber core area, A_s is the active area of the source, D_{core} is the core diameter, and D_s is the source diameter. Equation 11–1 is valid only when A_{core} is less than A_s and gives the loss in decibels.

For the case where the fiber core is larger than the source area, no area mismatch loss takes place. For multimode fibers with core diameters of 50 μm and larger, area mismatch loss is small because most communication LEDs and lasers have small active areas, typically less than 50 μm in diameter.

The N.A. mismatch loss is a result of a small fiber acceptance angle and hence low N.A. in comparison with the radiation angle of the source. To clarify this effect, this chapter will examine the coupling of an unlensed LED to a single-mode fiber with N.A. of 0.1. Figure 11–3 shows how little of the source radiation is actually coupled into the fiber. For Figure 11–3, where the source is assumed to be Lambertian, the ratio of total emitted power P_{tot} to power coupled P_{coup} into the fiber is, approximately,

$$P_{coup}/P_{tot} = (\text{N.A.})^2 \qquad \textbf{(11–2)}$$

(A Lambertian source has a radiation pattern such that $I = I_0 \times \cos\theta$. I is the intensity at angle θ, and I_0 is the maximum intensity at $\theta = 0$, perpendicular to the emitting surface.) The loss in decibels is then

$$\text{Loss}_{\text{N.A.}} = 20 \times \log(\text{N.A.}) \qquad \textbf{(11–3)}$$

For the single-mode fiber with N.A. of 0.1, the coupling loss is $20 \times \log(0.1)$, or -20 dB. That is, only one-hundredth of the radiated power is coupled to the fiber.

This problem is partially corrected by using various optical methods to narrow the radiated beam. Lensed LEDs and edge emitting LEDs have narrower radiation beams (about $\pm 35°$ in comparison with $\pm 60°$ for the Lambertian source) and thus couple more efficiently. It is clear that the

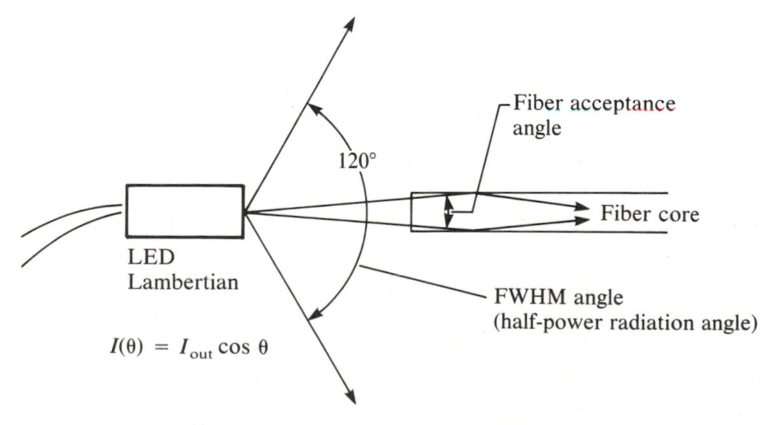

FIGURE 11–3 Coupling a LED to a single-mode fiber.

single-mode fiber should be used in conjunction with a laser, which has a very narrow beam (about 5°–10°), and not with an LED.

EXAMPLE 11–1

A step-index multimode fiber with N.A. of 0.22 and core diameter of 50 μm is coupled to an LED (Lambertian) with an active diameter of 100 μm. The total power radiated by the LED is 2 mW (3 dBm). Find the loss due to (1) area mismatch and (2) N.A. mismatch. Also, (3) find the power coupled into the fiber.

Solution

1. The area mismatch loss is

$$\begin{aligned}\text{Loss}_{\text{area}} &= 10 \times \log(A_{\text{core}}/A_s) = 10 \times \log(50^2/100^2) \\ &= 20 \times \log(1/2) = -6 \text{ dB}\end{aligned}$$

2. N.A. mismatch loss is

$$\text{Loss}_{\text{N.A.}} = 20 \times \log(0.22) = -13 \text{ dB}$$

3. The total loss is

$$\text{Loss}_{\text{tot}} = -6 - 13 = -19 \text{ dB}$$

The power coupled in dBm is

$$P_{\text{coup}} = 3 \text{ dBm} - 19 \text{ dB} = -16 \text{ dBm} = 25\ \mu\text{W}$$

The problems at the receiving end (the photodetector) are similar to those at the transmitting end. Here, the fiber (which is the effective light source) must be smaller in area than the active area of the photodetector. The acceptance angle of the photodetector must be large enough to accept all (as much as possible) of the light energy radiated by the fiber end. Typically, photodetectors have large active areas and large acceptance angles.

Reflection loss (see Chapter 1) is relatively small for both transmitting and receiving ends and is usually ignored. Eliminating or reducing the air gaps between the source and the fiber and between the fiber and the photodetector lowers the reflection loss almost to zero.

11–3 MODULATION, MULTIPLEXING, AND CODING

The previous section discussed design choices involving fiber optic technology. This section turns to some communication technology choices such as the type of modulation used, whether digital or analog multiplexing is required, and the kind of code format that is suitable. It analyzes the various communication technology choices by way of an example.

EXAMPLE 11–2

Draw a block diagram of a fiber optic communication system with the following requirements:

1. Eight digital channels (serial data), each operating at 1250 8-bit words per second.
2. Two analog channels with 5-kHz BW each. If PCM is used, each sample must be 6 bits and there must be at least 10,000 samples per second per channel.
3. Provision to allow for doubling total data transmission at a later date (eight additional digital channels and two additional analog channels).
4. Total distance between transmitter and receiver of 5 km.
5. Transmitter to use a single LED or laser (excluding expansion part).

Solution

This system is a hybrid. It contains both analog and digital channels. To accommodate both, you have two choices.

One choice is direct analog and digital transmission using frequency division multiplexing. Each subcarrier will carry a channel, analog or digital. Note that FDM is a linear system, not a pulse or digital one, even though it is used here for digital transmission.

The other choice is to convert the analog signals to digital ones, preferably PCM and to use TDM. (FDM could also be used.)

Both alternatives are shown here, the first using FDM (Figure 11–4) and the second using TDM (Figure 11–5) will be discussed. Figure 11–4 is a block diagram of the FDM system. Ten subcarriers are used in the FDM scheme, eight for the digital channels and two for the analog channels. Because FDM requires analog transmission and the subcarrier signal may not be distorted (otherwise there will be severe interchannel interference and cross-coupling), you have to use a LED, which is essentially a linear device (linear relation between current input and optical power output).

The input digital signal is code converted, that is, converted from the input code format (not specified) to the desired code, a Manchester (biphase) code, commonly used in fiber optic systems. The converted digital code is used to modulate a subcarrier (f_1 to f_8 for the eight digital channels). Here, either AM or FM can be used. Of course, the demodulators on the receiver side must match the type of modulation used. In digital FM (also called frequency shift keying), the demodulator used is often a **phase-lock loop** (a circuit that synchronizes with an incoming signal) circuit, available as an integrated circuit (IC) package.

The particular frequency allocations and BW requirements are discussed in Section 11–5. The analog channels use subcarriers f_9 and f_{10}. The modulated subcarriers are mixed to form a compound signal, which drives the LED. The

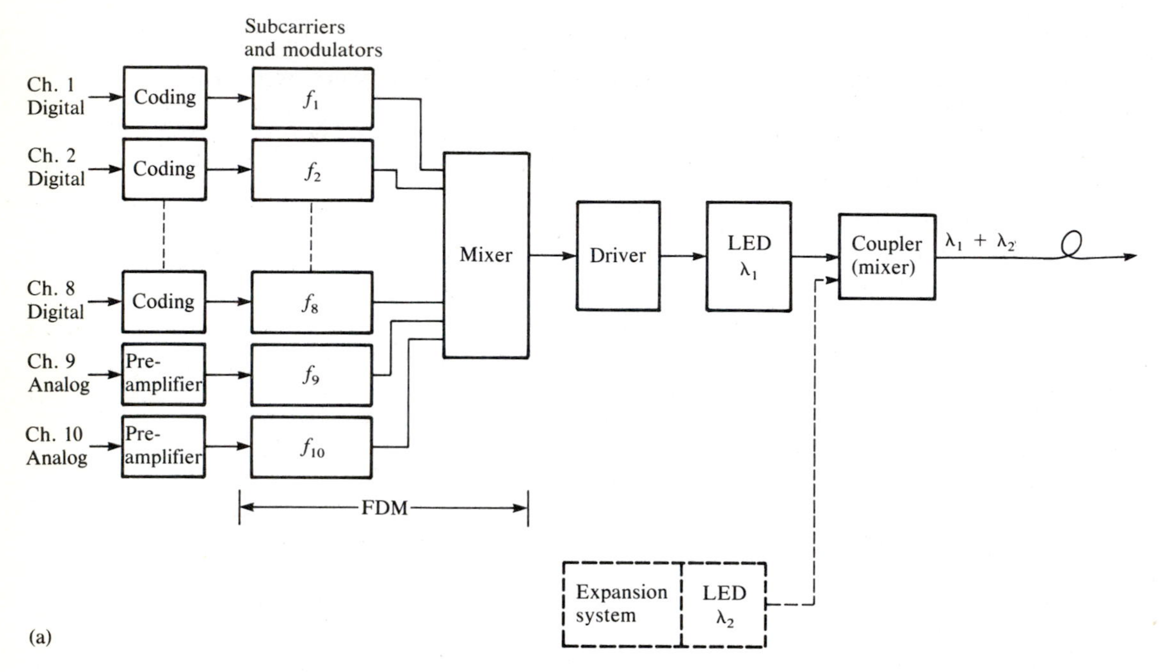

FIGURE 11–4 Frequency division multiplexing (FDM) system for Example 11–2. (a) Transmitter. (b) Receiver.

light intensity of the LED is modulated by the compound signal. This is intensity modulation.

Note that the requirement allowing a later expansion of the system is accomplished by using wavelength division multiplexing (WDM). The present system uses wavelength λ_1 (of, say, 0.82 μm), and the expansion system (identical to the existing system) uses λ_2 (of, say, 1.3 μm). The provisions for a later expansion could have been made by enlarging the mixer, allowing for an additional 10 subcarriers to be mixed. This would imply a much larger BW requirement for the single-wavelength system.

On the receiver side, provisions are made to separate λ_1 and λ_2 using a filter, such as a dichroic filter. The compound signal carried by λ_1 is photodetected. Any suitable photodetector can be used. After amplification, the signal is separated into subcarriers by 10 band pass filters. These pass a band of frequencies centered on the particular subcarrier frequency, f_1 to f_{10}. The BW of each filter is designed to accommodate the BW of the modulating signal. For the eight digital channels, each with a 16-kb/s rate, a BW of twice the bit rate would be reasonable. Similarly, each analog channel BW can be about 5 kHz. The demodulator extracts the analog or digital signals from the modulated subcarrier. Some signal reshaping and digital decoding, as required, complete the system.

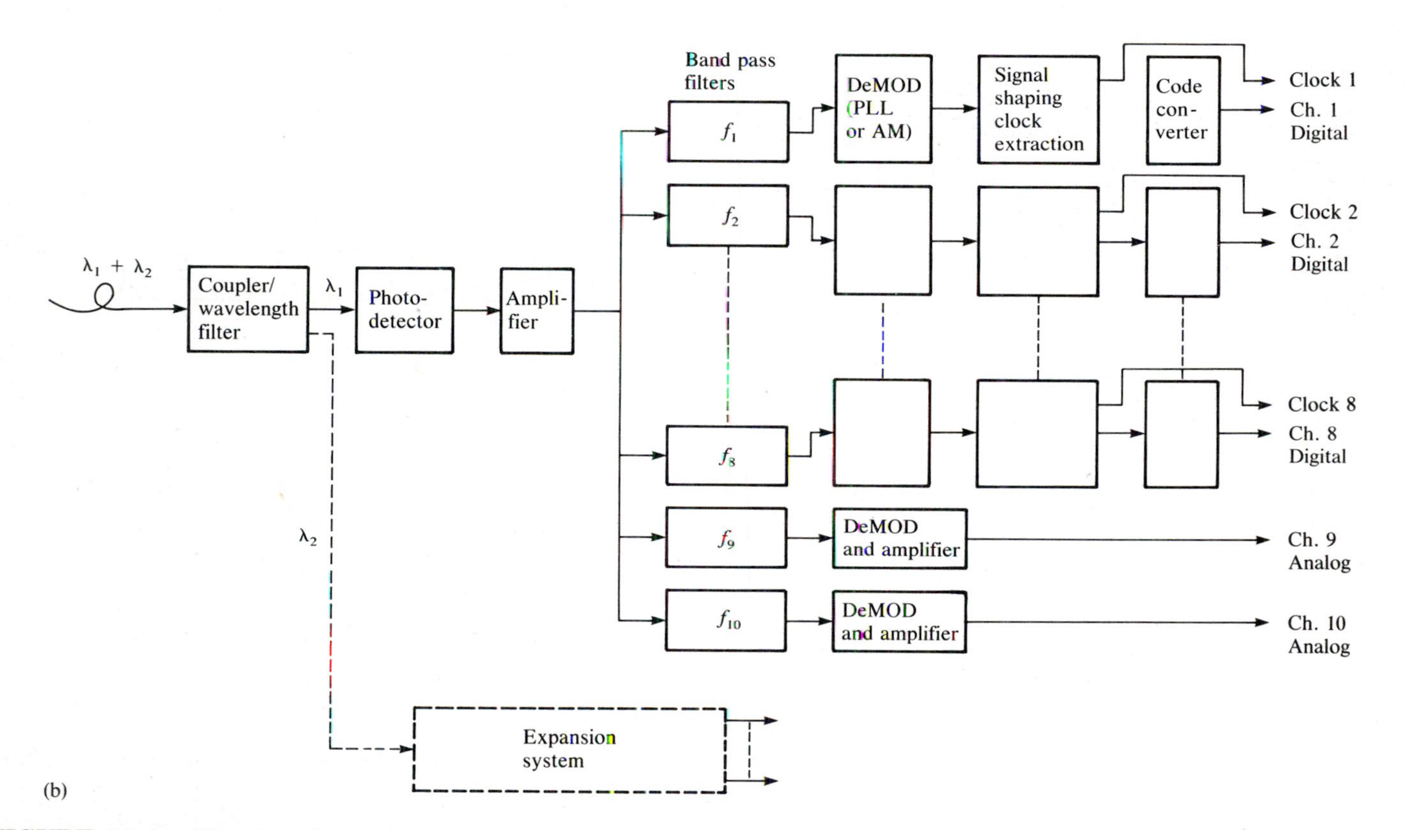

(b)

FIGURE 11–4 *(Continued)*

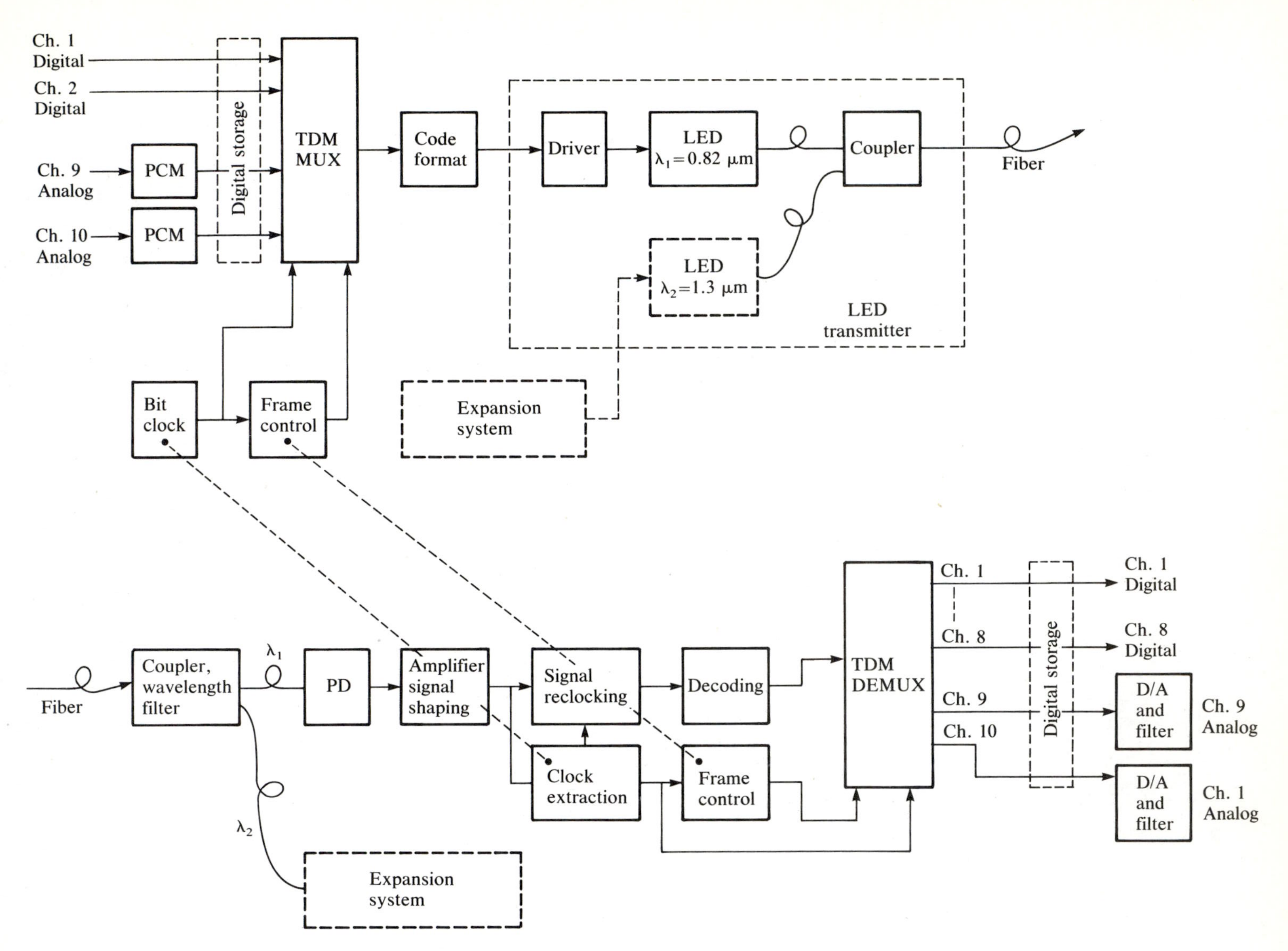

FIGURE 11–5 Time division multiplexing system for Example 11–2 (all-digital approach).

An alternative design is based on TDM, with the analog inputs converted to digital. The block diagram of such a system is shown in Figure 11–5. As with the FDM system, expansion needs are provided by using WDM (λ_1 and λ_2).

With TDM, you need synchronization between transmitter and receiver. You must make sure that when channel 1 is transmitted, the receiver MUX is set to deliver the data to the channel 1 output line only. The synchronization is accomplished by assuming a fixed frame structure and by transmitting the clock signal to the receiver. This clock transmission could be accomplished using a separate fiber or, as shown in Figure 11–5, by using a Manchester (or equivalent) code format. This format contains transitions for 1 and 0. Each bit time contains a transition. The bit clock is extracted from these transitions and is used to produce frame synchronization signals similar to those used in the transmission. The data storage at the input of the MUX allows the MUX to operate at a much higher rate than the incoming data.

In a fixed TDM system, the MUX switching rate must be the highest rate needed. Because the digital data consists of 8-bit words, choose the same word length for the PCM data. Each of the PCM channels transmits 10,000 8-bit words per second (6 bits of data and 2 bits blank) or one word every 100 μs. To accommodate 10 channels, the MUX must switch every 10 μs, yielding a **frame rate** (all 10 channels multiplexed) of 100 μs per frame, as shown in Figure 11–6. The bit rate transmitted becomes 8 bits per 10 μs, or 800 kb/s (neglecting synchronization time). Because it takes each digital input channel 800 μs to produce one 8-bit word (the bit rate of the digital channel is 10 kb/s),

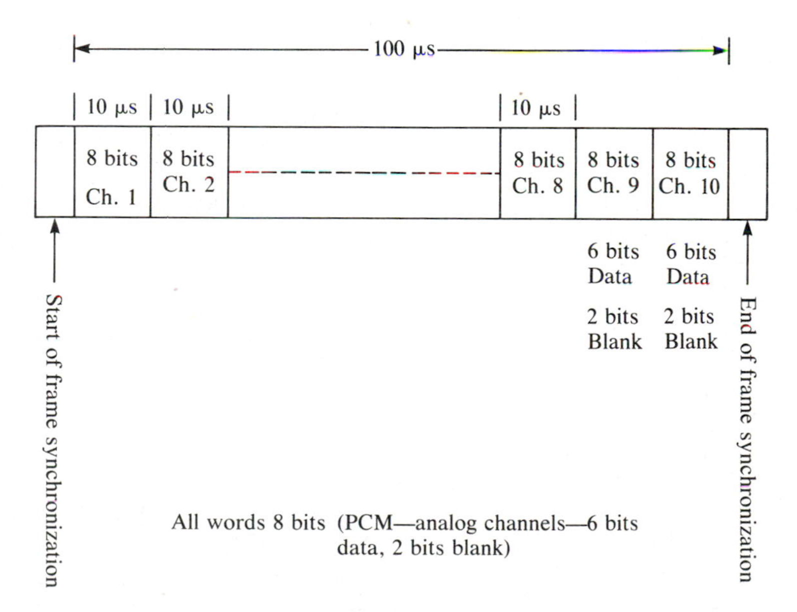

FIGURE 11–6 Frame structure for time division multiplexing solution of Example 11–2.

the 10 μs per channel is much too fast for the digital input rate. The result is that a large number of time slots will be wasted; that is, no data will be transmitted. To eliminate this problem, an **asynchronous TDM** multiplexing scheme (one in which higher data channels are visited more often) can be used. Using this method, the high-rate PCM channels are switched at a much higher rate than are the digital channels, and the transmitted rate is the sum of the individual channel rates. The transmitted bit rate becomes 240 kb/s, much lower than the 800 kb/s required for fixed TDM.

At the receiver, the DeMUX separates the incoming data stream into the individual channels. The 8-bit words of each channel (6 bits of the PCM channels) are stored to allow the receiving terminal to accept the data at a suitable rate.

The two approaches just discussed result in different BW and data rate requirements and affect the choice of optical fibers. The details of system data rate and BW considerations are discussed in Section 11–5.

Another important consideration is the required optical power (see Section 11–6). The particular length of fiber involved (5 km in Example 11–2) may cause excessive losses. It is often necessary to introduce repeaters along the fiber.

11–4 REPEATERS

Since repeaters are rarely used in analog systems, this text discusses only digital repeaters. The function of a repeater is to regenerate a deteriorated signal and retransmit it. Figure 11–7 demonstrates this function. The signal at the end of the fiber in Figure 11–7 is substantially distorted because of pulse spreading (pulse dispersion) and the power loss in the fiber. The repeater detects this signal and reshapes it into a clean, accurately timed pulse train, essentially a copy of the original pulse train. The reconstructed signal is then retransmitted with full power. The result is equivalent to reducing fiber

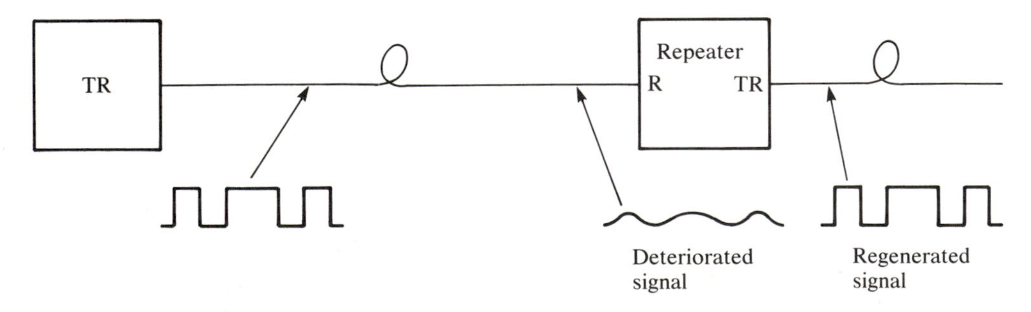

FIGURE 11–7 Typical function of repeater.

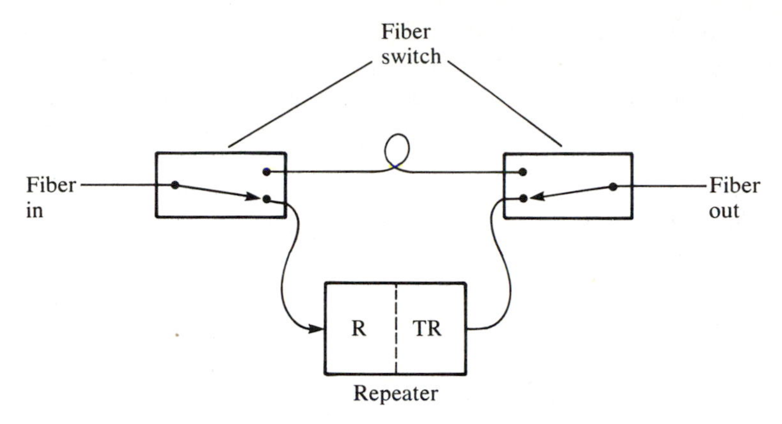

FIGURE 11–8 Repeater with bypass.

length. If a 10-km fiber has a repeater at the 5-km mark, the total dispersion and fiber loss is approximately that of a 5-km fiber, not a 10-km fiber. Note that both effective BW (data rate) and attenuation are improved.

The repeater consists of a receiver and transmitter similar to those used in transmitting and receiving terminals (see Figure 11–5). Some repeaters contain error-checking circuitry that can be used to initiate a retransmission of the original data. Most repeaters are constructed as self-contained, rugged packages, which allow the repeaters to work in remote, not easily accessible locations. The transatlantic fiber system, for example, has repeaters placed underwater. Of course, you hope that they will not need to be repaired or replaced for many years.

Some repeaters are designed with an automatic **bypass switch.** In case of repeater failure, the repeater is bridged by a fiber switch. This maintains the communication link, although only at a degraded performance level. A schematic arrangement of a bypass switch is shown in Figure 11–8. The switch is automatically actuated to bypass the repeater in case of failure. The use of bypass switches is most prevalent in local area network applications (see Section 12–3).

11–5 BANDWIDTH AND RISE TIME BUDGETS

The all-digital solution of Example 11–2 indicated a transmission data rate of 800 kb/s. This means that the rate-limiting effects of all system components, considered together, must allow operation at this rate. In other words, the rise time of the various components, such as amplifiers and LEDs, and the dispersion of the fiber must be small enough in comparison with the bit period. (Total rise time includes the effects of fiber dispersion.)

This can be stated in terms of BW requirements. Because each system component is limited in the BW that it can handle, the cumulative effect of all components should not limit the BW of the system below what is required. The 800-kb/s data rate of Example 11–2 translates into a 400-kHz BW requirement if NRZ coding is used. This means that each component must have a BW that is larger than 400 kHz, so that the combined effect will yield at least a 400-kHz BW. The rise time and bandwidth are related by

$$\mathrm{BW} = 0.35/t_r \qquad \textbf{(11–4)}$$

In practice, you can evaluate rise time requirements and translate them into BW or vice versa. Care must be taken to distinguish between optical BW and electrical BW when using Equation 11–4 (see Section 6–3–4). It is easier, particularly for digital systems, to deal with rise times and fiber dispersions rather than with BW, so you often choose to analyze system and component rise times rather than BW requirements.

11–5–1 Analog Systems

Analog systems are usually specified in terms of the required electrical BW. In the analysis or design of the system, use Equation 11–4 to obtain the required system rise time and then select system components to suit it. The relation between total system rise time and component rise time is given by

$$t_s = (t_{r1}{}^2 + t_{r2}{}^2 + \ldots)^{1/2} \qquad \textbf{(11–5)}$$

where t_s is the total system rise time and the t_{r1} and t_{r2}, . . . are the rise times associated with the various components.

To simplify matters, divide the system into five groupings:

1. Transmitting circuits t_{tc},
2. LED or laser t_L,
3. Fiber dispersion t_f,
4. Photodiode t_{ph}, and
5. Receiver circuits t_{rc}.

Hence, the system rise time can be expressed as

$$t_s = (t_{tc}{}^2 + t_L{}^2 + t_f^2 + t_{ph}{}^2 + t_{rc}{}^2)^{1/2} \qquad \textbf{(11–6)}$$

Its BW can be calculated using Equation 11–4.

EXAMPLE 11–3

A 2-km fiber with a BW × length product of 25 MHz × km (optical BW) is used in a communication system. (This BW is based on the total dispersion of the fiber, intermodal and intramodal. See Section 4–3.) The rise time of the other components are t_{tc} = 10 ns, t_L = 2 ns, t_{ph} = 3 ns, and t_{rc} = 12 ns. Calculate the electrical BW (electrical 3-dB frequency) for the system.

Solution

Because you are looking for electrical BW, first calculate the electrical BW of the 2-km fiber and then calculate its rise time. With BW × length of 25 MHz × km, a 2-km fiber has a BW of 25/2, or 12.5 MHz. The electrical BW of the 2-km fiber is given by Equation 6–4:

$$BW_{el} = 0.707 \times BW_{opt} = 8.8375 \text{ MHz}$$

The fiber rise time is

$$t_f = 0.35/(8.8375 \times 10^6) = 39.6 \text{ ns}$$

The system rise time is

$$t_s = (10^2 + 2^2 + 39.6^2 + 3^2 + 12^2)^{1/2} = 42.7 \text{ ns}$$

System BW_{el} is

$$BW_{el} = 0.35/(42.7 \times 10^{-9}) = 8.2 \text{ MHz}$$

Note that the fiber itself has the highest dispersion; t_f is 39.6 ns. The total BW is due almost completely to the fiber. (Compare fiber BW of 8.837 MHz to system BW of 8.2 MHz.) Improving fiber BW (using different fiber) is the key to a higher BW system. In analyzing and designing for a high BW, the worst component should always be improved first.

11–5–2 Digital Systems

Digital systems are analyzed on the basis of rise time rather than on BW. The system rise time is determined by the data rate and code format. Figure 11–9 shows the effect of system rise time on the transmitted pulse. System rise time represents the best rise time that you can expect. As was noted earlier, it depends on the rise time characteristics of the various system components.

In Figure 11–9(a), the system rise time is low (fast response), and there is little distortion. In Figure 11–9(b), with high system rise time (slow response), the distortion introduced affects amplitude, rise time, and fall time. Under these conditions, it becomes difficult to separate logic 1 from logic 0, and substantial **intersymbol interference** takes place; that is, the first pulse is still partially present as the following pulse is received.

Figure 11–10 demonstrates the effect of system rise time on the reception of data. At high pulse rate, the amplitude is so small that it becomes impossible to distinguish between logic 1 and logic 0. The signal received at the high rate does not cross the threshold between 0 and 1.

Because some deterioration in rise time (and fall time) is unavoidable, you must establish a criterion for an acceptable waveform. Note that an increase in pulse rise time is directly related to increases in system bit errors. (This topic is covered in more detail in the discussion of eye patterns in Section

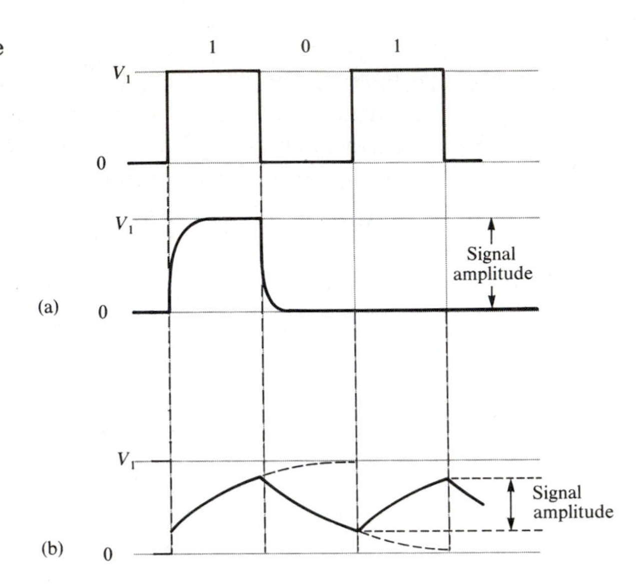

FIGURE 11–9 Effect of rise time. (a) Short rise time. (b) Long rise time.

11–7.) An acceptable criterion is to require a system rise time t_s of no more than 70% of the pulse width T_p:

$$t_s \leq 0.7 \times T_p \quad \textbf{(11–7)}$$

For a return to zero format, where T_p occupies half the bit time T (Figure 11–11), you find that

$$t_s \leq (0.7 \times T)/2 \quad \textbf{(11–8)}$$

$$t_s \leq 0.35/B_r \quad \textbf{(11–8a)}$$

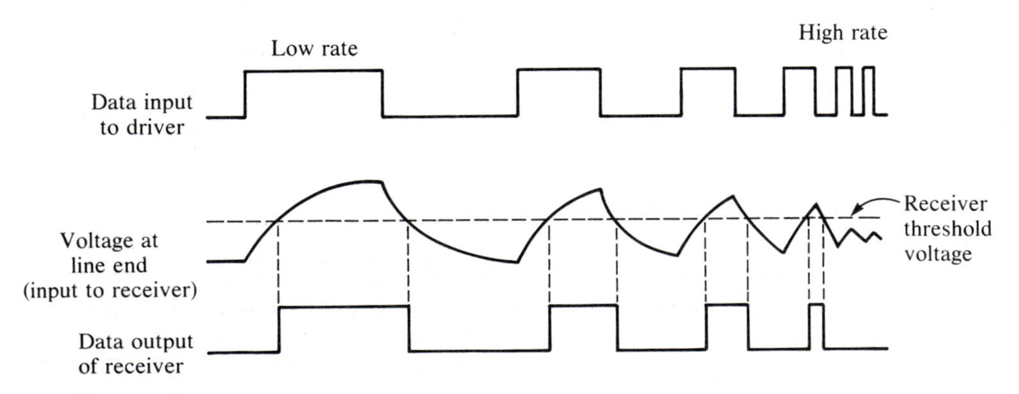

FIGURE 11–10 Distortion of data bits for varying data rates.
Source: The TTL Application Handbook, August 1973f, p. 14–7. Reprinted with permission of National Semiconductor.

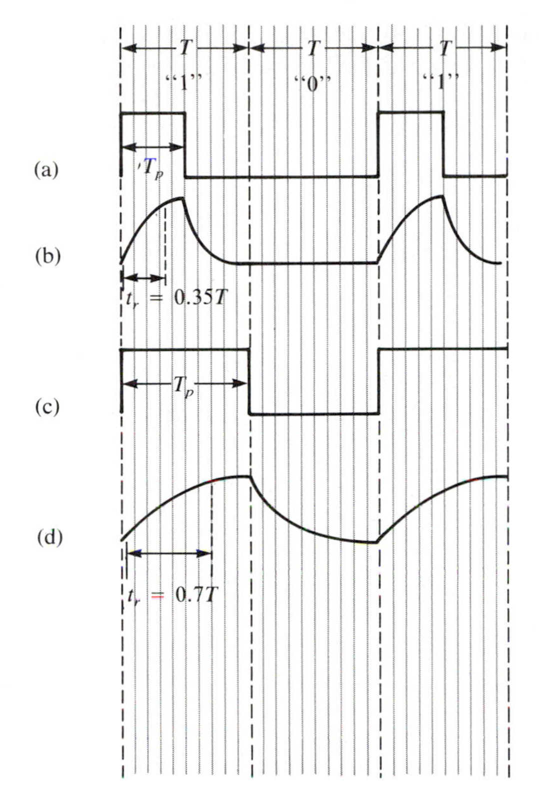

FIGURE 11–11 Effects of system rise time for return to zero (RZ) and nonreturn to zero (NRZ). (a) Transmitted RZ pulse train. (b) Received RZ signal with allowable t_r. (c) Transmitted NRZ pulse train. (d) Received NRZ signal with allowable t_r.

where $B_r = 1/T$ is the operating bit rate. For an NRZ format, where $T_p = T$, you get

$$t_s \leq 0.7/B_r \tag{11–8b}$$

For the same data rate, note that t_s for RZ must be half that for NRZ. In other words, RZ transmission requires a faster system.

EXAMPLE 11–4

A system uses components with the following specifications:

LED rise time t_L = 6 ns
Intermodal fiber dispersion = 20 ns/km
Intramodal fiber dispersion = 2 ns/km
PIN photodiode rise time t_{ph} = 8 ns

Assume that the effects of both transmit and receive circuits can be ignored. With NRZ format, find the maximum data rate (1) when a 5-km fiber is used and (2) when 500 m of fiber is used. Also, (3) find the dispersion per kilometer suitable for a 5-km link with a data rate of 50 Mb/s and a NRZ format.

Solution

1. For a 5-km length, you have dispersion as follows:

$$\text{Intermodal: } 10 \text{ ns} \times 5 = 50 \text{ ns}$$
$$\text{Intramodal: } 2 \text{ ns} \times 5 = 10 \text{ ns}$$
$$t_f = (50^2 + 10^2)^{1/2} = 51 \text{ ns}$$

   Total system rise time

$$t_s = (6^2 + 51^2 + 8^2)^{1/2} = 52 \text{ ns}$$
$$B_r = 0.7/t_s = 0.7/(52 \times 10^{-9}) = 13.46 \text{ Mb/s}$$

2. For a 500-m fiber

$$t_f = [(0.5 \times 10)^2 + (0.5 \times 2)^2]^{1/2} = 5.1 \text{ ns}$$
$$t_s = (6^2 + 5.1^2 + 8^2)^{1/2} = 11.22 \text{ ns}$$
$$B_r = 0.7/(10.22 \times 10^{-9}) = 68.5 \text{ Mb/s}$$

3.

$$t_s = 0.7/(50 \times 10^6) = 14 \text{ ns}$$
$$= (6^2 + t_f^2 + 8^2)^{1/2}$$
$$t_f = (14^2 - 6^2 - 8^2)^{1/2} = 9.8 \text{ ns}$$

On a per kilometer basis, the fiber must have a total dispersion of 9.8/5 = 1.96 ns/km. Only a single-mode fiber with no intermodal dispersion can have such a low total dispersion.

Note that in Example 11–4 part 3, using repeaters every kilometer would be enough to allow the use of the specified multimode fiber.

11–6 POWER BUDGET

The power arriving at the detector must be sufficient to allow clean detection, with few errors. This usually means that the signal power must be larger than the noise power present at the receiver. (Sources of noise are discussed in Section 11–8.) For clarity, simply assume that the power at the detector P_r must be above the threshhold level P_s. This level is called the **receiver sensitivity,** and it is related to the BER. Depending on the type of detector, and on specific transmission details, the receiver sensitivity can be as low as -45 dBm (the lower the better) for a BER of 10^{-9} (see Section 11–8). This means that the receiver can function with an input power of 31.6 nW (equivalent to -45 dBm) and have a probability of error of one in 10^9.

The power received is a function of

1. Power emanating from the light source P_L,
2. Source to fiber loss L_{sf},
3. Fiber loss which depends on fiber length L and loss per kilometer A,
4. Connector or splice losses L_{con}, and
5. Fiber to detector loss L_{fd}.

This text neglects secondary effects such as bend, waveguide, and surface losses (roughness and reflections). The allocation of power loss among system components is the **power budget.**

For reliable system operation, it is necessary to provide a **power margin.** Received power must be larger than receiver sensitivity by some margin L_m. Formulated mathematically, you have

$$L_m = P_r - P_s \tag{11–9}$$

L_m is the loss margin in dB, P_r is the receiver power, and P_s is the receiver sensitivity in dBm. With more details, Equation 11–9 becomes

$$L_m = P_t - L_{sf} - (A \times L) - L_{con} - L_{fd} - P_s \tag{11–10}$$

All units are dB or dBm.

EXAMPLE 11–5

A system has the following characteristics:

LED power = 1 mW (0 dBm)
LED to fiber loss = 10 dB
Fiber loss = 8 dB/km
Fiber length = 2 km
Total connector loss = 4.5 dB
Fiber to detector loss = 1.0 dB
Receiver sensitivity = −38 dBm

Find the loss margin.

Solution

$$L_m = 0 - 10 - (8 \times 2) - 4.5 - 1 - (-38 \text{ dBm}) = 6.5 \text{ dB}$$

The received power P_r is 6.5 dB above the receiver sensitivity.

The solution to Example 11–5 can be presented graphically, as shown in Figure 11–12. Here, assume the presence of connectors at the 0.5-, 1-, and 1.5-km points, each with 1.5-dB loss. The slope of the line represents loss per kilometer of the fiber and the sharp drops represent various **coupling losses.** The graphical representation serves to visualize the various losses. It permits easy manipulation of the various parameters to obtain a desired loss margin.

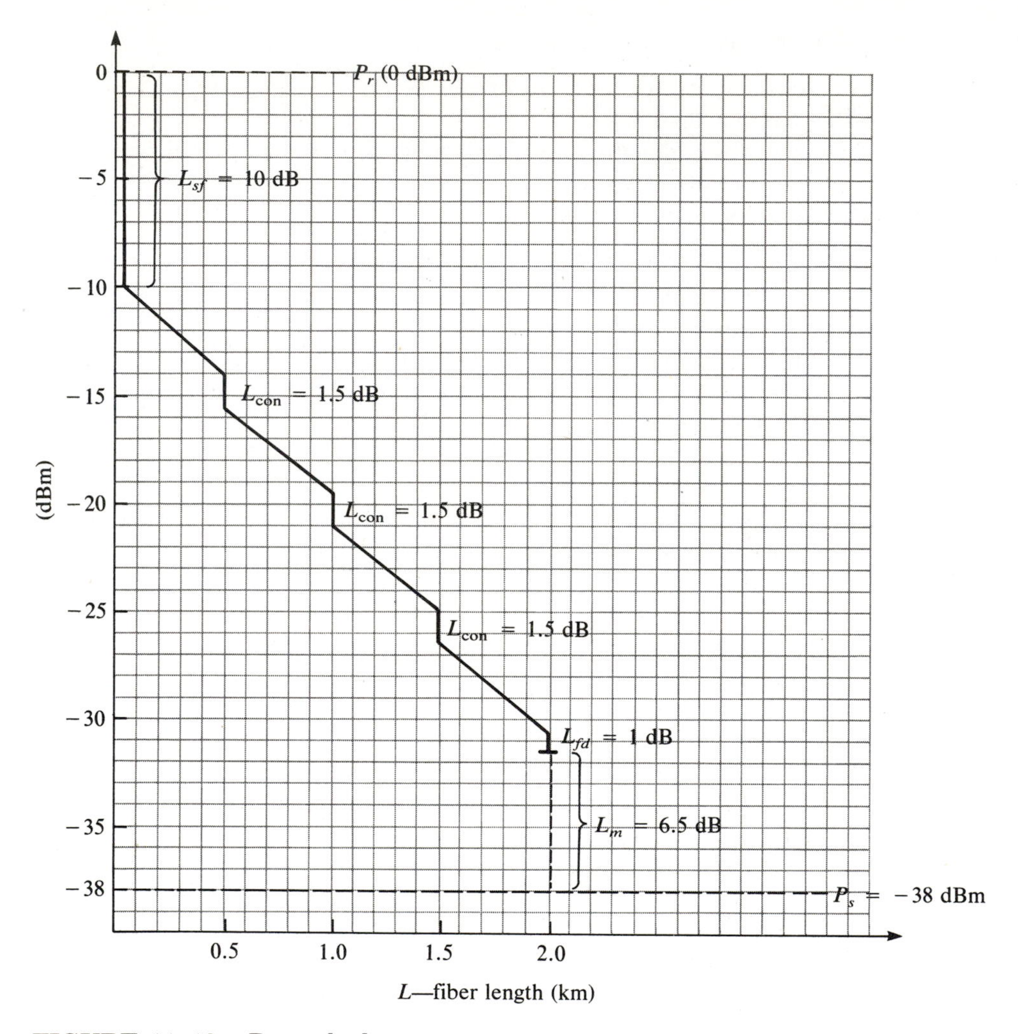

FIGURE 11–12 Power budget.

11–7 DYNAMIC RANGE

An important consideration in communication systems is the range of power variation that the system can tolerate. Input to a detector must not be too high to avoid saturation nor too low to ensure detection. Dynamic range (D.R.) of a receiver is defined as the ratio of maximum to minimum input signal power that the receiver can handle. The D.R. is usually given in decibels as

$$\text{D.R.} = 10 \times \log(P_{max}/P_{min}) \tag{11–11}$$

or

$$\text{D.R.} = P_{max}\ (\text{dBm}) - P_{min}\ (\text{dBm})$$

The larger the D.R. of a receiver, the better it is able to handle large variations in input power. These variations may be due to changes in line length, the addition of splices, couplers, or connectors, or other line losses. In systems where a number of transmit-receive terminals share a single fiber, the received power varies with the number of transmitting terminals as well.

EXAMPLE 11–6

A laser transmitter and a receiver are connected by an optical fiber with 5-dB/km loss. The fiber may vary in length from 100 m to 5 km. The laser power coupled into the fiber varies between 0.8 and 1.6 mW. Find the required D.R. of the receiver.

Solution
First convert the laser power to dBm.

$$0.8 \text{ mW} = -0.97 \text{ dBm} \qquad 1.6 \text{ mW} = 2.0 \text{ dBm}$$

P_{min} can be found by using the minimum laser power with the maximum fiber length (maximum loss):

$$P_{min} = -0.97 - (5 \text{ km} \times 5 \text{ dB/km}) = -25.97 \text{ dBm}$$

P_{max} can be found by using the maximum laser power with the shortest fiber:

$$P_{max} = 2.0 - (0.1 \text{ km} \times 5 \text{ dB/km}) = 1.5 \text{ dBm}$$
$$\text{D.R.} = 1.5 - (-25.97) = 27.47 \text{ dB}$$

11–8 NOISE, BIT ERROR RATE, AND EYE PATTERN

The receiver sensitivity is a function of the amount of electrical **noise** present at the detector. The term "electrical noise" refers to random or spurious electrical current, unrelated to signal current. The **noise current** interferes with the signal and makes detection difficult.

In analog systems, the noise distorts the signal waveform. What you consider acceptable noise depends on the strength of the signal and the extent of distortion that you are willing to tolerate. In digital systems, the noise may result in detection errors. A 1 signal may be detected as a 0 or a 0 as a 1. The larger the noise relative to the signal, the greater the probability that a detection error will occur.

11–8–1 Noise and Signal-to-Noise Ratio

There are two primary sources of electrical noise at the detector. (Noise from optical phenomena in the fiber is excluded.) **Thermal noise** results from random motion of electrons; all electrons are constantly in motion because of heat energy. Only at absolute zero temperature will this motion stop. This random motion is also present in the photodetector load resistor R_L. R_L is used to convert the detected signal current to a voltage for further processing. The random current in R_L generates a corresponding voltage, which is processed with the signal. The mean (average) of the square of this noise current i_t^2 is given by

$$i_t^2 = (4 \times k \times T \times \mathrm{BW})/R_L \quad \textbf{(11–12)}$$

where T is the absolute temperature of the system; BW is the system electrical bandwidth; R_L is the load resistor; and k is Boltzman's constant.

Shot noise is the result of light detection being a discrete process. Each generated electron-hole pair is like a small current pulse. The signal current is the average value of many such current pulses. A more careful examination of the signal shows that the pulse nature of the detection is still apparent. If the average detected current is I, there will be fluctuations above and below I related to individual (or groups of) electrons generated at particular instants. The mean square of this noise current i_s^2 is

$$i_s^2 = 2 \times q_e \times \mathrm{BW} \times I_p \quad \textbf{(11–13)}$$

q_e is the charge of the electron and I_p is the average detector current. Equations 11–12 and 11–13 show that the square of the noise current is proportional to BW. As a result, the noise current is related to $(\mathrm{BW})^{1/2}$ and is often given in $\mathrm{nA/(Hz)}^{1/2}$.

The significance of the noise is always relative to the signal. For this reason, **signal-to-noise ratio** (S/N) is defined as the signal power divided by the noise power. Because S/N is a ratio of powers, it can be given in decibels. A 20/1 S/N is equivalent to $10 \times \log(20)$, or 13 dB, signal to noise.

11–8–2 Miscellaneous Noise Sources

Although thermal and shot noise are the primary noise sources, there are a number of other sources of noise, both optical and electrical. Lasers are prone to random modulation of their intensity, which is a form of optical noise. Multimode fibers, particularly with misaligned splices and connectors, introduce optical noise because of the interference patterns created by the different propagation modes, known as **speckle noise.** Background light that falls on the detector is another source of optical noise.

The amplifier that is used to process the detected signal introduces electrical noise. The S/N at the output of the amplifier is smaller (worse) than that at the input because of the noise introduced by the amplifier.

11–8–3 Bit Error Rate

The performance of the communication system in terms of errors or signal distortion is directly related to the S/N of the receiver. A lower S/N results in more distortion for analog system and more errors for digital systems. Errors in detection, detection of 1 when 0 is transmitted, and vice versa, are given in statistical terms. The BER is defined as the probability of an error occurring for a number of detections.

For example, if you anticipate that you will have one error out of 1000 detections, the BER is 1/1000, or 10^{-3}. A BER of 10^{-9} means that the probability that an error will occur at any detection is 10^{-9} or that in 10^9 detections, you anticipate only one error. More precisely, out of a large number of detections, you expect that one out of every 10^9 will be an error.

The relation between S/N and BER is given, approximately, in Figure 11–13. In digital signals, the signal power is considered the power of the peak signal (typically logic 1). Figure 11–13 shows that a large improvement in BER is achieved by small changes in S/N for S/N above about 15 dB. To obtain a BER of 10^{-9} (a common requirement for digital systems), you need S/N of 21.6 dB. The received electrical power must be more than 100 times the noise power at the receiver.

Characteristics of receiver noise are often given in terms of optical power. The **noise-equivalent power** (NEP) is the optical signal power required to make S/N equal to 1, with a 1-Hz system BW. It is the optical signal power that produces an electrical signal equal to the noise signal. NEP is given in

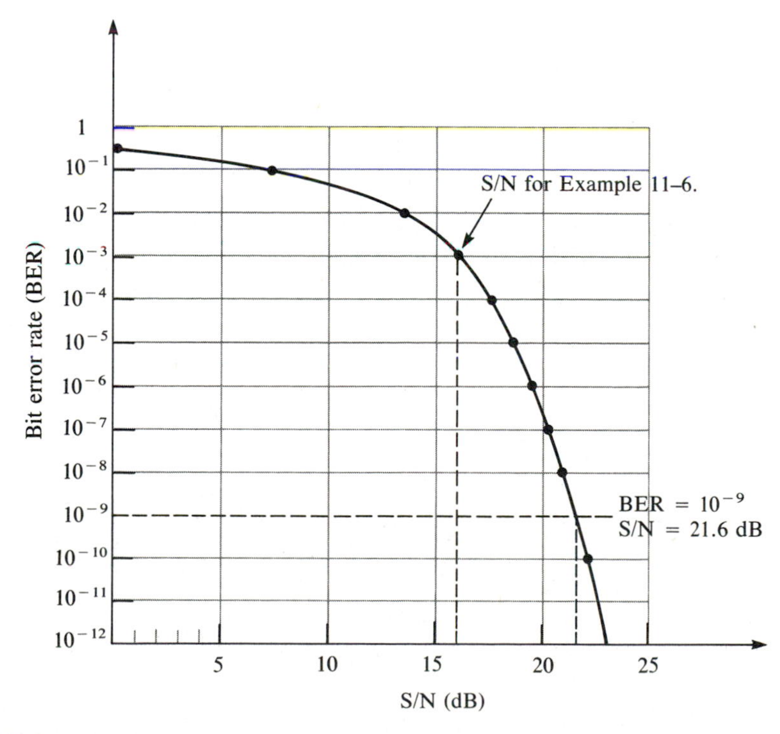

FIGURE 11–13 Relation between signal to noise ratio (S/N) (electrical) and bit error rate (BER).

W/(Hz)$^{1/2}$. For quality reception, the optical power must be many times the NEP. The **minimum detectable power** P_{min} is often referred to as the power required to produce S/N of 1 (0 dB); however, not all systems can operate properly at this signal level.

Example 11–7 demonstrates the relationships among S/N, BER, and NEP.

EXAMPLE 11–7

A simplified version of a photodetector receiver is shown in Figure 11–14. The system has the following characteristics:

BW = 1 MHz
P_s = 20 nW electrical signal power (related to P_{in})
P_n = 0.5 nW electrical noise power
R = 0.8 A/W photodiode responsivity

Find (1) S/N in decibels, (2) BER (use Figure 11–13), and (3) NEP.

Solution

1. S/N = $10 \times \log(P_s/P_n) = 10 \times \log(20/0.5) = 16$ dB
2. From Figure 11–13, a S/N of 16 dB, BER = 10^{-3}.
3. To find NEP, you have to find the optical power equivalent to the 0.5 nW of noise in the 100-kΩ load R_L. First find i_n, the noise current that yields 0.5 nW noise power:

$$P_n = i_n^2 \times R_L$$
$$i_n = (P_n/R_L)^{1/2} = [(0.5 \times 10^{-9})/10^5]^{1/2} = 0.707 \times 10^{-7} \text{ A}$$

Next, find the P_{in} required to yield this i_n. Use the responsivity to convert from current to input optical power:

$$P_{in} = (0.707 \times 10^{-7})/0.8 = 0.884 \times 10^{-7} \text{ W}$$

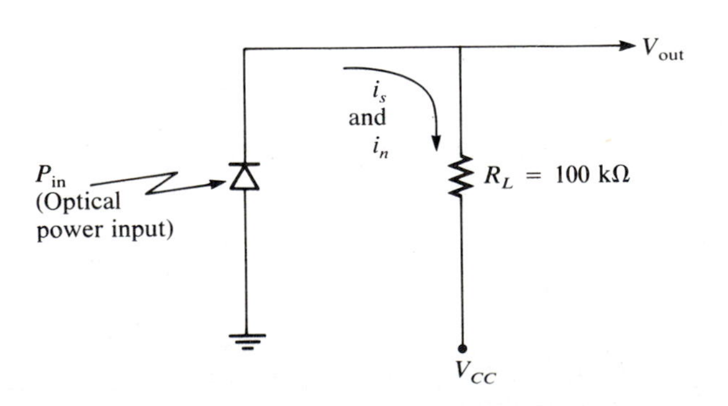

FIGURE 11–14 Circuit for Example 11–7.

This P_{in} is equivalent to the noise power for a 1-MHz BW. Per cycle, you have

$$\text{NEP} = (0.884 \times 10^{-7})/(10^6)^{1/2} = 0.884 \times 10^{-10}\ \text{W/Hz}^{1/2}$$

The BER of a receiver depends on the S/N, which in turn depends on the system BW (the data rate of the system). Figure 11–15 gives a typical relation between the received signal power and the BER for particular data rates. For example, if the received power is -36 dBm, the BER, when operating at 10 Mb/s, is about 2×10^{-10}, while at 25 Mb/s, the BER is about 3×10^{-7}. To maintain a 10^{-10} BER for a 25-Mb/s rate, the input power must be increased to approximately -35 dBm.

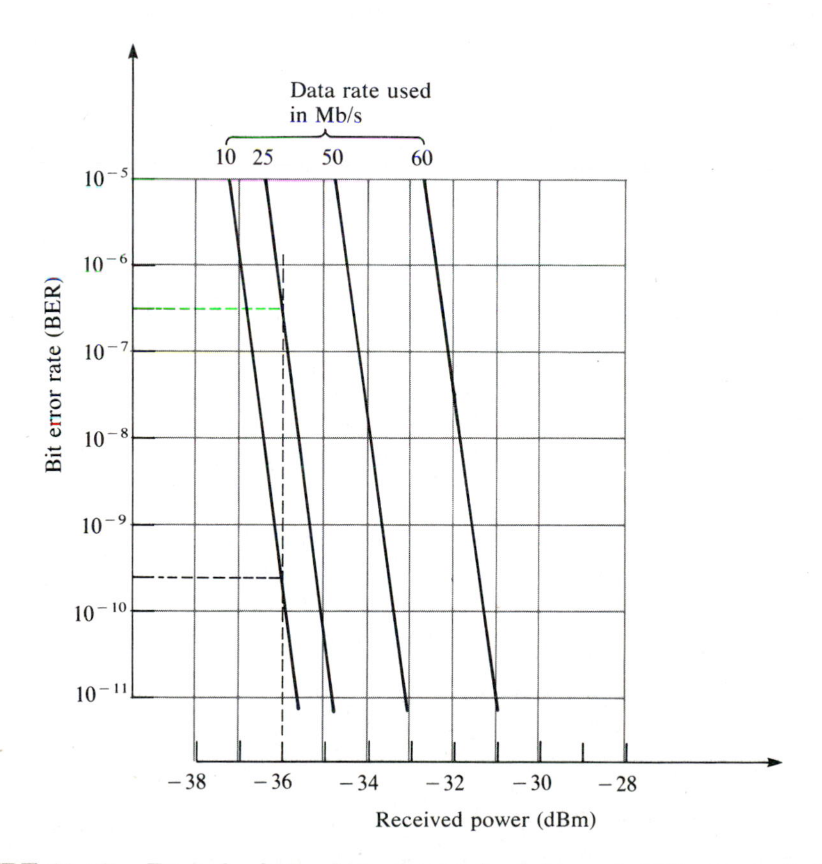

FIGURE 11–15 Typical relation between received power and bit error rate for various data rates for receiver designed for 50-Mb/s operation.
Courtesy of AMP, Inc., Harrisburg, Pa. Reprinted with permission.

11–8–4 Eye Pattern

The effects of noise and system rise time on the performance of a digital system can be visualized by the **eye pattern.**

An eye pattern is obtained when displaying a random pulse train on an oscilloscope, using fixed trigger intervals. The process is shown in Figure 11–16. The oscilloscope is triggered at fixed intervals as shown. The first and second traces are both displayed in the same time span because the trace always starts by the trigger signal. The result is the superposition of the traces, as shown in Figure 11–16(c). The waveforms have been assumed to be ideal, that is, zero rise and fall times. As a result, the eye pattern is perfectly square.

The area of the square represents the time interval during which the bit can be detected. As Figure 11–16(c) shows, the distinction between a 0 and a 1 is very clear. Figure 11–17 shows the effects of noise and deterioration in rise time and fall time. Plate 8 shows an eye pattern with the rise time about one-quarter of the bit time and no noise. Plate 7 has a rise time of one-half the bit time and no noise. Note that as the rise time increases, the eye closes. The area inside the eye pattern decreases, effectively shortening the detection interval.

Plates 6 through 1 (of Figure 11–17) show eye patterns with progressively higher noise and worse rise times. The noise is measured in time **jitter** rather

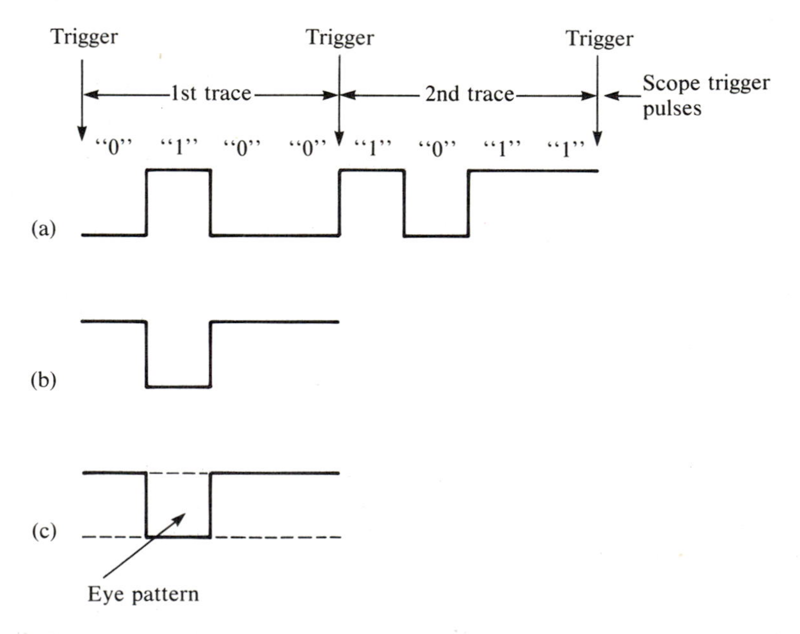

FIGURE 11–16 The eye pattern. (a) Data pulse train, nonreturn to zero format first and second traces. (b) Second trace as displayed on scope. (c) Superposition of first and second trace, resulting in the eye pattern.

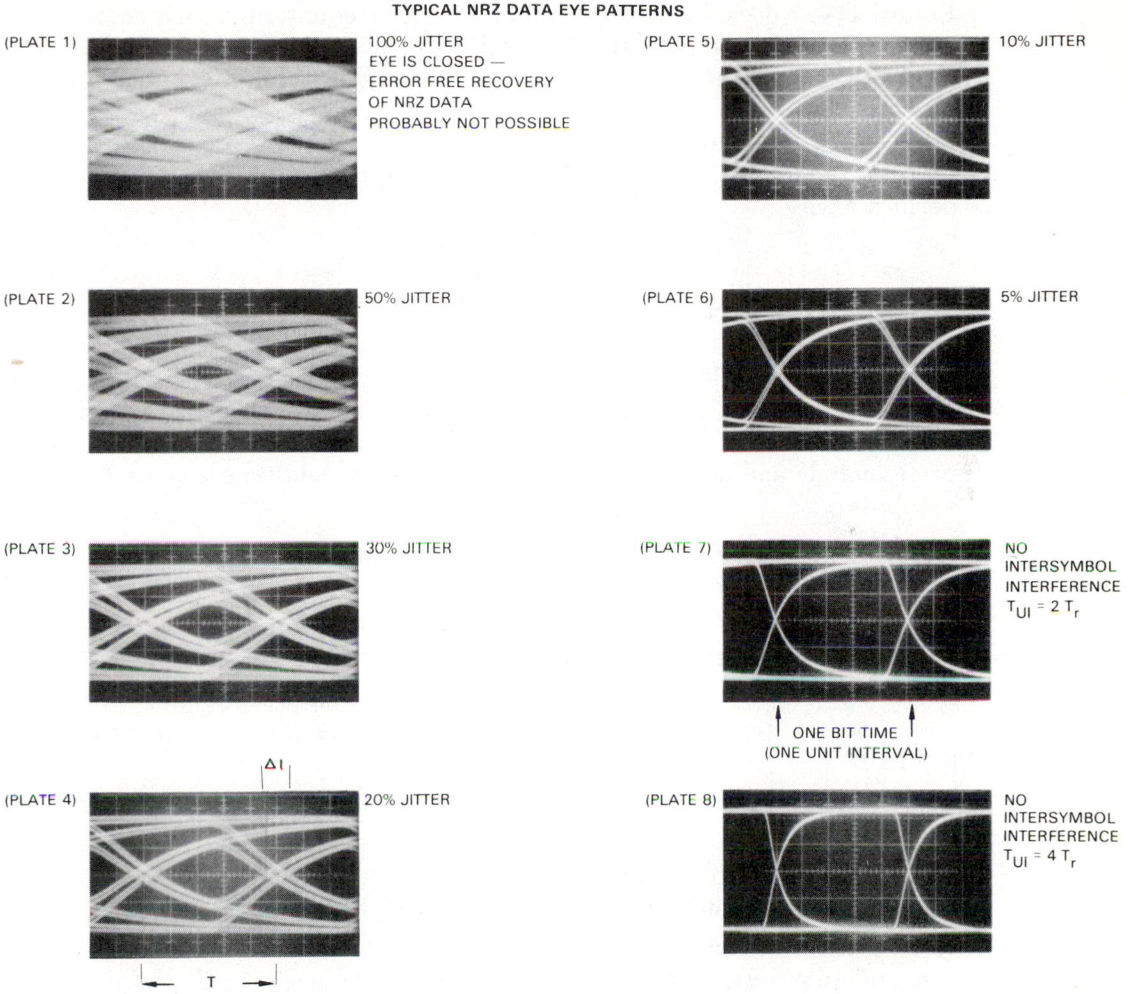

TYPICAL NRZ DATA EYE PATTERNS

(PLATE 1) 100% JITTER
EYE IS CLOSED —
ERROR FREE RECOVERY
OF NRZ DATA
PROBABLY NOT POSSIBLE

(PLATE 2) 50% JITTER

(PLATE 3) 30% JITTER

(PLATE 4) 20% JITTER
Δt
T

(PLATE 5) 10% JITTER

(PLATE 6) 5% JITTER

(PLATE 7) NO
INTERSYMBOL
INTERFERENCE
$T_{UI} = 2\ T_r$
ONE BIT TIME
(ONE UNIT INTERVAL)

(PLATE 8) NO
INTERSYMBOL
INTERFERENCE
$T_{UI} = 4\ T_r$

FIGURE 11–17 Eye patterns showing effects of noise and rise time limitations.
Source: The TTL Application Handbook, August 1973, p. 14–10. Reprinted with permission of National Semiconductor.

than in volts. The time jitter is a random, unpredictable change in bit time caused by noise. A 10% jitter, for example, means that the peak time change is 10% of bit time. You see that the combination of noise and poor rise time makes it increasingly difficult to distinguish 0 from 1. The area of the eye is practically zero in Plate 1, with 100% jitter.

The amount of jitter can be measured as shown in Plate 4 of Figure 11–17. Δt is the amount of jitter, and T is the bit time:

$$(\Delta t/T) \times 100 = \%\text{jitter} \qquad \textbf{(11–14)}$$

Figure 11–17 does not include the effects of attenuation, which reduces the amplitude of the signal. This effect will be displayed as a drop in the vertical size of the eye pattern, reducing the eye area and thus making detection more difficult.

The eye pattern test is often used to test the performance of a digital data link. It gives both a qualitative and a quantitative measure of link performance.

SUMMARY AND GLOSSARY

Most of the terms listed here are new ones introduced in this chapter. The explanations given here serve as a review of the material covered in this chapter. They also represent the technical language used by communication professionals and as such, they are an important addition to your technical vocabulary.

Asynchronous TDM. A TDM MUX scheme in which the higher data rate channels are visited more often.

Bandwidth budget. The allocation of BW limitation among the system components. (See "rise time budget.")

Bit error rate. The expected or probable error rate in a digital system. If one error is statistically expected to occur in 10^9 bits, the bit error rate is 10^{-9} (1 in 10^9).

Bypass switch. A switching arrangement that automatically bridges a failed repeater.

Coupling loss. The power loss introduced by the connection between the light source and the fiber or the fiber and the photodetector.

Eye pattern. A pattern displayed on an oscilloscope, giving visual qualitative and quantitative representation of the performance of a digital data link. (See Section 11–8–4.)

Frame rate. The reciprocal of the time it takes for a TDM MUX to complete one scan of all input channels.

Intersymbol interference. The interference of 1 bit with the next succeeding bit. (A 1 affecting the following 0.)

Jitter. Random variation in phase or bit time in a digital pulse train.

Minimum detectable power. The input signal power that produces a signal-to-noise ratio of 1.

Noise. See "noise current," "shot noise," "speckle noise," and "thermal noise."

Noise current. Random motion of electrons, not related to the signal.

Noise-equivalent power. The amount of optical power that produces a current in a photodetector of the same magnitude as the noise current.

Phase-lock loop. A circuit, available in an integrated circuit package, that synchronizes a local oscillator with an incoming signal.

Power budget. The allocation of power loss among system components such as fiber, coupling, and connectors.

Power margin. The incident power at the receiver minus the receiver sensitivity, usually in decibels.

Receiver sensitivity. The signal power, in dBm, at the receiver that results in a particular bit error rate (BER). Typically, the BER chosen is 10^{-9}.

Repeater. An integrated receiver-transmitter that is used to reamplify and reshape the digital signal.

Rise time budget. The allocation of rise time among system components. Each component contributes to system rise time (and system bandwidth). The rise time for the components determines the rise time for the system.

Self-clocking code. A digital format that contains clock information. For example, a transition during each bit time, as in the Manchester code, allows the extraction (generation) of a clock signal.

Shot noise. Noise caused by the discrete nature of electrons.

Signal-to-noise ratio. The ratio of the signal power to the noise power.

Speckle noise. Noise introduced by variations in interference patterns of the different modes that propagate in a fiber. It can occur only with highly coherent transmission.

Thermal noise. Random motion of electrons induced by temperature.

FORMULAS

$$\text{Loss}_{\text{area}} = 10 \times \log(A_{\text{core}}/A_s) \tag{11–1}$$

Coupling loss between source and fiber due to area mismatch.

$$\text{Loss}_{\text{area}} = 20 \times \log(D_{\text{core}}/D_s) \tag{11–1a}$$

Coupling loss given in terms of the source and fiber diameters.

$$P_{\text{coup}}/P_{\text{tot}} = (\text{N.A.})^2 \tag{11–2}$$

Ratio of coupled power to total power for a source, related to the N.A. of the fiber.

$$\text{Loss}_{\text{N.A.}} = 20 \times \log(\text{N.A.}) \tag{11–3}$$

Loss due to fiber N.A.

$$\text{BW} = 0.35/t_r \tag{11–4}$$

Relation between BW and rise time.

$$t_s = (t_{r1}{}^2 + t_{r2}{}^2 + \ldots)^{1/2} \tag{11–5}$$

The system rise time related to the rise times of its components.

$$t_s = (t_{tc}{}^2 + t_L{}^2 + t_f^2 + t_{ph}{}^2 + t_{rc}{}^2)^{1/2} \tag{11–6}$$

System rise time, in terms of the rise times of
- t_{tc}, *transmitter circuitry*
- t_L, *light source*
- t_f, *fiber dispersion*
- t_{ph}, *photodiode*
- t_{rc}, *receiver circuitry*

$$t_s \leq 0.7 \times T_p \tag{11–7}$$

Criterion for system rise time t_s in terms of the pulse width T_p for NRZ.

$$t_s \leq (0.7 \times T)/2 \tag{11–8}$$

System rise time t_s for RZ transmission, in terms of the bit time T.

$$t_s \leq 0.35/B_r \tag{11–8a}$$

Relation between system rise time and the bit rate B_r for RZ.

$$t_s \leq 0.7/B_r \tag{11–8b}$$

Same as Equation 11–8a, but for NRZ.

$$L_m = P_r - P_s \tag{11–9}$$

Power margin (at receiver):
- P_r, *received power*
- P_s, *receiver sensitivity*

$$L_m = P_t - L_{sf} - (A \times L) - L_{con} - L_{fd} - P_s \qquad \textbf{(11–10)}$$

Power margin in terms of component losses and receiver sensitivity.

$$\text{D.R.} = 10 \times \log(P_{max}/P_{min}) \qquad \textbf{(11–11)}$$

Dynamic range given in decibels.

$$i_t^2 = (4 \times k \times T \times \text{BW})/R_L \qquad \textbf{(11–12)}$$

The mean thermal noise current in terms of the absolute temperature T, BW, the load resistance R_L, and Boltzman's constant k.

$$i_s^2 = 2 \times q_e \times \text{BW} \times I_p) \qquad \textbf{(11–13)}$$

Shot noise related to the charge of an electron q_e, BW, and the average photocurrent I_p.

$$(\Delta t/T) \times 100 = \%\text{jitter} \qquad \textbf{(11–14)}$$

The ratio of the change in bit time to the bit time as a measure of the amount of jitter.

QUESTIONS

1. What is the function of a repeater? When is it necessary?
2. Compare multimode step-index, multimode graded-index, and single-mode fibers, in terms of cost and data rate.
3. A LED is unsuitable for use with single-mode fibers and a laser is not recommended for use with multimode fibers. Why?
4. What are the basic loss mechanisms in coupling from source to fiber?
5. Why do you expect a lensed LED to be more efficient than an unlensed LED?
6. What is the purpose of multiplexing?
7. What other approaches to the design of the system in Example 11–2 can you suggest?
8. Manchester code is self-clocking. What does this mean?
9. Can a laser be used in the FDM approach to the solution of Example 11–2? Give your reasons.
10. What is a bypass switch?

11. Give the general relation among system BW, the data rate, and rise time.
12. To improve system rise time, you first try to improve the weakest link. What does this mean?
13. With too large a rise time (for a particular data rate), intersymbol interference may occur. What does this mean?
14. What five factors affect the received power?
15. What is meant by receiver sensitivity?
16. What are the dominant sources of noise in a fiber optic communication system?
17. What is the S/N?
18. What is BER?
19. How is the BER related to the NEP of the receiver?
20. How is the BER related to the system BW?
21. If the power margin is reduced because of additional line losses, what will happen to the BER? Explain.
22. What is the eye pattern?
23. What is the significance of the size of the eye opening?
24. What are the factors that affect the size of the eye?
25. How can percent jitter be measured using the eye pattern?

PROBLEMS

1. The active area of the NDL4103P LED (a NEC Corporation product) is about 960 μm^2. What is the smallest core diameter that can be used to keep area mismatch loss below 1 dB?
2. Find the loss from area mismatch when the fiber in Problem 1 is coupled to a single-mode fiber with core diameter of 6 μm.
3. Find the area mismatch loss when a 60-μm-diameter LED is coupled to a 50-μm core.
4. A LED (unlensed) is used with a fiber with N.A. of 0.2. Find the coupling loss. (The source is Lambertian.)
5. Find the coupling loss from both N.A. and area mismatch when a Lambertian source with active diameter of 60 μm is coupled to a fiber with core diameter of 50 μm and N.A. of 0.22.
6. For Example 11–2, draw a block diagram using multiplexing at the optical level. Show all couplers.
7. Find the bit rate for a PCM system for a 4-kHz signal with three samples per cycle and 8 bits per sample.
8. A TDM MUX is used to multiplex 16 channels. Each channel operates at 10,000 characters per second, with each character consisting of 8 bits. The MUX output transmits one character at a time. Assume the input data can be stored before transmission. Find

a. The channel switching rate (input channels per second)
b. The output data rate in bits per second
c. The output data rate in characters per second

9. An FDM MUX is used to multiplex eight analog channels, each with a BW of 5 kHz. If 20% of the total BW is devoted to guard bands (unused frequency bands between the subcarriers),
   a. Find the total BW.
   b. Assume that the lowest subcarrier is 1 MHz. Give the frequency of all subsequent subcarriers.
10. What is the system BW with LED rise time = 15 ns, fiber dispersion = 30 ns, and photodiode rise time = 10 ns? (Ignore all other components.)
11. A 10-km fiber with a specified 50-MHz × km bandwidth (optical BW) is used with t_L of 20 ns and t_{ph} of 15 ns. (Ignore all other components.) Find
   a. System BW
   b. System BW if a perfect repeater is inserted at the 5-km mark
12. Find the digital data rate (NRZ format) for a system using 2 km of fiber, with

    $t_L = 12$ ns
    $t_f = 24$ ns
    $t_{ph} = 10$ ns
    $t_{tc} = 5$ ns
    $t_{rc} = 5$ ns

13. Repeat Problem 12 for RZ format.
14. Find the fiber dispersion per kilometer that will allow the system in Problem 12 to operate at 30 Mb/s
15. A system has a BER of 10^{-10}. How many bit errors do you expect in a transmission of 10^{11} bits?
16. A fiber system has a LED with P_t of 2 mW (total radiated power) and

    Coupling to fiber loss = 17 dB
    Fiber loss = 5 dB/km
    Connector loss = 0.6 dB per connector
    Coupling to photodiode loss = 1.2 dB
    Receiver sensitivity = −38 dBm

   a. If a 3-km length of fiber with three connectors is used in a system, what is the power margin? Draw the power budget diagram. Assume connectors are located at the LED, the 1-km mark, and at the photodiode.
   b. Assume that Figure 11–15 is applicable. Obtain the BER when operating at 60 Mb/s.
17. With the same parameters as those in Problem 16, what is the maximum fiber length that can be used? (Receiver power must be at least −38 dBm.)
18. A system has total noise power of 10^{-10} W and signal power of 10 nW. Use Figure 11–13 to find the BER.
19. The total noise current in a photodetector is 1.8×10^{-13} A/(Hz)$^{1/2}$ (model

FND–100, made by EG&G). The responsivity is 0.62 A/W, and the operating BW is 1 MHz.
   a. Find the NEP.
   b. Find the total noise power at the 1-MHz BW.
   c. With a signal power at the receiver of 100 nW at the 1-MHz BW, find the S/N.

20. Repeat Problem 19 with a BW of 10 kHz.

21. A photodiode has an NEP of 10^{-13} W/(Hz)$^{1/2}$ with a signal power of 18 nW. Find the S/N in decibels with a BW of
   a. 1 MHz
   b. 10 MHz
   Also, estimate BER for step a, using Figure 11–13, and BER for a 10-MHz BW, using Figure 11–13.

12 Applications

CHAPTER OBJECTIVES

This chapter presents some applications of fiber optic technology. Two important applications are long-haul communication such as telephone intercity trunk lines, and local area networks, used to connect relatively close computer stations. The important features of these systems are discussed to give you an appreciation of the similarities and differences between the two.

The discussion of community antenna television (CATV), closed circuit television (CCTV), and local telephone loop are intended to give you a broader view of fiber optic communication.

The use of fiber optic technology in measuring systems is keyed to the development of fiber optic sensors. You will learn the principles of fiber optic sensors and some particular fiber optic sensor devices.

The discussion of medical applications, which includes specific devices and methods, will further broaden your view of this technology.

Sections 12–7 and 12–8 cover coherent optical communication and integrated fiber optics and electro-optics, respectively, as future technology. You will see a part of what the technology is expected to achieve in the future.

12–1 INTRODUCTION

Fiber optic technology finds applications in a large variety of areas. It is prominent in communication. It is used in medicine, both as a diagnostic tool and in medical procedures. It is used as a sensor in industrial applications and as a source of light in various consumer-oriented areas. This chapter covers just a few of the applications of fiber optics and emphasizes communication, both long- and short-haul, because these are the mainstays of fiber optics.

12–2 LONG-HAUL COMMUNICATION

One of the most effective motivations for developing fiber optic communication is its application to **long-haul communication.** Typically, long-haul systems are used for long distances, 10 km and more, in **point-to-point communication** with very high data rates (see Figure 12–1). The point-to-point feature implies a direct fiber route with no intervening user stations. A telephone connection between New York and Los Angeles, for example, is a typical long-haul application for fiber optics. Because both cities are large, the connection is likely to carry a large number of conversations (or computer data channels) and hence to have high data density. The fiber optic system offers the required high data rate at economical prices and with high reliability.

Typically, the point-to-point long-haul communication system used in telephone trunk lines requires the following features:

1. *Low losses*. Because a long distance is involved, it is important to minimize the number of repeaters necessary. A low-loss fiber will allow repeaters to be spaced as much as 40 km apart. This spacing reduces maintenance and installation costs.
2. *High data capacity*. High data capacity will allow a large number of conversations or digital channels to use a single fiber. This is highly cost effective, both in terms of installation and maintenance. The high capacity implies the use of very low dispersion fiber (single mode) and a very narrow line width source, probably a single-frequency laser.
3. *Long continuous fiber*. Manufacturing fiber is a complex process, and producing a continuous length of fiber requires special care. Typically, the unbroken length is about 5 km. A long continuous fiber requires fewer splices and hence, keeps power loss to a minimum.
4. *Structural reliability*. Because the cable is unattended, it must withstand harsh environmental conditions. It should also be structurally sound to allow easy field handling and installation.
5. *Field repairability*. The cable and other components must be field repairable. The cable cannot be brought to a laboratory for repairs.

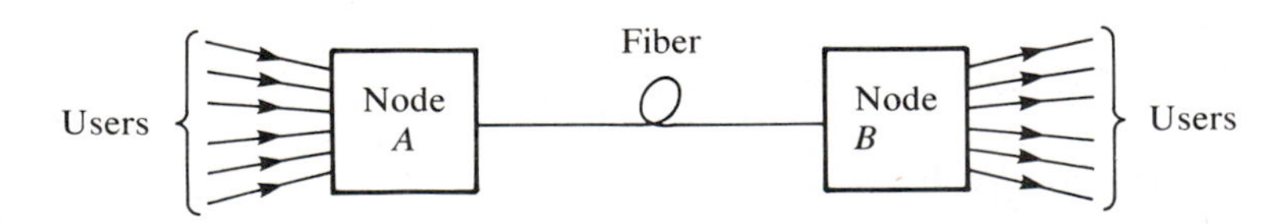

FIGURE 12–1 Point-to-point link.

6. *Security and privacy.* The cable must provide secure and private transmission. Secure data transmission means that it is difficult to tap illegally into the line.

Long-haul fiber optic systems have been constructed with nearly all of the features listed. The single mode fiber operating at 1.3 and 1.5 μm has been used at 560 Mb/s over unrepeatered distances of more than 50 km. (A system installed in Taiwan by the AT&T Corporation in 1988 operates at 417 Mb/s over an unrepeatered length of 104 km. Most of the 104 km—87 km—are installed underwater.) Compare this with the 1.5-km spacing of a typical coaxial installation. Fiber splicing techniques, with splice losses of 0.1–0.2 dB, have been developed that allow field splicing of a multifiber cable.

Repeaters in long-haul systems are often installed in remote areas (e.g., underwater in transoceanic installations), and it is necessary to supply electrical power to the repeater. This is accomplished by including copper wires as part of the fiber optic cable. Typically, it is these copper wires that deteriorate first, particularly where the cable is slightly damaged and the wires and fibers are exposed to the environment.

Most telephone systems use pulse code modulation (PCM), operating at a variety of standardized rates. In the United States, the standard rates are based on a 64-kb/s rate per voice channel. Table 12–1 gives the designation for each data rate, the number of voice channels accommodated (using PCM in a time division multiplexing transmission) and the standard rates. Although the rates are based on the 64-kb/s basic voice channel rate, they are not exactly multiples of 64 kb/s. Word and frame synchronization bits affect the total data rate. Higher rates, such as 500 Mb/s and above, have recently been used, as well.

The European standards vary slightly from those used in the United States. These are given in Table 12–2.

A rather sophisticated long-haul **network** (a communication line intended to serve many stations, or nodes) which is not point to point was installed in July 1988 under the auspices of the National Science Foundation. It is called NSFNET. It links six supercomputer centers throughout the United

TABLE 12–1
U.S. Telephone Transmission

Designation	No. of Voice Channels	Data Rate (Mb/s)
T1	24	1.544
T1C	48 (2 × T1)	3.152
T2	96 (4 × T1)	6.312
T3	672 (28 × T1)	44.736
T3C	1344 (56 × T1)	91.053
T4	4032 (168 × T1)	274.176

TABLE 12–2
European Telephone Pulse Code Modulation Rates

No. of Channels	Data Rate (Mb/s)
30	2.048
120	8.448
480	34.368
1920	139.364
7680	565.000

States. It is also tied in with seven regional academic computer networks that serve hundreds of users in the major regions of the country.

12–3 LOCAL AREA NETWORKS

The **local area network** (LAN) is a computer-oriented communication system. In contrast to the long-haul systems such as telephone trunk lines, the LAN operates over short distances such as 1 or 2 km. It is multiuser oriented; many devices communicate through the LAN. The following are applications of the LAN:

1. Communication between computers, allowing each computer to use data and programs from any other.
2. Computer-to-peripheral communication, replacing copper wires connecting the computer to a printer, monitor, work station, or other device.
3. Connection of a number of users to a single (usually expensive) peripheral device, for example, a sophisticated laser printer or a plotter. (In this way, several computers can share the peripheral device.)

All these uses require multiple access to the optical fiber. Typically, couplers, star or T, are used to connect the various communicating stations.

The LAN system must have a number of features that were not needed in the long-haul system and can do without other features that are important for the long haul system.

Dispersion (ns/km) and loss (dB/km) are not of major importance of the LAN system. This does not imply that it operates at lower data rates. Because the data rate of a fiber is inversely proportional to its length, short fiber length can operate at high data rates, even with relatively high dispersion per kilometer. Total fiber loss is directly related to length; short fibers have low losses.

Long continuous fiber, easy field repairability, and security and privacy are only of secondary importance for the LAN. Structural reliability is easy to guarantee because the cable is usually installed in a protected area and can easily be maintained. One of the most important features of the LAN is the ease with which stations can be added to the system. This refers to the methods used to splice in or connect additional computer stations to the existing network.

There are numerous ways of interconnecting computers (or any communicating nodes).

The topology of the **ring network** is shown in Figure 12–2. Here, the data pass through all nodes, and only the node for which the data are intended use those data. The bidirectional ring (Figure 12–2(b)) provides an extra communication path to increase reliability. The failure of a repeater or a stretch of fiber between repeaters can be compensated for by reconfiguring the network, as shown in Figure 12–2(c). With the reconfiguration, all operating nodes can still communicate with all others, using a part of the information being sent (the message) to indicate the intended node (address) and using a number of timing and control bits. The nodes in the ring are often provided with **bypass** switches to allow the data to bypass a failed node.

Figure 12–3 shows details of a coupler that may be used in a ring structure. This structure, often called a passive ring, does not need bypass provisions because a failed node will not interfere with the data flow. The coupler, however, introduces a relatively large loss (typically, 3 dB) into the data path.

Another topology used is the **star network** structure. Figure 12–4 shows a star topology using a transmissive star coupler (Figure 12–4(a)) and a reflective star coupler (Figure 12–4(b)). With the star coupler, the signal from any input is shared with all outputs, allowing any node to communicate with all other nodes.

Note that the process of adding a new node to the network is somewhat more complicated for the ring network than for the star network. For the latter, you can use an unused input/output line, if available, or replace the coupler with a larger one. For the ring structure, you must splice in a new node and make sure the node is programmed appropriately to receive and to transmit data on the network.

A **bus** configuration is shown in Figure 12–5. In this configuration the T couplers must allow communication to and from (two-way communication) the bus fiber.

12–4 OTHER COMMUNICATION SYSTEMS

Fiber optic communication has been used in **community antenna television,** residential intercoms, industrial communication and control, local subscriber loops, and more. In the CATV system, a common antenna is used to receive the

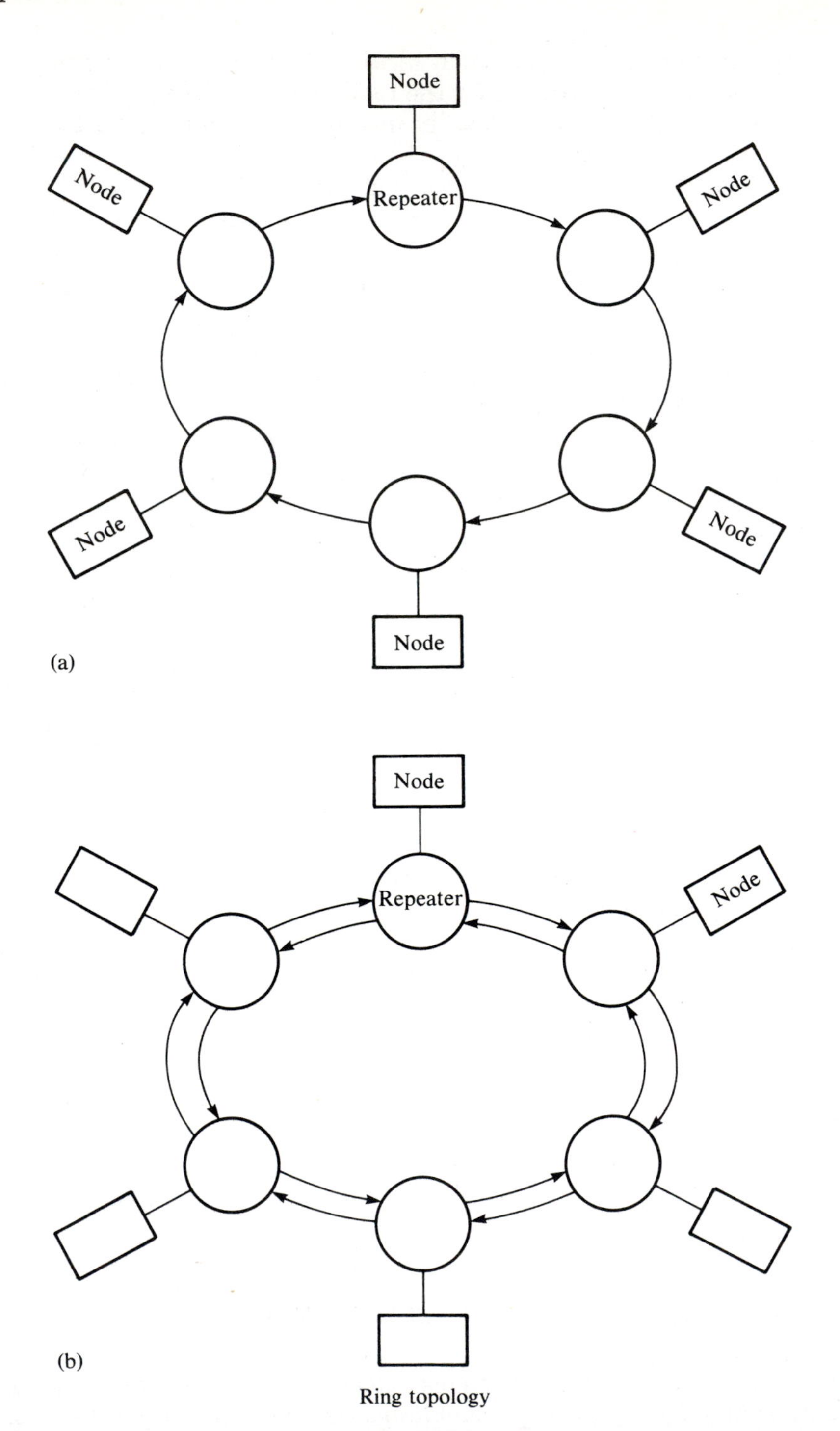

FIGURE 12–2 Topology of ring network. (a) Unidirectional. (b) Bidirectional.

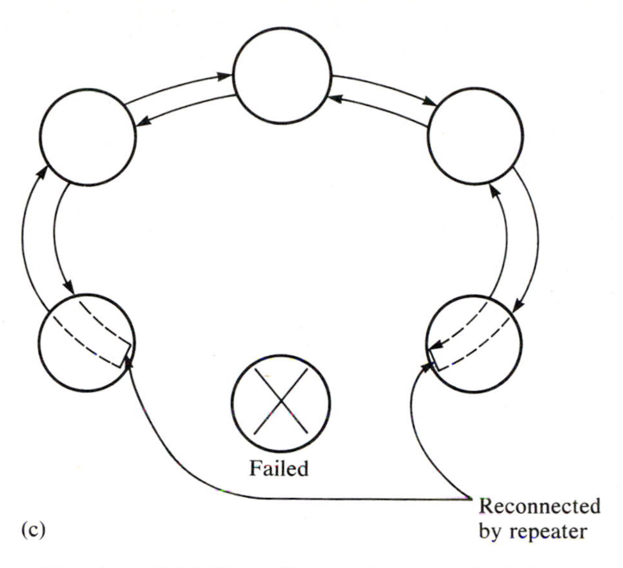

FIGURE 12–2 *(Continued)* (c) Reconfigured due to node failure.

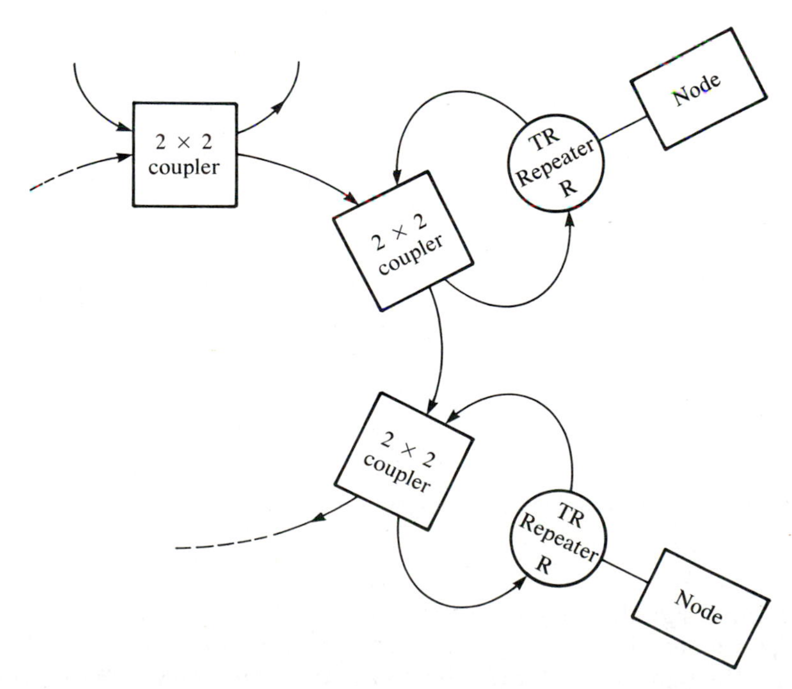

FIGURE 12–3 2 × 2 coupler used in a ring network.

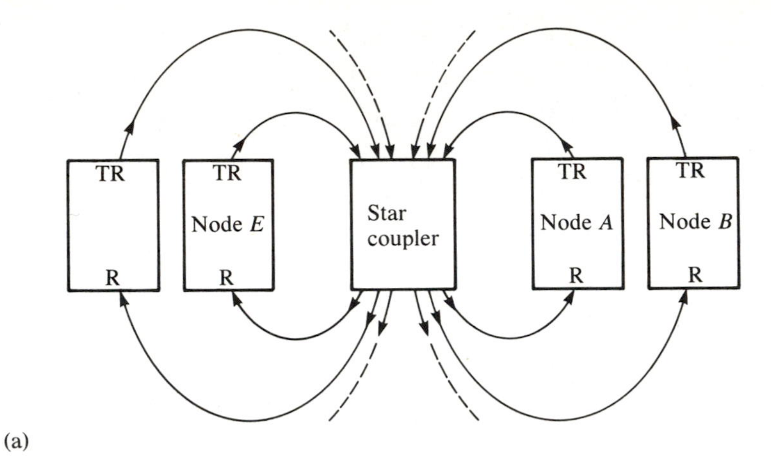

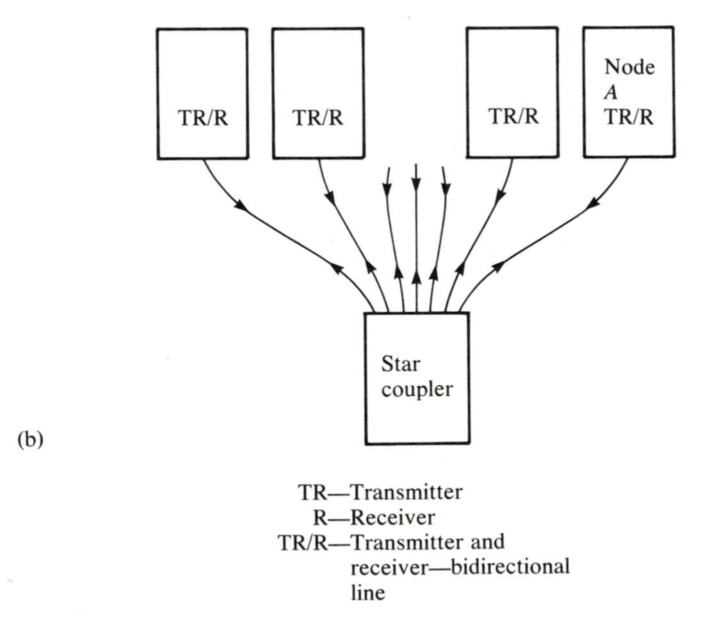

FIGURE 12–4 Star coupler in a network. (a) Transmissive star. (b) Reflective star.

multichannel signal, and a fiber system is used to distribute the signal to local users. The video signals of the various channels either are used directly to modulate the light source for the fiber optic transmission or are first converted to digital form and then distributed to the user. In either method, a suitable multiplexing scheme must be used.

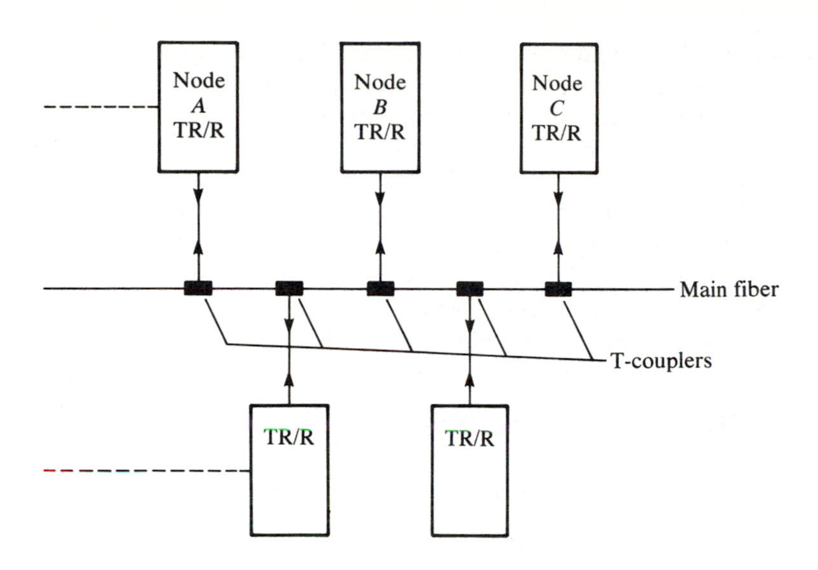

FIGURE 12–5 Bus structure.

For the analog system, the bandwidth (BW) of the transmitted multiplexed signal (representing all the received television channels) is approximately the sum of the individual video channel BWs. A 100-channel system would have a total BW of about 800 MHz (5 MHz per channel plus guard bands).

A digital system requires digitization of the video signals. A 5-MHz video BW requires a 10-MHz sampling rate and 7–9 bits per sample, resulting in a bit rate per channel of 70–90 Mb/s. (The exact sampling rate is 10.74 MHz, yielding bit rates of 75.18–96.66 Mb/s.) In spite of the enormous bit rate required for the digital system, a large number of CATV (and **closed-circuit television**) systems use it because multiplexing (TDM) is simpler for digital than for analog systems.

A product introduced by Alcatel/STR Corporation of Connecticut, the OVID-4, is capable of transmitting 4 standard video channels and up to 24 audio channels on a single fiber using digital transmission. The system can operate over an unrepeatered distance of 40 km.

Fiber optic systems are often used in television studios for CCTV. The fiber cable used is a fraction of the weight of an equivalent copper cable, making it particularly suitable for connecting portable television cameras to fixed studio monitors.

A novel use of fiber optic communication is in its replacement of the standard copper wire intercom systems in large buildings. Again, the small size and low weight of the fiber cable are major advantages.

In recent years, telephone companies have begun to connect fiber optic cables to homes, in what is called the **subscriber loop,** providing a variety of services in addition to voice communication. The fiber optic subscriber loop is expected to replace most of the copper cabling presently in use. An experimental 500-home fiber optic subscriber loop was installed in 1980 in the town of Biarritz, France. By 1986, the experiment had been expanded to serve more than 150,000 homes in Paris and Montpellier, France. In 1986, the New York Telephone Company began a subscriber loop to serve users in the Rockefeller Center in New York City. The French loop is an analog system, and the New York loop is digital. The loops offer what is called an **integrated service digital network,** which includes voice lines, teletype, direct stock exchange data, and other services (a mix of digital and analog communication).

12–5 FIBER OPTIC SENSORS

The optical fiber can be used as a sensor in the measurement of pressure, displacement, absence or presence of an object, temperature, and more. It can also be used as a link between a remote sensor and a monitoring station. The latter is like any fiber optic communication system. It carries the electrical signal from a sensor to the monitoring station.

The concepts underlying the operation of an optical fiber sensor can be divided into three groups:

1. Variations of light intensity,
2. Changes in phase, and
3. Changes in diffraction patterns **(moire fringes).**

(These represent the most used techniques. Variations in modal distribution and frequency changes can also be used.)

12–5–1 Intensity-Modulated Sensor

The most popular sensors, because of their simplicity, are those that use intensity variations. Figure 12–6 shows this idea. Here the displacement of an object is related to the intensity of the light arriving at the detector. As ΔL (the displacement) increases, the detected light intensity decreases. To avoid the effects of spurious light at the detector, the object is illuminated by a relatively narrow line light source, and the detector contains a filter matching the source wavelength. Also, note that sleeves extend beyond the end of fibers to reduce illumination of the fibers by spurious light.

Figure 12–7 is a diagram of a device that uses the same principle presented in Figure 12–6. Here, it is used to measure the maximum displacement of a speaker assembly under dynamic conditions. The speaker is driven by a signal in the kilohertz range, and the displacement of the assembly is monitored instantaneously by the fiber optic transceiver.

Figure 12–8 shows another intensity sensor. A length of fiber (bare fiber with no plastic jacket) is placed between two corrugated blocks. The force on

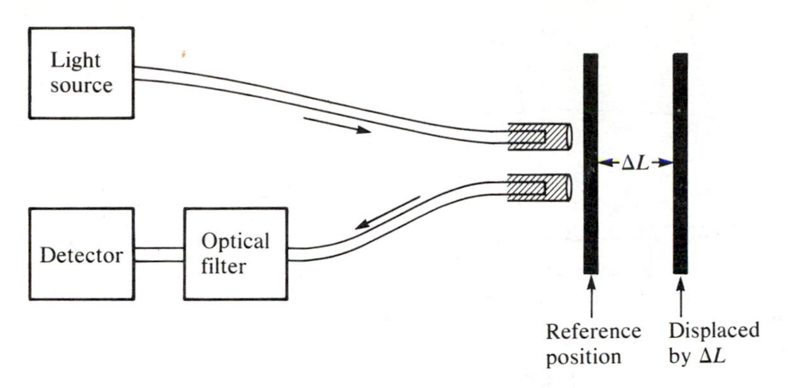

FIGURE 12–6 Displacement measurement.

the top block presses it against the fiber, introducing microbend losses in the fiber. These losses increase with increased force. The changes in intensity induced by the force are measured with respect to a direct unmodulated signal from the light source, eliminating source variations as a source of error.

12–5–2 Phase Sensors

Figure 12–9 shows a temperature-sensing system that relies on phase variations. As the sensor element is heated, the fiber elongates, increasing the length of the light path. This process changes the phase of the light that

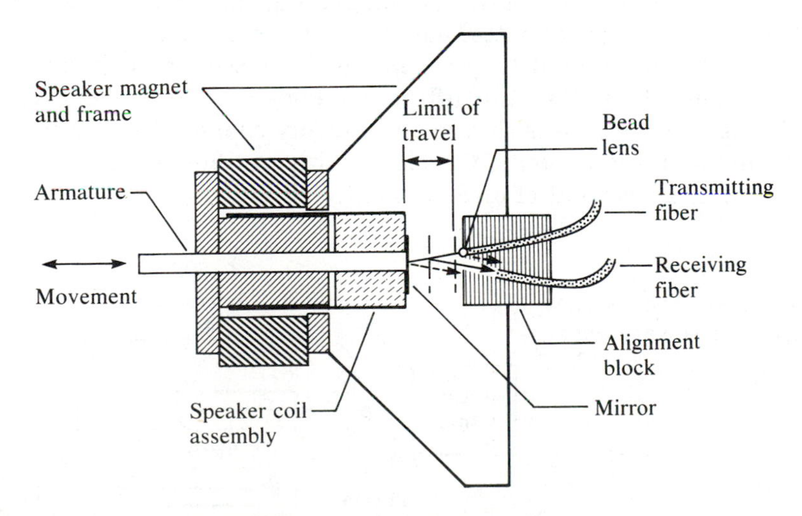

FIGURE 12–7 Dynamic displacement measurement.
Source: J. E. Moulton, III, "Low-Cost Optical Out-of-Range Detection for Microcontrollers," *Fiber Optic Products News,* June 1988.

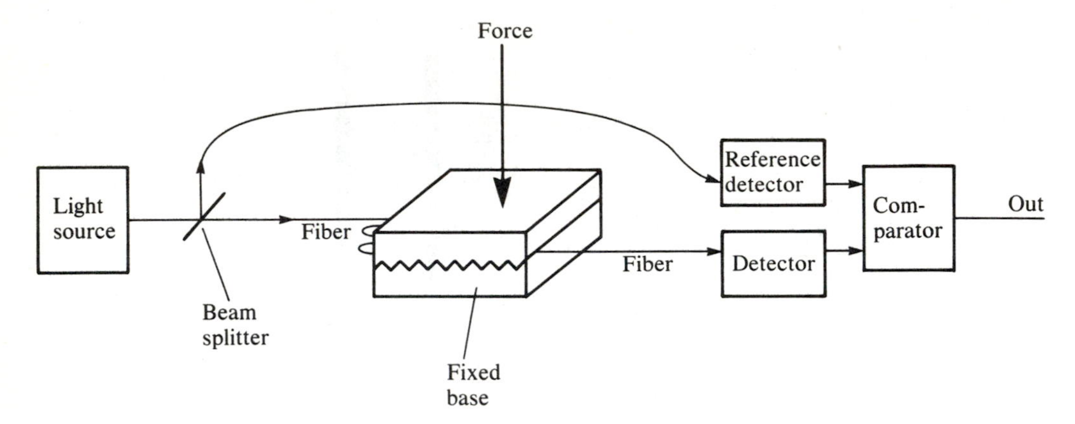

FIGURE 12–8 Force sensor using microbend losses.

reaches the detector. The reference fiber provides a phase reference for the phase comparator. This process is very complex, because it involves measuring the phase of a sine wave at the frequency of light. In addition, the light source must be highly coherent (single frequency and constant phase). (The detection is based on combining the reference light with the light passing through the sensing element to produce diffraction-like patterns [moire fringes].)

12–5–3 Diffraction Grating Sensors

Figure 12–10 shows the use of diffraction gratings for sensing displacement. Two diffraction gratings are butted against each other, forming the equivalent of a new grating. As one of the gratings is displaced, the effective grating distance is changed and thus the diffraction angle (shown by dashed lines) is changed. The detector senses this as an intensity change. Here, you digitally count the high- and low-intensity cycles as the gratings move. This yields a direct digital measurement of displacement.

FIGURE 12–9 Temperature sensor using phase variations.

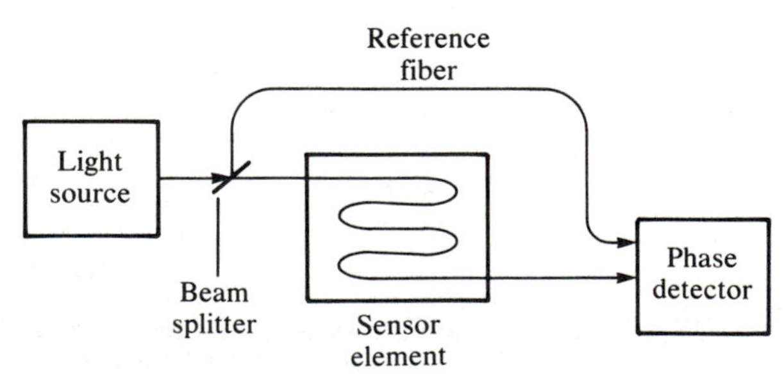

FIGURE 12–10
Displacement sensing using diffraction patterns.

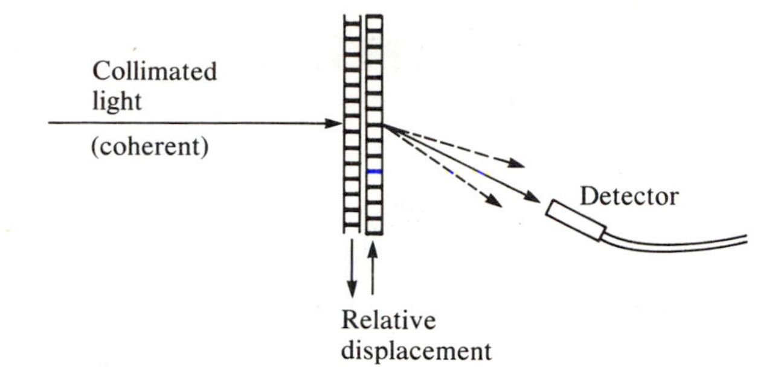

12–5–4 The Fiber Advantage

The sensors discussed here represent only some of the technology being developed. Some of the reasons for using fibers rather than other materials are related to the immunity of fibers to the environment, electrical interference, and high temperatures. The fiber systems allow the monitoring station to be placed at a safe distance from the hazardous environment where the sensing must take place. (Consider monitoring various parameters at or near a nuclear explosion.)

12–6 MEDICAL APPLICATIONS

Fiber optic technology can be used in medical **diagnostics** as well as in medical procedures. The fiber optic endoscope shown in Figure 12–11 is used to inspect internal organs for diagnostic purposes. In industrial applications, an endoscope (sometimes called a borescope) may be used to inspect parts where direct visual inspection is impossible.

In Figure 12–11, end B of the endoscope is the end of the fiber that views the object. End A is connected to a light source to provide proper illumination. End C is the viewing lens. A video borescope is shown in Figure 12–12. (Typically, the term "borescope" is used in industrial rather than in medical applications.) Here, the viewing lens is connected through a fiber to a video camera and monitor, providing a video display of the viewed area.

Fiber optic use in medical procedures is best demonstrated by the LASTAC system. (This system is registered by GV Medical, Inc., of Minneapolis, Minn. LASTAC stands for laser enhanced transluminal angioplasty.) The procedure using LASTAC is shown in Figure 12–13. Here, optical energy, transmitted through the fiber, is used to evaporate built-up plaque that is blocking an artery. The first such procedure was performed by Dr. L. A. Nordstrom of the University of Minnesota on March 8, 1988. Final evaluation of the procedure is not complete.

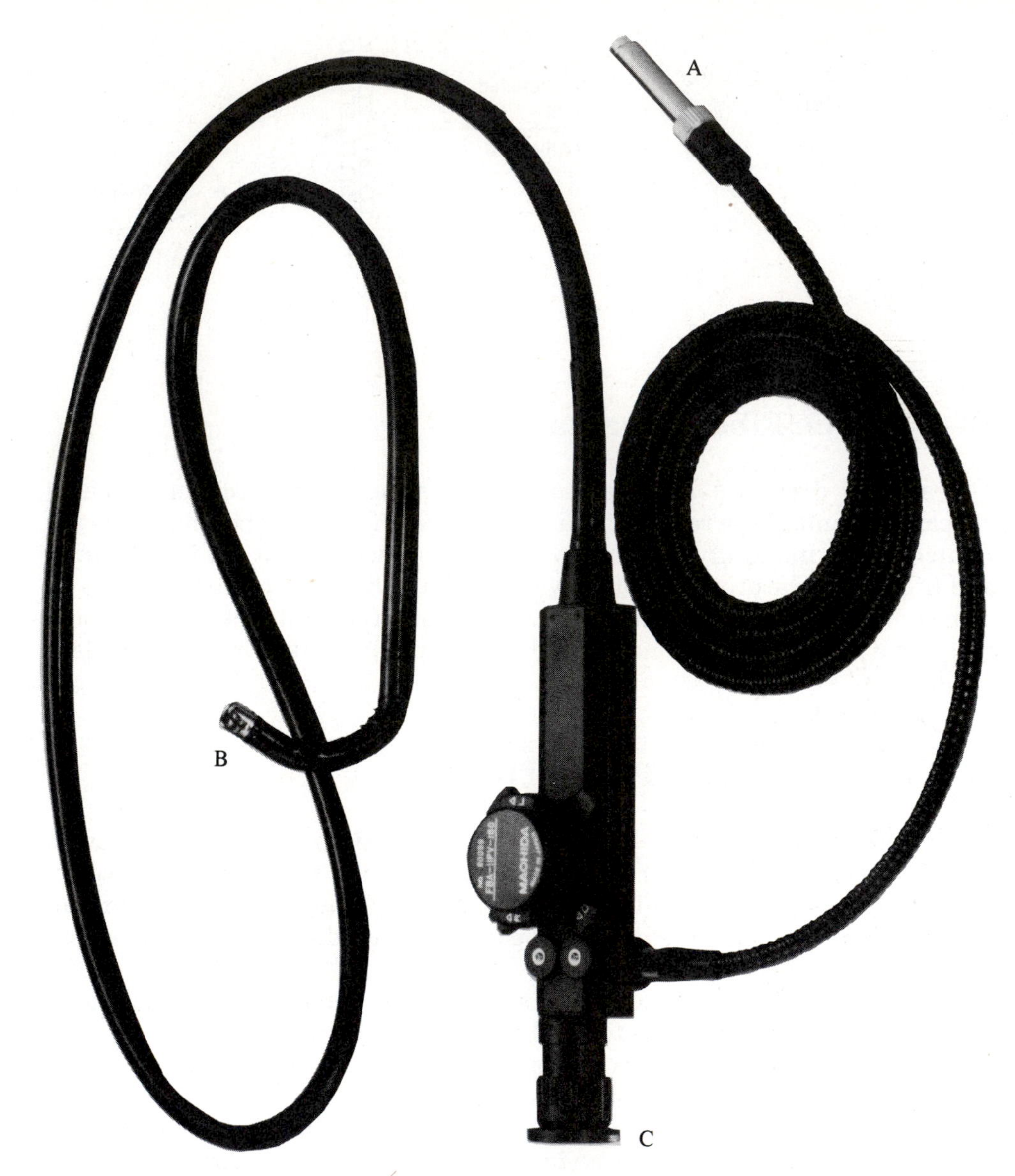

FIGURE 12–11 Photograph of endoscope (borescope).
Source: Machida, Inc., Orangeburg, N.Y.

Fiber optic systems have been used to treat cancer. This process involves the injection of special chemicals that penetrate only the cancerous cells. Optical energy (heat energy) transmitted via the fiber illuminates the affected area and is absorbed by the special chemical in the cancerous cells. The generated heat destroys the cancerous cells.

BORESCOPIC VIDEO SYSTEM

Machida Incorporated introduces a state of the art, borescopic video inspection system with the versatility to adapt to the users needs. It is now possible to combine in one compact unit, a color or black & white camera, 7" high resolution monitor, video tape recorder, character generation and/or video micrometer.

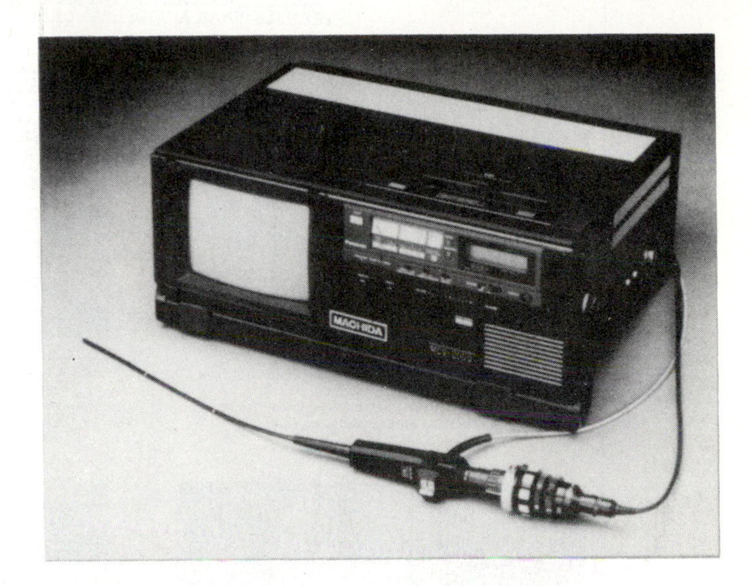

FEATURES:

- **Compact**
- **Light Weight**
- **Versatile**
- **Simple to Set Up**
- **Eliminates Messy Wiring**

COLOR VIDEO SYSTEM
Model MCV-2000

Consists Of:
- Color Miniature Camera
- 7" High Resolution Monitor with integrated processor
- Machida Borescope Adapter
- All Cables
- Carrying case

Price: $5842.00

BLACK 7 WHITE VIDEO SYSTEM
Model MCV-1000

Consists Of:
- Black & White Miniature Camera
- 7" High Resolution Monitor with integrated processor
- Machida Borescope Adapter
- All Cables
- Carrying Case

Price: $5165.00

THE FOLLOWING OPTIONS ARE AVAILABLE IN ANY COMBINATION WITH EITHER SYSTEM

OPTION #1	1/2" VHS Video Recorder\Playback Unit	$1417.00
OPTION #2	Video Micrometer	$2750.00
OPTION #3	Character Generator	$667.00
OPTION #4	Microphone	$33.00
OPTION #5	Adapter To Couple System To Olympus Fiberscope	$500.00

MACHIDA INCORPORATED
CORPORATE HEADQUARTERS
East Coast Service Center
40 Ramland Road South Orangeburg, NY 10962-2698 914-365-0600
West Coast Service Center
1440 South State College Blvd., Suite 2A-1, Anaheim, CA 92806 714/533-3530

CUSTOMER SERVICE
800-431-5420

Prices Effective 10-88

FIGURE 12–12 Data sheet for borescopic video system.
Source: Machida, Inc., Orangeburg, N.Y.

STEPS IN THE DIRECT LASER-ENHANCED TRANSLUMINAL ANGIOPLASTY PROCEDURE

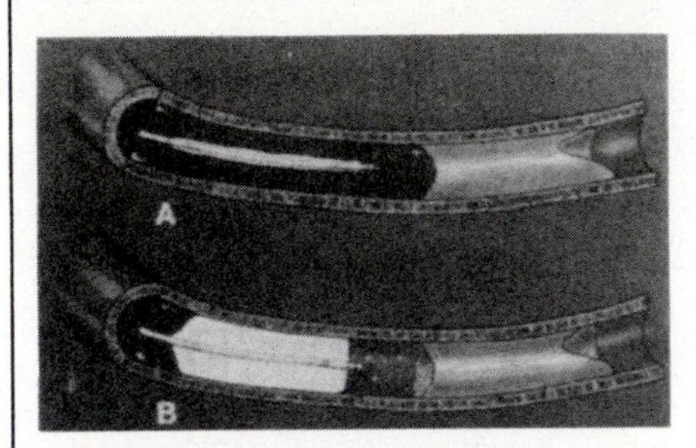

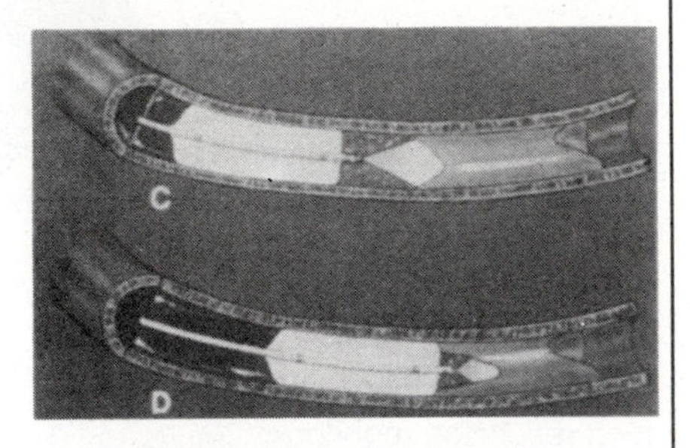

A - The balloon catheter is advanced close to the lesion of interest.
B - The balloon is inflated, blocking blood flow. Irrigating fluid clears the lasing field.

C - Laser light is emitted from the optical fiber, vaporizing plaque.
D - The balloon is deflated, advanced, and reinflated. Laser light is emitted, lengthening the new channel through the lesion.

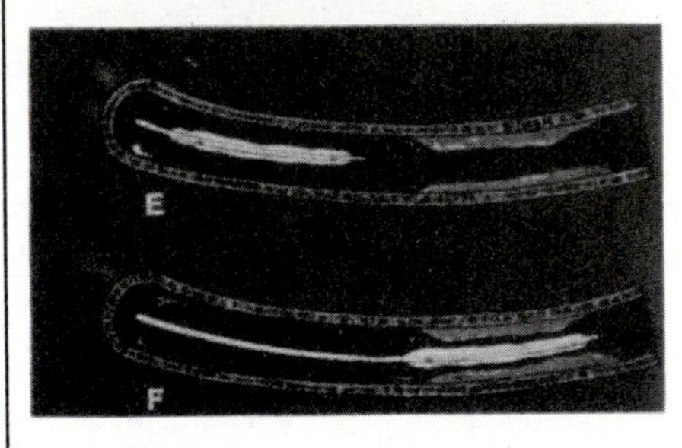

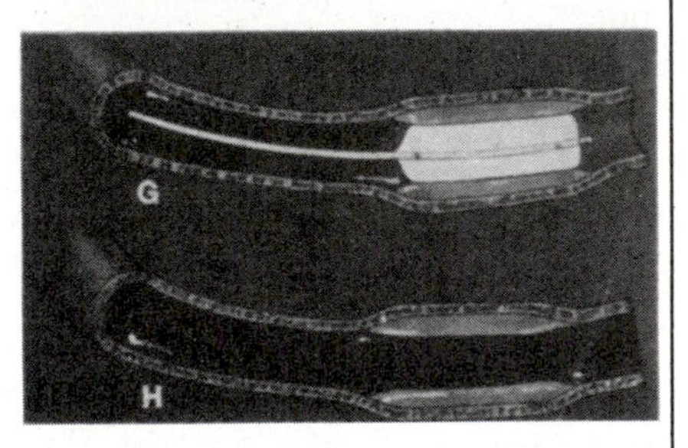

E - After the lased channel is created through the lesion, the balloon is totally deflated.
F - The balloon is passed through the channel.

G - The balloon is inflated, opening the lesion.
H - The balloon catheter is removed to complete the process.

*Caution—Investigational device. Limited by Federal (or United States) law to investigational use.

FIGURE 12–13 LASTAC procedure.
Source: L. K. Franklin, "Laser Enhanced Transluminal Angioplasty—The Lastac System," *Fiber Optic Product News,* April 1988.

12–7 COHERENT OPTICAL COMMUNICATION

Coherent optical communication, which is said to be the wave of the future, is, in principle, the same as the standard broadcast system in which a carrier is modulated by the information to be transmitted. However, conventional circuit techniques used in modulation and demodulation cannot be used in coherent optical communication. The major difficulty is the frequency of the carrier—light frequency. The light frequency is too high.

To modulate the amplitude of a light source, that is, to vary the amplitude of a sinusoidal carrier at 10^{12}–10^{13} Hz, special technology is required. Similarly, receivers can use neither conventional amplification nor **heterodyne reception** technology. (Heterodyne refers to the technology of beating the incoming carrier signal against a local oscillator, yielding an intermediate frequency that is constant and easily amplified.) The principles of these techniques apply; however, the details do not follow regular circuit technology.

A regular photodetector cannot be used in **coherent communication** because it responds to light intensity only. Coherent communication treats the light as a sine wave, where amplitude, frequency, and phase are important parameters. These parameters do not play a role in standard intensity-modulated systems.

The power of coherent communication lies in the enormous BW that it affords. If you assume that a carrier of 2.3×10^{12} Hz (λ of 1.3 μm) has a modulation BW of 1% of the carrier frequency, you obtain an information BW of 2.3×10^{10} Hz, or 23 GHz. This is far beyond anything possible in other systems and is compatible with the capabilities of single-mode, low-dispersion, optical fibers.

In addition to the BW, the coherent system offers the potential of fiber links operating over long unrepeatered distances because of the high receiver sensitivity attainable in coherent communication systems.

Some coherent systems under experimentation during 1987 and 1988 are listed in Table 12–3.

Field trials of a coherent system were conducted by British Telecom Company in mid-1988. These trials involved the use of an existing single-mode fiber link of 67.4 km in the Cambridge telephone exchange, over a loop to Saint Neots. Another system was operated over a 108.6-km single-mode fiber loop to Bedford via Saint Neots. The wavelength used was 1.55 μm. The loss was 19.4 dB for the 67.4-km link and 29.6 dB for the 108.6-km link. The transmission rate was 565 Mb/s. No errors were detected for a duration of 5 hours in the 67.4-km link and for 4.5 hours in the 108.6-km link. Because these experiments were conducted in existing links and not in a laboratory environment, they bring coherent communication much closer to practical use.

TABLE 12–3
Experimental Coherent Systems

Company	Bit Rate (Mb/s)	Wavelength (μm)	Distance (km)
AT&T	2000	1.53	170
Fujitsu	1200	1.54	190
NTT	4000	1.55	155

12–8 INTEGRATED FIBER OPTICS AND ELECTRO-OPTICS

The operating speed of today's fiber optic systems is ultimately limited by the capabilities of the electronic circuitry used. Integrated optics attempts to replace some of the electronic data processing with direct optical processing. The latter is literally limited only by the speed of light. It should operate at rates of thousands of gigahertz. The integrated electro-optic (and fiber optic) approach envisions an optical integrated circuit (similar to the electronic integrated circuit [IC] chip) that will contain optical and electro-optical devices such as optical amplifiers, filters, switches, couplers, multiplexers, modulators, bistable circuits, detectors, and lasers. The copper wires will be replaced by optical waveguides, which are planar optical fibers. Because the size of the optical IC is measured in micrometers, it is limited to single-mode operation, where waveguides are usually a few micrometers in diameter.

A typical integrated optic waveguide, which can be constructed by the same technology used in manufacturing electronic ICs, is shown in Figure 12–14. A small layer of optical material with refractive index n_1 is embedded in a substrate with refractive index n_2. Here, $n_3 = 1$ is the refractive index of air above the waveguide. The n_1 and n_2 are selected so that $n_1 > n_2$ and clearly, $n_1 > n_3$. This structure forms an optical waveguide, similar to an optical fiber. The n_2 and n_3 materials serve as the cladding and the n_1 material as the core. Because one side of the waveguide is air (n_3), expect the waveguide to have relatively high losses. This is insignificant, however, because these waveguides are only a few millimeters long.

Optical components such as switches can be built from materials that exhibit changes in their refractive index when subjected to an electric field. This allows the refractive angle to be changed by an electrical signal, directing the light to one of the outputs. A schematic representation of such a switch is shown in Figure 12–15. The structure is compatible with optical integration because it can be constructed from planar optical guides.

Other components such as optical amplifiers, modulators, and filters can also be constructed using the IC technology. The final product may be an optical and electro-optical data processing system contained in a single IC

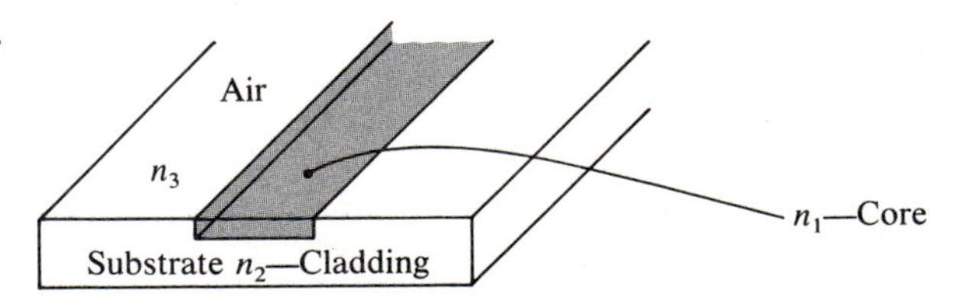

FIGURE 12–14 Planar integrated circuit optical waveguide.

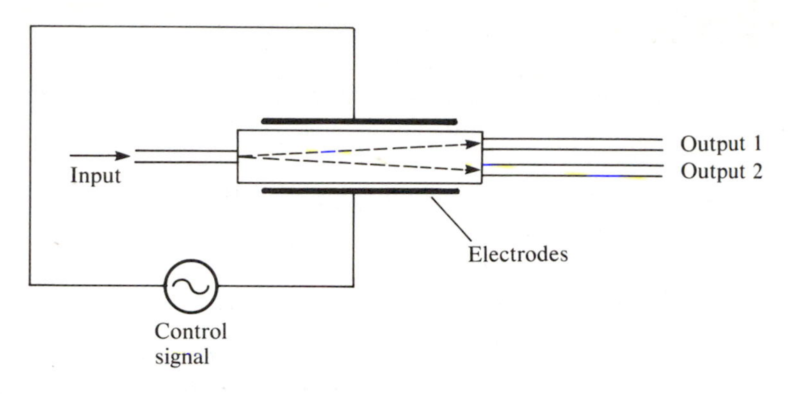

FIGURE 12–15 Electro-optical switch.

(similar to the digital computer ICs). This will provide a high-speed front-end processor for optical signals. The output of the optical processors will consist of lower-speed signals, which can then be converted to electrical signals for further processing. The integrated electro-optic technology should also bring about a substantial reduction in costs, just as the electronic IC has.

12–9 FUTURE OF FIBER OPTICS

It is unlikely that developments in the next 10 or 20 years will bring as many advances to fiber optics as the past 10 years have brought. Nevertheless, you can expect the development of coherent optical communication applications, with data rates increasing from 5 to 10 Gb/s and with unrepeatered lengths of over 150 km. This development will entail a major thrust (1) in the area of integrated optics to replace parts of the drive and processing circuitry and (2) in super reliable, highly stable, single frequency lasers (extremely narrow line width). To complement the high data rate, faster multiplexers and demultiplexers will be developed, probably using integrated optical technology.

In the medical field, you can expect some experimental laser-enhanced procedures to become widely used. This will require techniques for precise control of laser output as well as precise positioning of the fiber used. Fiber optic diagnostic tools will include instruments for precise measurement of blood flow and video displays of various internal organs.

Home telephone service, using fiber cables connected to the telephone (subscriber loop), will provide a wide range of services and equipment, including CATV, fire detection and burglar alarms, video conferencing equipment, financial information services, and other consumer services.

The last 10 years have witnessed an explosion in fiber optic technology. Over the next 20 years, you will see this technology applied and a new technology evolve.

SUMMARY AND GLOSSARY

The list provided here contains keywords to serve as a summary of the material covered. Understanding these terms and how they are related to the subject is equivalent to reviewing the chapter. Some of these terms are also part of the technical jargon and useful for any practitioner.

Bus. Data distribution method used in digital systems.

Bypass. A method of short circuiting a failed node on a network by allowing data to bypass the node.

Closed-circuit television. Use of television cameras and monitors on local premises, for example, safety monitors in banks and supermarkets.

Coherent communication. Communication system using a sinusoidal carrier, modulated by information. In optical coherent communication, the sinusoidal carrier is light (10^{13} Hz).

Community antenna television. The use of a common antenna to receive video channels, which are then distributed by wire or fiber to many subscribers.

Diagnostics. The act of analyzing and determining a particular medical problem.

Heterodyne reception. The use of a local oscillator to produce an intermediate frequency signal that is substantially lower than the received signal and that contains all the transmitted information of the carrier.

Integrated service digital network. The use, typically by a public utility company such as the telephone company, of a digital network to provide a large variety of customer services. (These services include voice, community antenna television, and computer communication.)

Local area network. A digital network that spans only a few kilometers and usually serves a single facility, building, or plant.

Long-haul communication. A long-distance communication link, typically, point to point.

Moire fringes. Interference pattern created by diffraction.

Network. A communication link intended to serve many customers. Connecting to or disconnecting from the network is similar to connecting or disconnecting telephone service.

Point-to-point communication. A communication link connecting only two nodes, typically, used in long-haul systems.

Ring network. A particular network topology (structure) in which data flow in a circular fashion between nodes.

Star network. A network topology in which all nodes communicate via a control distribution node.

Subscriber loop. The telephone link that connects the subscriber to the main telephone system.

QUESTIONS

1. What are the fundamental differences between long-haul and LAN communication systems?
2. What are the important features of a long-haul system?
3. Why is it not important to require the use of a high BW × km fiber in LANs, while it is important in long-haul systems?
4. What is a ring topology?
5. Compare the ring, the star, and the bus networks.
6. How does the BW of an analog multiplexed, 10-channel video system compare to the same system when a digital MUX is used?
7. What is meant by the subscriber loop?
8. Why is it almost impossible to provide an integrated service digital network (ISDN) using copper wires? (Consider the BW required.)
9. What are the concepts used in fiber optic sensors?
10. What is an example of a sensor that uses light intensity variations?
11. Figure 12–8 uses what is essentially a mode mixing block (Chapter 3). Compare the sensor application to mode mixing.
12. In a phase-sensitive sensor, why is a reference light used?
13. How are the diffraction gratings used to measure displacement?
14. What are the advantages of using fiber optic technology in remote sensors?
15. What is an example of the use of fiber optics in medical diagnostic applications?
16. What is a fiber optic application in surgery?
17. What is coherent communication?
18. What are the advantages of coherent communication over intensity-modulated technology?
19. What is integrated electro-optics?
20. Why is integrated electro-optics (and fiber optics) important?
21. What are some of the devices that can be manufactured using electro-optics?
22. Eighty years ago, nobody thought that flying was possible, not to mention reaching the moon. Now that fiber optic and related technologies are here, what do you expect from them over the next 30 years? (Be specific.)

APPENDIX I
Table of Constants and Units

Constant/Unit	Symbol	Value
Angstrom	Å	10^{-9} m
Boltzman's constant	k	1.381×10^{-23} J/K
Electron charge	q_e	1.602×10^{-19} C
Degrees kelvin	K or T	°C + 273.15
Electronvolt	eV	1.602×10^{-19} J
Joule	J	6.242×10^{18} eV
Planck's constant	h	6.626×10^{-34} J-s
		4.136×10^{15} eV $\times$ s
Speed of light in free space	c	2.998×10^{8} m/s
		($\sim 3 \times 10^{8}$ m/s)
$(k \times T)/q_e$ at room temperature		0.026 V

APPENDIX II
Laser Safety

Safety concerns are primarily related to operating lasers rather than to the lower power light-emitting diodes (LEDs). The most important factor affecting laser hazards is the intensity of the optical energy at the exposed surface. A highly concentrated beam is more dangerous than a diffused beam. The intensity is also related to the total optical power emitted. As a result, the low power and relatively diffused LED source is not considered a hazard. Nevertheless, it would not be smart to look directly into a LED source, particularly in light of the daily arrival of new higher power LEDs and of the optical systems that often concentrate optical energy.

NEVER LOOK INTO THE SOURCE DIRECTLY! You may not see anything because the wavelengths used often are not in the visible range; however, your eyes may be affected.

Safety regulations are intended to minimize (it is hoped, to eliminate) possible damage to the user of the laser or the laser system. The safety regulations are related to the degree of potential hazard of the laser. These regulations are characterized by three major aspects enumerated in standard Z136.1–1986 section 3.1 of the American National Standards Institute (ANSI).

1. The laser or laser system's capability of injuring personnel
2. The environment in which the laser is used
3. The personnel who may use or be exposed to laser radiation.

So that unnecessary restrictions are not placed on lasers, they are grouped into classes based on the first aspect listed here. Thus, the low-hazard laser is not subject to the same regulation as the potentially dangerous laser.

The laser classes range from class 1 to class 4 (with two levels of class 3), with the lower numbers indicating lower hazard levels. The class descriptions quoted next are described in a bulletin titled "Laser Safety Guide," issued by the Laser Institute of America.

> Class 1 denotes exempt lasers or laser systems that cannot, under normal operating conditions, produce a hazard.
>
> Class 2 denotes low power visible laser systems which, because of the normal human aversion responses, do not normally present a hazard, but may present some potential for hazard if viewed directly for extended periods of time (like many conventional light sources).
>
> Class 3a denotes lasers or laser systems that normally would not produce a hazard if viewed with the unaided eye. They may present a hazard if viewed using collecting optics.
>
> Class 3b denotes lasers or laser systems that can produce a hazard if viewed directly. This includes intrabeam viewing of specular reflections. Except for the higher power class 3b lasers, this class laser will not produce a hazardous diffuse reflection.[1]
>
> Class 4 denotes lasers or laser systems that can produce a hazard not only from direct or specular reflections, but also from a diffuse reflection. In addition, such lasers may produce fire hazards and skin hazards.

The class of a laser or laser system is usually determined by the manufacturer. (In special cases, the user may choose to perform certain measurements to establish the class.) However, the evaluation of the potential hazard of a laser or laser system is not determined solely by the classification. Other factors to consider are the specific use and the application environment. For example, any completely enclosed laser system with emission from the enclosure below a certain level is classified as class 1, which means it is exempt from special regulations. Such a system, however, may contain a higher class laser requiring safety controls during servicing, when the enclosure is removed.

In addition to ANSI standards, the Bureau of Radiological Health of the U.S. Department of Human Services has developed laser product safety standards that are limited to performance characteristics and apply specifically to laser manufacturers.

---

1. A diffuse reflection is one where the reflected radiant energy follows Lambert's law where, in essence, the radiation is reflected over a wide angular range.

All lasers require special labeling. A typical label is shown here, giving

1. Peak power of the laser,
2. Operating wavelength, and
3. Classification.

Typical laser label.

Before starting to work with lasers (and future high-power LEDs), you must first protect your eyes. Never look directly at a laser. Do not expose your skin to the light beam. A single exposure may not harm you; however, the effects of repeated exposure are not well known.

BE CAUTIOUS. Establish safe working habits.

APPENDIX III
Energy Gap E_g for Various Substances

Substance	Symbol	E_g (eV)
Aluminum antimony	AlSb	1.6
Aluminum arsenide	AlAs	2.16
Aluminum phosphide	AlP	2.45
Cadmium sulfide	CdS	2.42
Gallium antimony	GaSb	0.70
Gallium arsenide	GaAs	1.43
Gallium phosphide	GaP	2.26
Germanium	Ge	0.67
Indium arsenide	InAs	0.36
Indium phosphide	InP	1.28
Lead sulfide	PbS	0.37
Silicon	Si	1.11
Zinc sulfide	ZnS	3.60

Answers to Selected Problems

Chapter 2

1.a. $\lambda = 10^5$ m
1.b. $\lambda = 297.03$ m
1.c. $\lambda = 5.26$ m
1.d. $\lambda = 0.496$ m
1.e. $\lambda = 300\ \mu$m
1.f. $\lambda = 0.857\ \mu$m
1.g. $\lambda = 0.6\ \mu$m
3. $f = 4.83 \times 10^{14}$ Hz
5. $f = 2.57\ \mu$s
7. $t = 0.2$ ms
9. $\theta = 45°$
11. $\theta_2 = 15.2°$
13.a. $t_1 = 5$ ns
13.b. $t_2 = 5.3$ ns
13.c. $\Delta t = 0.3$ ns
15. $\theta_x = -16.44°$
17. Loss = 0.028 (2.8%)

Chapter 3

1. $\theta < 13.7°$
3.a. $n_{\text{core}} = 1.43$
3.b. $\theta_c = 11.8°$
3.c. N.A. = 0.291
3.d. $\theta_a = 16.94°$ (acceptance angle = 33.9°)
5. $n_1 = 1.51$
7. $\theta_c = 44.4°$
9. $\Delta f = 2.81 \times 10^{13}$ Hz
11. $\Delta f = 2.31 \times 10^{13}$ Hz

Chapter 4

1. $\text{loss}|_{\text{dB}} = -13$ dB
3. loss = 1.85 dB/km
5. $\Delta t = 0.67\ \mu$s
7. $\Delta t = 28.8$ ns
9. output pulse width = 32 ns
11.a. $\Delta t = 0.4$ ns
11.b. output pulse width = 2.04 ns
13. $\Delta t = 1.5$ ps/km × nm
15. output pulse width = 540 ns

Chapter 5

1.a. $M_N = 14{,}858 \approx 15{,}000$
1.b. $M_N = 1750$
1.c. $M_N = 1570$
3.a. $V = 172.4$, $M_N \approx 7400$
3.b. $V = 59$, $M_N \approx 900$
3.c. $V = 56$, $M_N \approx 800$
5. $n = 1.434$
7. $a \leq 2.97\ \mu$m
9. $n_2 = 1.3953$
11. $n_2 = 1.3986$

Chapter 6

1.a. sampling rate = 3×10^4 samples/s
1.b. data rate = 240 kb/s
3.a. BW = 10 MHz
NRZ B_r(max) = 20 Mb/s

3.b. BW = 1 GHz
B_r = 2 Gb/s
5. BW = 40 MHz
7. BW = 100 MHz × km
9. B_r(max) = 83.3 Mb/s
11. BW_{op} = 212.2 MHz

Chapter 7

1. 6000 samples/s
3. BW = 1500 Hz
5. output rate = 512 kb/s

Chapter 8

1. $loss_{lat}$ = 1.61 dB
3.a. $loss_{ang}$ = 0.58 dB
3.b. $loss_{ang}$ = 0.4 dB
total loss = 0.64 dB
3.c. difference = 0.64 − 0.58 = 0.06 dB
5. $loss_{lat}$ = 0.44 dB
7.a. loss = 0.2 dB
7.b. $loss_{Fr}$ = 0.3 dB
9.a. $loss_{area}$ = 1.53 dB
9.b. no loss
11. $loss_{Fr}$ = 0.005 dB
13.a. PD = 14.3 dB
13.b. PD = 9.5 dB
15. Power ratio = −0.97 dB (80% fiber)
Power ratio = −7.0 dB (20% fiber)
17. P = 63 μW (−1 dB fiber)
P = 16 μW (−6.86 dB fiber)

Chapter 9

1. P_{out} = 3.2 nW
3. P_{out} = 1 mW
P_F = 50 μW
5. n_c = 0.165
P_F = 0.132 mW
7. λ = 8.86 μm
9. R_1 = 200 Ω
R_B = 5.75 kΩ
11. I_F = 71.7 mA
13. I_F = ±20 mA

Chapter 10

1. λ_C = 0.92 μm
3. I_P = 0.35 μA
5. R = 0.51 A/W
7. I_P = 0.56 μA
V_{out} = 28 mV
9.a. I_P = 160 nA
V_{out} = −16 mV
9.b. V_{out} = 21 mV
11.a. R = 0.91 A/W
11.b. V_{out} = −72.8 mV
13. I_{B1} = 0.2274 μA
P_{in} = 0.38 μW
15. i_t = 4.11 pA
17. i_n = 57.1 pA
19. i_n = 0.9×10^{-14} A
21. NEP = 17.77×10^{-10} W

Chapter 11

1. R = 15.6 μm
smallest diameter = 31.2 μm
3. L_{area} = −1.58 dB
5. total loss = −14.73 dB
7. bit rate = 96 kb/s
9.a. BW_{tot} = 50 kHz
9.b. ch. 1 = 1 MHz
ch. 2 = 1 μHz + 6.25 kHz = 1.00625 MHz etc.
11.a. BW = 3.4 MHz
b. BW = 6.3 MHz
13. B_r = 11.9 mb/s
15. Expected: 10 errors
17. Added length = 1.2 km
19.a. NEP = 2.9×10^{-13} W/(Hz)$^{1/2}$
b. NEP (1 MHz) = 2.9×10^{-10} W
c. S/N = 344.8
$S/N|_{dB}$ = 25.4 dB
21.a. NEP (1 MHz) = 10^{-10} W
b. NEP (10 MHz) = 3.16×10^{-10} W
c. BER = 10^{-11} (NEP = 10^{-10} W)
d. BER = 10^{-4} (NEP = 3.16×10^{-10} W)

Glossary

ABSORPTION. A loss process in which impurities take up (absorb) some of the propagating light energy.

ACCEPTANCE ANGLE. The range of angles within which an injected light beam will enter and propagate in the fiber. Related to the numerical aperture.

AMPLITUDE MODULATION (AM). A scheme involving changing the amplitude of a carrier in accordance with the amplitude of the input signal. Input information is ultimately carried by side bands.

ANGULAR MISALIGNMENT. The angle by which one of the fibers to be joined deviates from the axis of the other fiber.

ANGULAR RESPONSE. The relationship between the generated photo current in a photodetector and the angle of incidence. The angle is measured with respect to a line perpendicular to the sensitive area.

APD. Avalanche photodiode. A photodiode that operates with a large enough reverse bias to cause avalanche multiplication.

AREA MISMATCH. The fact that the two fiber ends to be joined, by connector or splice, may not have the same area.

ASCII CODE. American Standards Committee for Information Interchange Code. A 7 bit binary code used to represent a variety of symbols, including the digits 0 through 9, the letters A through Z and a through z, and so forth.

AVALANCHE MULTIPLICATION. The increase in emitted electrons (generated electron-hole pairs) due to the avalanche phenomenon in the diode, given as a multiplicative factor. (A photogenerated free electron collides with an atom, producing another free electron, and so on.)

BACK SCATTER. Light scattered back in the direction of the light source, typically caused by an obstruction, micro or macro.

Bandwidth budget. The allocation of bandwidth limitation among system components. (*See also* Rise time budget.)

Bandwidth, electrical. Standard definition of bandwidth. For systems with dc response ($f_1 = 0$), it is the frequency for which the output power drops by 3 dB or the amplitude drops to 0.707 of its maximum value.

Bandwidth, optical. The frequency for which the optical output power drops to one-half its maximum value

Baud. A line-signaling rate. Signal transitions per second.

Beam angle. The angle of the radiating beam as determined by the half-power points. (*See* Full angle at half maximum.)

Beam angle, horizontal. The beam angle parallel to the emitting surface.

Beam angle, vertical. The beam angle in a direction perpendicular to the emitting surface.

Biphase. A coding format for digital transmission (also called the Manchester code.)

Bit error rate, BER. The expected or probable error rate in a digital system. If one error is statistically expected to occur in 10^9 bits, the BER is 10^{-9} (1 per 10^9 bits).

Bit time. The time slot in a transmission during which the data bit is present.

Bus. Data distribution method used in digital systems.

Bypass. A method of short circuiting a failed node on a network by allowing data to bypass the node.

Capacity. When used in connection with a digital system or line, capacity refers to the highest data rate with which the system can operate. Maximum bit rate.

CATV. Community antenna television. The use of a common antenna receiving video channels, which are then distributed by wire or fiber to many subscribers.

CCTV. Closed circuit television. Use of TV camera and monitor in local premises, such as safety monitors in banks and supermarkets.

Cladding. The outer coating of an optical fiber (not the structural covering) usually made of glass or plastic.

Coherence, spatial. An electromagnetic wave with fixed, nonrandom phase at a point in space is said to have spatial coherence.

Coherence, temporal. An electromagnetic wave of very narrow range of frequencies is said to have temporal coherence. Ideally, a single frequency (single wavelength) wave.

Coherent communication. Communication system using a sinusoidal carrier, modulated by information. In optical coherent communication, the sinusoidal carrier is light ($= 10^{14}$ Hz).

Core. The center portion, usually made of glass or plastic, of an optical fiber.

Coupler. A means of coupling optical power among different fibers.

Coupler, star. A star-like structure that allows coupling of optical power between a large number of fibers.

Coupler, T. Typically, a coupler with 3 ports (sometimes 4 for a dual T structure).

Cross coupling. The power coupled between fibers that are not expected to be coupled at all, usually given in decibels.

Cross section mismatch. *See* Profile mismatch.

Cutoff frequency, f_c. The lowest optical frequency that will produce a photo current. Frequencies below f_c do not have enough energy to produce a photo current for a particular photodetector.

Cutoff wavelength, λ_c. Similar to cutoff frequency, but expressed in wavelength. Wavelengths larger than λ_c will not produce photo current.

Dark current. The photodiode leakage current, with no incident light.

Data rate (bits/sec). The highest rate of digital binary information that the system can handle (related to bandwidth). (The unit of digital information is the bit.)

Demultiplexing (DeMUX). The retrieval of the original data from a multiplexed channel.

Depressed index cladding fiber, DIC. *See* Refractive index profile: W profile.

Dichroic filter. Optical device that attenuates, or "passes," light signals, depending on their wavelength.

Diffraction. The process that causes an electromagnetic wave to bend as it passes by a sharp edge (for example, the edge of a razor blade) obstructing the wave.

Diffraction grating. A set of opaque, or reflective slits (about the size of λ) spaced a constant distance S. (This grating produces an efficient diffraction effect.)

Directivity. A measure of how well radiating energy, light, coupled from one fiber to the other, is directed. Typically given in terms of relative power in the reverse direction (in a T coupler).

Dispersion. Pulse broadening effect. The effect that causes the output pulse to be wider than the input pulse.

Dispersion, chromatic. *See* Intramodal dispersion. It is called chromatic (color dependent) since it depends on line width (which is related to color).

Dispersion, intermodal. Dispersion caused by the delay between different modes. Typically, we consider the extremes: the delay between the shortest path (0 mode) and the longest path (the critical mode).

Dispersion, intramodal. Dispersion that is independent of modes, related to the line width of the source, and caused by variations in the refractive index as a function of wavelength.

Dispersion-shifted fiber. A fiber designed to have its minimum intramodal dispersion at about 1.5 μm, coinciding with the minimum loss window.

Duplex (full duplex). A communication channel that allows *simultaneous* transmission in both directions.

Efficiency. Generally defined as the ratio of output power to input power.

- **coupling.** The ratio of power inside (coupled to) the fiber to the power of the source connected to the fiber.
- **external.** The ratio of radiated power to the input electrical power, for a light source.
- **internal.** *See* Quantum efficiency.
- **quantum.** The number of photons generated divided by the number of electrons injected into the light source.

Emission. Emission of photons; optical energy.

spectrum. The range of wavelengths between the half-power points.

spontaneous. Photons emitted as a direct result of injected electrons, with no secondary stimulation.

stimulated. Mode of radiation in which generated photons collide with electrons, stimulating more emission of photons. The multiplication effect.

Emitter, edge. Light emmiter with radiation coming from the edge of the active region, sometimes abbreviated as ELED.

Emitter, surface. Emission emanates from the active area surface (SLED).

Endoscope. An instrument used to inspect (view) the inside of internal organs such as intestines.

End separation. Represents the distance between fiber edges, *S,* usually given in relation to the fiber core diameter, *d,* as *S/d.*

Energy gap. The energy difference between a conduction electron and a valence electron.

Energy pump. To sustain stimulated emission, an external energy source is needed, an energy pump.

Error detection. Digital techniques used to *detect* the presence of digital errors (bit errors).

Error detection and correction. Digital techniques that provide identification of the bit error, and, hence, allow error correction.

Excess loss. Loss in a coupler in excess of the power division. The ratio of total output power, at all output ports, over the total input power, at all input ports.

Eye pattern. A pattern displayed on an oscilloscope, giving visual qualitative and quantitative representation of the performance of a digital data link.

Fabry Perot Cavity. A cavity with reflectors at both ends, causing a light beam to bounce back and forth, producing standing waves. The cavity is said to resonate when standing waves are at a maximum.

Fall time, t_f. The time it takes for a waveform to change from 90% to 10% of its maximum amplitude.

Far field pattern. To determine the radiation pattern of a light source, intensity measurements must be taken at some distance from the radiating surface.

Fiber loss (dB/km). The loss of energy of light as it travels along the fiber, expressed in decibels per kilometer. (The decibel is a logarithmic unit relating the input power to the output power [input to the fiber and output from the fiber.])

Frame. One data set of multiplexed channels.

Frame rate (for MUX). The reciprocal of the time it takes for a TDM MUX to complete one scan of all input channels.

Frequency modulation (FM). A scheme in which the carrier frequency is changed in accordance with the input signal.

Fresnel reflection (losses). The reflection from the boundary between materials with differing refractive indices, where refraction rather than reflection is the dominant effect.

Full angle at half maximum (half power width). In reference to radiation patterns, the width, in degrees, between the two points at which radiated power is one-half its maximum intensity.

Graded-index. *See* Refractive index profile, graded-index.

GRIN lens. A lens made like a graded-index fiber (but with a substantially larger diameter) with carefully selected length.

Guard band. The range of frequencies separating the different subcarriers in an FDM system.

Guard ring. An extra diode structure (also reverse biased) surrounding the photo-diode-active area. It reduces the dark current.

Half duplex. A communication channel that allows transmission in both directions, one direction at a time (not simultaneously).

Heterodyne reception. The use of a local oscillator to produce an intermediate frequency signal that is substantially lower in frequency than the received signal, and which contains all of the transmitted information of the carrier itself.

Heterostructure. A LED or LASER structure with multiple P and N layers.

Index matching. The use of special materials with suitable refractive indices to eliminate refractive index discontinuities at the junction of two fiber edges in a connector, or splice.

Intensity modulation. In connection with an optical system, it refers to changes in the intensity of the light used to represent data carried by the light.

Interference pattern. Bands of high and low light brightness caused by interference between coherent light beams.

Intersymbol interference, ISI. The interference of one bit with the next succeeding bit. (A 1 affecting the following 0.)

IRED. A light emitting diode that operates in the infrared region. Infrared emitting diode.

ISDN. Integrated Service Digital Network. The use, typically by public utility companies such as the telephone company, of a digital network to provide a large variety of customer services. These include voice, CATV, computer communication, and so forth.

Jitter. Random variation in phase or bit time in a digital pulse train.

Laser. Light amplification by Stimulated Emission of Radiation. Refers to a device that produces light, typically of a single color (nearly a single frequency) by stimulated emission. Some such devices are the gas laser, the ruby laser, the semiconductor laser, and the diode laser.

Lasing. Operation in a sustained stimulated emission mode.

Lateral misalignment. The distance between the centers of the two fiber edges butted together in a connector, or splice.

LED. Light emitting diode. A semiconductor diode that emits light in response to a forward current.

Line width, $\Delta\lambda$. The range of wavelengths between the two points (two wavelengths) of half-power emission of a light source.

Local area network (LAN). A communication system that interconnects computers or other systems located within a relatively short distance of 1 or 2 mi.

Line width, relative. The ratio of line width to the center wavelength of the source, $\Delta\lambda/\lambda$.

Loose buffer. A fiber cable construction where the fiber is free to move (loose) inside a buffer jacket.

Loss. The ratio of output power to input power expressed in dB (decibels).

Manchester code. *See* Biphase code.

Meridional ray. A ray that passes through the fiber center line.

Minimum bending radius. Represents the sharpest possible bend the cable or bare fiber can tolerate before it breaks. It is the radius of the smallest possible circle the fiber can be formed into.

Minimum detectable power. The input signal power that produces a S/N of 1.

Mode(s). The propagation of light energy in an optical fiber takes place at distinct angles of propagation called modes of propagation, or simply modes.

Mode conversion. The transfer of light energy from one *mode* to another (mode coupling).

Mode distribution. The amount of energy carried by each mode in an optical fiber, usually given in relative terms.

Mode distribution, steady state. After a certain length of fiber, the power carried by each mode does not change any more. The mode distribution is said to be in the steady state distribution.

Moire fringes. Interference pattern created by diffraction.

Monochromatic. Of single color or single frequency. A light source that has a very narrow line width is said to be monochromatic.

Multiplexing. Techniques used to transmit data from many sources via a single channel (a single fiber or wire). (The sources may be analog or digital.)

Multiplexing, frequency division (FDM). The bandwidth of the common channel is shared among a number of transmitting sources. Each source is allocated a certain range of frequencies.

Multiplexing, space. The use of different physical media (different optical fibers) for the different transmitting channels.

Multiplexing, time division (TDM). *Time* on the common channels is shared among a number of transmitting sources. Each is allocated a certain time slot.

Multiplexing, wavelength division (WDM). Different wavelengths (colors) are used for the different transmitting sources.

Network. A communication link intended to serve many customers. Connecting to or disconnecting from the network is similar to connecting or disconnecting phone service.

Noise current. Random motion of electrons, not related to the signal.

Noise equivalent power, NEP. The amount of optical power that produces a current in a photodetector of the same magnitude as the noise current.

Noise, external. Electrical noise generated outside the device or system (line interference, transients).

Noise, internal. Electrical noise generated by the device itself (internal thermal, resistor noise).

Noise, shot. Noise caused by the discrete nature of electrons.

Noise, speckle. Noise introduced by variations in interference patterns of the different modes that propagate in a fiber. It can occur only with highly coherent transmission and is caused by fiber discontinuities.

Noise, thermal. Random motion of electrons (random current) related to the device temperature.

Normalized frequency. The V number, where V is a function of fiber radius, N.A., and the wavelength of operation.

Numerical aperture, N.A. The sine of the critical propagation angle ($\sin \theta_c$).

Numerical aperture mismatch. The fact that the N.A. of the two fiber edges to be joined may not be equal.

Period, *T*. The period of a repetitive waveform.

Phase lock loop (PLL). A circuit available in an IC package that synchronizes a local oscillator with an incoming signal. (The phases of the two signals are locked together.)

Photoconductive mode. A reverse biased photodiode where the conduction of the diode varies with incident light.

Photocurrent. Current that is due solely to incident optical power.

Photodetector. A device that converts optical energy to electrical energy.

Photogeneration. The generation by optical energy of electron-hole pairs, the electron having moved to the conduction band, leaving a mobile hole in the valence band.

Photon. The packet of light energy, analogous to the basic atomic parts (electron, proton, etc.).

Phototransistor. A transistor whose base current is produced by incident light.

Photovoltaic mode. A photodiode operation with zero bias, in which current (voltage) is generated as a result of incident light.

Pigtail mounting. A LED or LASER with a fiber permanently attached to it.

PIN. A photodiode with three regions: a positive region, an intrinsic material region, and a negative region.

PIN-FET. An integrated package containing a PIN at the input of a FET.

Planck's constant, *h*. The constant h that relates the energy of the photon to its wavelength ($h = 6.626 \times 10^{-34}$ joule-seconds).

P-N. A photodiode with P and N regions.

Population inversion. The condition in which there is a high concentration of electrons (and holes) in a region, where recombination is likely.

Port. A point of input or output to a system or a circuit. Fiber ends allowing access to a system.

Power budget. The allocation of power loss among system components, such as fiber, coupling, and connectors.

Power margin. The incident power at the receiver minus the receiver sensitivity, usually in decibels.

Power ratio. The ratio of power between two output fibers in a T coupler.

Profile mismatch. The cross-sections of two fibers may not have the same shape. For example, one may be elliptical, the other circular.

Propagation angle. The angle a beam inside a fiber makes with the fiber axis.

Propagation angle, critical. Rays with propagation angles larger than the critical angle are not confined to the fiber. They exit the fiber.

Pulse amplitude modulation (PAM). The amplitude of a pulse train is modified in accordance with an input signal amplitude.

Pulse code modulation (PCM). The input signal is sampled. Each sample is converted to a digital code used in the transmission.

Refraction. A phenomenon that causes light to change its direction as it travels from one material to another (described by Snell's law of refraction).

Refractive index. The ratio of the speed of light in a vacuum to the speed of light in a material n, where n is greater than or equal to 1.

Refractive index profile. The way the refractive index varies as a function of the distance from the fiber center.

Refractive index profile: graded-index (parabolic). The refractive index changes as the square of the ratio r/a, highest for $r = 0$, lowest for $r = a$. r is the distance of an arbitrary point in the core from the center; a is the core radius.

Refractive index profile: triangular. The refractive index changes linearly as the point moves from the center, $r = 0$, to the core edge, $r = a$.

Refractive index profile, *W*. The cladding has two values of n: low close to the core and at the cladding outer edge, and higher in between.

Relative refractive index difference, Δ. The ratio of the refractive index difference over the core index.

Repeater. A device or system used to reshape (amplify) the signal along a transmission cable. (As a minimum, it consists of a receiver, an amplifier, and a transmitter. The signal is received, reshaped, amplified, and retransmitted along the cable; typically, analog signals are amplified and digital signals are reshaped.)

Responsivity, *R*. *See* Radiant sensitivity.

Ring network. A particular network typology (structure) where data flows in a circular fashion between nodes.

Rise time, t_r. The time it takes the waveform parameter (voltage, current, power) to change from 10% to 90% of its maximum amplitude.

Rise time budget. The allocation of rise time among system components. Each component contributes to system rise time (and to the reduction of system bandwidth.) The rise time for the components determines the rise time for the system.

Sample rate. The number of samples taken per second.

Sampling. To examine (measure) a continuous analog signal at specific intervals of time.

Scattering. The process that causes light to be transmitted in all directions.

Scattering, Mie. Scattering loss caused by imperfections and obstructions larger than the wavelength.

Scattering, Rayleigh. Scattering caused by very small obstructions, the size of λ and smaller. Scattering caused by the molecular structure.

Self clocking code. A digital format that contains clock information. For example, a transition during each bit time, as in the Manchester code, allows the extraction (generation) of a clock signal directly from the data.

Signal to noise ratio, S/N. The ratio of signal power to noise power in an electronic system.

Simplex channel. A communication channel allowing transmission in one direction only.

Single-mode fiber. A fiber designed to allow light propagation in one mode only, the lowest order mode.

Skew rays. Rays that propagate in the fiber without crossing the fiber center line. (spiral rays).

Spectral response. The relation between generated photo current and the incident optical wavelength for a particular detector.

Splice. A method of permanently connecting two fibers.

Splice, elastomeric. A type of mechanical splice developed by GTE.

Splice, fusion. A splice accomplished by fusing the fiber ends, using heat.

Splice, mechanical. A splice where the two fiber ends are held together mechanically.

Standing waves. When a transmitted wave is reflected at the far end of a cavity, the wave in the cavity may seem to be standing still. The troughs and the crests of the wave are stationary. This standing wave situation occurs only when the reflected wave has the appropriate phase. It is dependent upon the length of the cavity measured in wavelengths.

Star network. A network topology where all nodes communicate via a control distribution node, in a star-like structure.

Step-index fiber. A fiber made of a core and cladding with two refractive indices, n_{core} and n_{clad}.

Subcarrier. When using FDM, each portion of the shared common channel bandwidth is a subcarrier.

Subscriber loop. The telephone link that connects the subscriber to the main telephone system.

Synchronization. Keeping a stream of data locked to a specific rate.

Threshold current. The minimum LASER current at which lasing takes place.

Tight buffer. A fiber cable where the outside jacket is tightly wrapped around the fiber itself.

Total internal reflection. Rays traveling at shallow angles, below the critical propagation angle, from a high-index material to a low-index material, undergo total internal reflection and do not cross into the low-index material. This behavior is the same as that of a reflected ray.

Index